# 199/I Current Topics in Microbiology and Immunology

Editors

Springer
Berlin
Heidelberg
New York
Barcelona
Budapest
Hong Kong
London
Milan
Paris
Tokyo

# The Molecular Repertoire of Adenoviruses I

## Virion Structure and Infection

Edited by W. Doerfler and P. Böhm

With 51 Figures

Springer

WALTER DOERFLER
PETRA BÖHM
Institut für Genetik
der Universität zu Köln
Weyertal 121
50931 Köln
Germany

*Cover illustration: Confocal micrograph of an adenovirus-infected cell nucleus in which the 72K viral DNA binding protein and the cellular fibrillarin protein have been stained by immunofluorescence. Red spots show the position of the 72K protein while green spots show the location of the fibrillarin protein. Fibrillarin is released from its normal location in the nucleolus late during infection and becomes diffusely located throughout the nucleoplasm. (More information about the effect of adenovirus infection on nuclear organization is given in the chapter by E. Bridge and U. Pettersson.)*

*Cover design: Künkel+Lopka, Ilvesheim*

ISBN-13: 978-3-642-79498-8 e-ISBN: 978-3-642-79496-4
DOI: 10.1007/978-3-642-79496-4


Library of Congress Catalog Card Number 15-12910
Softcover reprint of the hardcover 1st Edition 1995



Typesetting: Thomson Press (India) Ltd, Madras
SPIN: 10479617 27/3020/SPS – 5 4 3 2 1 0 – Printed on acid-free paper

# Preface

When asked whether they wanted to contribute a chapter to these three volumes on the *Molecular Repertoire of Adenoviruses* almost all of my colleagues working in the field responded favorably and supplied fine chapters on their areas of expertise.

In 1995, adenoviruses continue to provide a major and highly developed experimental system for biochemical, cell biological, genetic, virological, and epidemiological investigations in mammalian molecular biology. As a considerable body of information has become available thanks to the continued efforts of many, this virus system has become particularly useful to those who want to address details of biological mechanisms and their relation to structure and function.

We have tried to cover as wide a field as possible in current adenovirus research and to encompass the entire gamut of adenovirology and adenovirus molecular biology. In spite of the tendencies that come from many corners to seduce researchers into directing their efforts towards applied molecular biology, which of course has its place and merits, there are still sturdy groups who pursue their interests in basic molecular biology and in particular in adenovirus research, an "eternal archetype," as one of the experts chose to name the virus.

In the first of the three volumes, we present an overview of adenovirus research and go on to cover the topics of the structure and assembly of adenoviruses, viral infection, and viral gene products. In the following two volumes, we turn our attention to topics such as DNA replication, recombination and integration, post-transcriptional control, transformation and E1A, adenovirus genetics, pathogenesis, and gene therapy.

The three volumes appearing now have also been initiated to mark with some, perhaps unavoidable, delay the occasion of the 40th anniversary of the discovery of adenoviruses by W.P. Rowe and R.J. Huebner and, independently, by M.R. Hilleman and J.H. Werner in 1953/1954.

I was encouraged to undertake the task of editing the current three volumes on adenovirus molecular biology by visits to many libraries in different countries on several continents during the last decade and by the observation that the three volumes on *The Molecular Biology of Adenoviruses*, which I had edited for Springer-Verlag's series *Current Topics in Microbiology and Immunology* in 1983 and 1984, had apparently been diligently worked through by many readers. The craftsmanship of the book covers had barely resisted the intellectual assault and showed that the books had been consulted with curiosity.

There are few viral genomes that have not been developed into useful vector systems. An increasing number of research reports have been devoted to adenoviruses as potential vectors for human somatic gene therapy. Should that virus system indeed prove its worth for this important application, researchers will have to appreciate the subtleties of the biology and the molecular biology of this virus system. As someone who has not been directly involved in research on gene therapy, it appears to me that considerable efforts will have to be expended before all the problems concerning the therapeutic application of the adenovirus system or, more likely, of even better systems will be solved. The importance of this goal justifies and will require a great deal of very active research.

We wish to thank all the contributors, the editors of the *Current Topics* series, and in particular Springer-Verlag for their help in getting these three volumes to press.

Cologne

Walter Doerfler
Petra Böhm

# List of Contents

# List of Contents of Companion Volume 199/II

# List of Contents of Companion Volume 199/III

# List of Contributors

(Their addresses can be found at the beginning of their respective chapters.)

# Adenovirus — An Eternal Archetype

L. Philipson



## 1 Introduction

Adenovirus constitutes a large group of DNA viruses that infect humans, animals, and birds (Pereira et al. 1963). The adenovirus genome containing 36 kb of DNA is organized in early and late transcription units (Ginsberg 1984; Horwitz 1990). Transcription of the latter is activated at the onset of viral DNA replication. From each of the six early transcription units, numerous messenger RNA are generated, and the encoded polypeptides are required to establish productive viral replication, transformation, and viral latency in infected cells. The first transcription units to be activated shortly after infection, regions E1A and E1B, encode proteins involved in cellular transformation and transactivation of viral and cellular transcription units (Flint and Shenk 1989). Regions E2A and E2B encode proteins involved in adenovirus DNA replication, including the viral polymerase terminal protein and the DNA-binding protein (Horwitz 1990). Finally, regions E3 and E4 are involved in other early viral functions leading to suppression of host defense mechanisms (Wold and Gooding 1991), to transcriptional activation of other promoters (Hardy et al. 1989), and to preferential transport of viral mRNA (Babiss and Ginsberg 1984; Halbert et al. 1985).

---

The Skirball Institute of Biomolecular Medicine, New York University Medical Center, 540 First Avenue, New York, NY 10016, USA

The products of the early transcription units comprise only a minor part of the total cellular mRNA and proteins. Therefore, the early phase is not associated with dramatic alterations in cellular metabolism. The late phase of adenovirus infection is triggered by the onset of viral DNA replication, which begins around 6–8 h after a high-multiplicity infection (GINSBERG 1984). Unlike early virus transcription, which is initiated and promoted at six different locations, initiation of DNA replication activates a major late promoter (MLP) located at 16.4 map units of the viral genome. Less abundant transcripts are generated from this site during the early phase of viral replication. A few other promoters are active during the late phase of adenovirus infection. The MLP generates five families of late transcripts (L1–L5) by differential splicing and polyadenylation of a large primary transcript that terminates at 99 map units at the right end of the genome (ZIFF 1980).

Every mRNA transcribed by RNA polymerase II (poll II) from the MLP contains an identical 5' noncoding segment which is 200 nucleotides in length and called the tripartite leader. It is derived from the splicing of three small exons located upstream of the L1 protein-coding family of the late transcript (BERGET et al. 1977; CHOW et al. 1977). Most of the late adenoviral mRNA encode structural polypeptides of the viral particle and proteins involved in packaging of the genomic DNA. A few mRNA, however, encode proteins that are required for the trimer formation, modification, translation or transport of viral structural proteins. During the late phase of adenovirus infection, large amounts of virus-encoded RNA pol III products are also synthesized, referred to as viral-associated (VA) RNA I and II (SÖDERLUND et al. 1976; WEINMANN et al. 1976). At least VA RNA I seems to be required for translation of viral mRNA at late times in the infectious cycle (THIMMAPPAYA et al. 1982).

Productive infection of cultured cells lasts for 30 h, typically beginning with early transcription within 1 h after virus attachment. During the late phase, dramatic alterations in cellular metabolism are observed, including the cessation of cell division. Second, there is an almost exclusive synthesis of late viral polypeptides; third, the rate of transport of cellular mRNA from the nucleus to the cytoplasm is strongly reduced without a concomitant inhibition of transcription from cellular genes. Thus, the machinery of the cell involved in synthesis of DNA, RNA, and protein is almost totally redirected towards the production and assembly of viral particles (HORWITZ 1990).

Due to the easy molecular dissection of the adenovirus and the early identification of all the mRNA and their products, this virus system provides an eternal archetype for several mechanisms involved in viral replication as well as in gene expression in mammalian cells.

I will attempt, in this synopsis, to review some of the events that made adenovirus the preferred model for several intracellular biological events. Among these are the mechanisms for receptor recognition and penetration, the transcriptional control mechanisms of the early genes, and the role of adenovirus in establishing the tumor suppressor concept. Equally important is the role of the E3 region in suppressing the host defense mechanisms, which probably facilitates establishment of a latent phenotype of the virus.

Of more general interest is the role adenovirus has provided in establishing models for gene expression in mammalian cells. The adenovirus was the first system to reveal that multiple mRNA can be derived from a defined large transcript. This finding led to the first and ultimate proof for the discontinuous gene concept, which applies also to the majority of mRNA in eukaryotic cells.

Still under investigation are the specific control mechanisms involved in preferential viral mRNA transport during the late phase as well as the intricate translational control mechanisms which allow preferential viral mRNA translation. The role of VA RNA and the importance of the tripartite leader in this process is now becoming clarified. In all these instances the adenovirus has provided a model for dissecting the molecular details involved in the control of viral and cellular gene expression. Adenovirus has provided additional models for macromolecular synthesis in mammalian cells, but because of space limitations, they could not be covered in this review. The first cell-free system for DNA replication using adenovirus DNA is one pertinent example (HORWITZ 1990). Another is the dissection and identification of the adenovirus oncogenes, which led to the introduction of the calcium–phosphate transfection technique (GRAHAM and VAN DER EB 1973) and culminated in the demonstration that the oncogenic potential of different adenovirus serotypes may reside in their capacity to influence expression of the transplantation antigens of the host cell (VAESSEN et al. 1986).

Before turning to the selected topics, it is appropriate to also recognize the groups who initiated the studies on adenovirus biochemistry (GREEN 1970; GINSBERG 1984). They provided the backbone on which everyone could build in establishing the virus as the preferred tool to dissect gene expression in mammalian cells.

## 2 Receptor Recognition and Penetration

Adenovirus attaches with high efficiency to cellular receptors via the fiber protein (PHILIPSON et al. 1968; DEFER et al. 1990). Although the identity of the primary receptor is still unknown, it may be shared with coxsackie B3 virus (LONBERG-HOLM et al. 1976). The vitronectin-binding integrins have recently been identified as the secondary receptor (WICKHAM et al. 1993). The latter may mediate virus internalization through attachment to the penton base protein. The complexed virus is internalized by receptor-mediated endocytosis (CHARDONNET and DALES 1970; VARGA et al. 1991). In the endosomes, it encounters an acidic pH, which may trigger the penetration of a modified virus into the cytoplasm.

Due to the recent emphasis on adenovirus as a vector for gene therapy and an effective carrier for introducing DNA into mammalian cells (COTTEN et al. 1993; ROSENFELD et al. 1991), the receptor recognition and the penetration steps have recently been reinvestigated. The advantage of utilizing adenovirus for gene delivery is its ability to be taken up by non-dividing cells and to enter several different cell types. In addition, stable recombinant and replication defective particles which carry large inserts can readily be purified from this human DNA virus.

It was already established in 1969 that adenovirus undergoes multiple sequential uncoating steps as it moves from the cell surface to the nuclear membrane (LONBERG-HOLM and PHILIPSON 1969). A careful recent analysis using biochemical methods demonstrated that the virus particle, which is stable as long as it is outside the cell, is dismantled upon entry by sequential elimination of structural proteins, leading to an efficient release of its DNA into the nucleus of the cell (GREBER et al. 1993). As shown in Fig. 1, the uncoating process already begins during endocytotic uptake at the plasma membrane or immediately thereafter. Further changes take place in the endosomes and in the cytoplasm. Thirty minutes after internalization half of the parental DNA has accumulated in the nucleus of the infected cells.

The first proteins to be eliminated from the penetrating virus are those located in each of the 12 vertices of the viral particle. It appears that the entire penton complex, consisting of the fiber and the penton base, dissociates at this early stage. The fiber is eliminated 10–12 min after initiation of penetration, and the penton base disappears 3–5 min later. The fiber release might be necessary to make the virus competent for penetration, and the penton base may have a specific role at this stage since, by itself, it might be responsible for the escape of the virus from the endosomes. A cytopathic effect has been associated with the

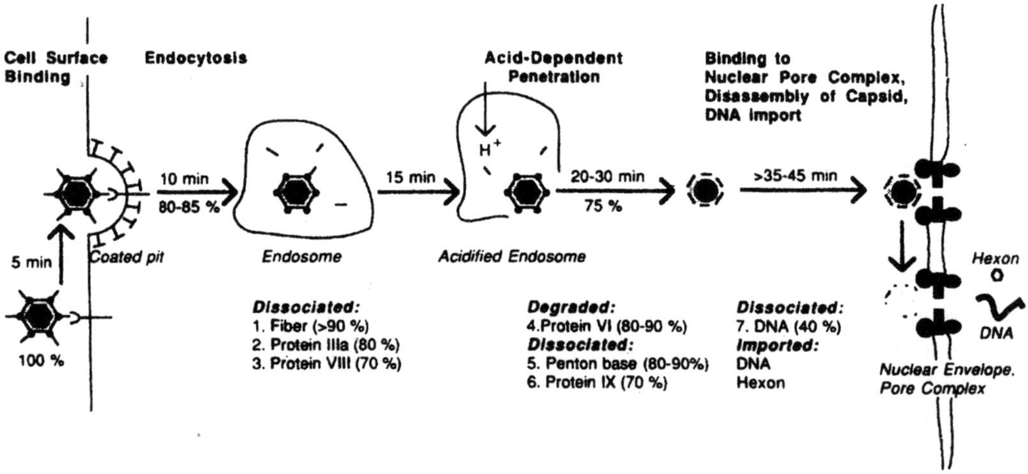


**Fig. 1.** The different steps involved in uncoating of adenovirus during cell entry. The virus binds to a high-affinity fiber receptor and is internalized via coated pits after a lag period of 5 min. Efficient penetration from the endosomes occurs after 15 min, and the virus is recovered at nuclear pore complexes after around 30 min. About 50% of the cell-bound viruses release DNA inside the nucleus, probably associated with the viral core proteins polypeptides V and VII. The efficiencies of the individual dismantling events are indicated in *brackets*. *Bars* represent the fibers, *filled circles* the vertex complex, and *filled hexagons* the virus and the *central black unit* depicts the viral chromosome (Reproduced with permission from GREBER et al. 1993)

penton base (Valentine and Pereira 1965), and acid treatment of the virus which removes the vertex region allows the virus to attach directly to liposomes specifically at the vertices (Blumenthal et al. 1986). All these early events of protein elimination may be due to protein degradation by a viral-coded protease, since a virus mutated in the viral protease (ad2*ts* 1; Weber 1976) does not lose its fiber during cell entry (Greber et al. 1993), nor does it penetrate into the cytoplasm subsequent to endocytosis (Miles et al. 1980). It may also be significant that a protein kinase, probably of cellular origin, is associated with the adenovirus particle (Akusjärvi et al. 1978). It would be interesting to analyze the effect of this kinase in the lytic events involved in the endosome escape mechanism.

After penetration to the cytoplasm, the virus loses proteins VI and VIII, which provide a cement between the core and the internal surface of the outer capsid (Greber et al. 1993). Shortly thereafter, the hexons are destabilized, including the externally located protein IX, which dissociates from the virus in the cytoplasm. The last group of proteins, V and VII, which bind directly to the viral DNA and form the viral core, are probably retained during entry of the DNA into the nucleus, although the DNA becomes highly sensitive to DNAse within the first 30 min after initiation of attachment. Somewhat later, a nuclease-generated ladder arrangement of DNA similar to the one observed for cellular nucleosomes may be discerned for the viral DNA, suggesting that it combines with cellular histones (Tate and Philipson 1979). More recently it has been claimed that protein V becomes dissociated from the core before the DNA enters the nucleus (Greber et al. 1993).

The uncoating pathway is efficient and rapid. At least 50% of the attached virus can release the DNA inside the nuclei, and only a very small proportion of the incoming viruses is routed to lysosomes. The uncoating process is probably guided by viral as well as cellular products. A full understanding of the uncoating and DNA delivery mechanism used by adenovirus is desirable if these viruses are going to be useful as a DNA vehicle in gene transfer. We may, therefore, expect a molecular dissection of several of these steps to fully take advantage of the adenovirus uptake, penetration, and delivery system for gene therapy.

## 3 Early Gene Reveals Transcriptional Control

When recombinant techniques and DNA sequencing became available, it was natural to analyze viral genes that could easily be amplified in vivo. The structure of several early adenoviral genes established that the viruses during evolution have combined in one and the same gene, a plethora of different functional domains necessary to enhance viral transcription or viral replication. When the adenovirus E1A region was characterized, it soon became evident that the E1A proteins had a fundamental role in adenovirus transcription and transformation. Viruses carrying mutations that impaired the expression of the E1A product produced reduced concentration of viral early mRNA from several other regions of the adenovirus genome (BERK et al. 1979; JONES and SHENK 1979). Soon thereafter, it was established that this effect was due to inefficient transcription from remote promoters (NEVINS 1981). Similar *trans*-activation is widespread among viral systems, and several viral proteins that enhance the rate of transcription or interact with DNA sequences have been identified among SV40 virus, herpes virus, and papilloma virus. Other viral proteins such as the adenovirus E1B 19-kDa protein obviously increase the level of target gene expression, but it has not yet been proven that this occurs at the level of transcription. The transcriptional activators among the early viral proteins fall in two categories: one is characterized by sequence-specific DNA binding, e.g., the SV40 T antigen (FRIED and PRIVES 1986) and the papilloma E2 protein (SPALHOLZ et al. 1988), whereas proteins in the second category, comprising the herpes α-*trans*-activation factor (αTF) (MICHAEL et al. 1988), the pseudorabies immediate early protein (ABMAYR et al. 1988), and the adenovirus E1A protein (MORAN and MATHEWS 1987), do not bind to DNA. Nevertheless, the latter are definitely involved in transcriptional control and must, therefore, function through protein–protein interaction. The E1A adenovirus protein is by far the most extensively studied among the viral *trans*-activators, mainly because it is not only a transcriptional regulatory protein but also an oncogene product. However, more recent results favor the notion that E1A-induced transformation and transcriptional activation can be separated as two distinct units in the two proteins generated from the E1A gene (MORAN and MATHEWS 1987; FLINT and SHENK 1989), again emphasizing that several functional domains have been juxtapositioned through evolution of the virus.

The *trans*-activation activity of the E1A protein was discovered mainly through its critical role in productive infection. It appears that all early adenovirus promoters can respond to the E1A protein irrespective of whether they are located in the adenovirus genome, provided as plasmids, or integrated at different sites in the cellular genome (BERK 1986). The *trans*-activatable promoters, whether viral or cellular, lack any common sequence elements, implying that *trans*-activation is not mediated through *cis*-DNA elements and suggesting that E1A binds to a family of different proteins rather than to DNA. *Trans*-activation by the E1A protein furthermore, not only involves transcription by RNA pol II, but also by RNA pol III (GAYNOR et al. 1985; HOEFFLER and ROEDER 1985), emphasizing that it acts by a mechanism that is indirect and of a general nature, since it activates several cellular promoters in addition to the viral promoters that use pol II and pol III.

During the early phase of productive adenovirus infection, two major differentially spliced E1A mRNA are generated. These alternatively spliced mRNA encode one 243- and one 289-amino acid protein, respectively. The proteins share N-terminal sequences and differ only in a 46-residue internal exon segment unique to the larger E1A protein, as shown in Fig. 2. Mutation in the CR1 and CR2 regions causes little change in *trans*-activation. Based on mutation in the CR3 region, this unique segment of the 289-residue protein appears to be involved in *trans*-activation (MORAN and MATHEWS 1987). A synthetic peptide comprising the 46-amino acid segment in the longer E1A protein and three adjacent conserved

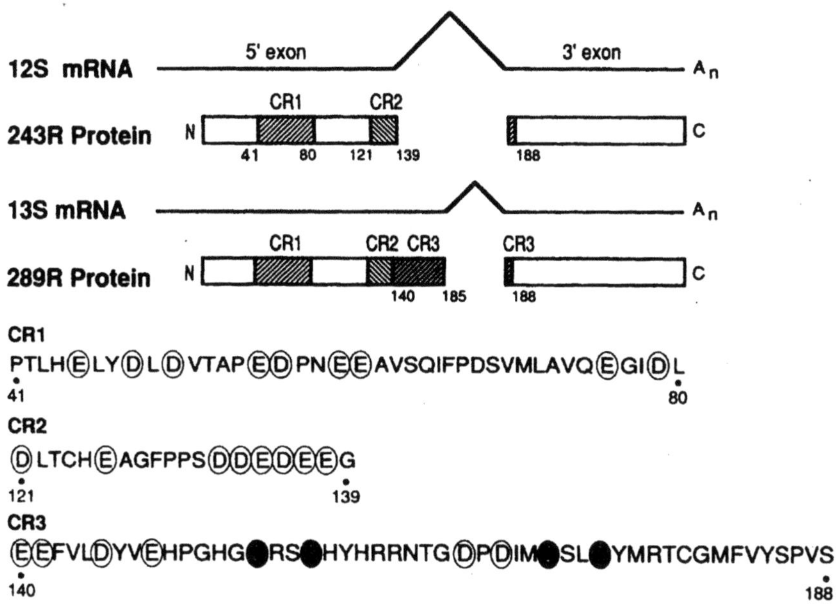


**Fig. 2.** The two most important mRNA and polypeptides originating from the E1A region. The sequences of conserved protein domains are specifically identified as the CR1, CR2, and CR3 regions and the amino acids in these regions have been identified below. Acidic residues within these sequences are marked by *open circles,* and cysteines comprising the metal-binding region in CR3 are marked by *shaded circles.* CR3 is the *trans*-activating domain (Reproduced with permission from the Annual Review of Genetics, Volume 23, (c) 1989, by Annual Reviews Inc; FLINT and SHENK 1989)

amino acids from the C-terminal exon of the E1A gene stimulates transcription from several adenovirus promoters, including those using pol II or pol III (GREEN et al. 1988). The peptide is, of course, less effective on a molar basis compared with the full-length E1A protein.

Normally, the E1A protein is heterogenous when isolated from infected cells, probably due to phosphorylation at several sites, but no site appears essential for *trans*-activation or oncogenic transformation (RICHTER et al. 1988). It appears that the E1A protein stimulates transcription from a diverse array of promoters. The mechanism of *trans*-activation has, therefore, been sought among the cellular transcription components mediating initiation at both pol II and pol III promoters. The most attractive hypothesis is that the E1A protein functions as an essential catalyst for the transcription initiation complex required for both classes of promoters. Whether this occurs by an enzymatic function of the E1A protein or by a stoichiometric binding to a shared protein is still unclear (FLINT and SHENK 1989). The TATA-binding protein (TBP), which is shared by pol II and pol III transcription systems, may be a good candidate for the latter mode of action (MEYERS and SHARP 1993). Since multiple proteins are required to form a transcription initiation complex, this model can only be proven by biochemical isolation of the protein complex from an E1A-induced promoter element that contains the E1A protein.

It may, however, be a laborious task, since numerous proteins are bound to each promoter site.

## 4 E1A Region Strengthens the Tumor Suppressor Concept

Evidence for negative regulation of growth of mammalian cells came from different quarters. Early results stemmed from somatic cell hybridization experiments showing that fusion of several different tumor cells with normal cells almost invariably results in the outgrowth of non-tumorigenic hybrids (HARRIS 1988), suggesting that the normal cells may donate protein(s) capable of suppressing the neoplastic phenotype. Human genetics provided a second clue to the existence of such tumor suppressors. More than 20 years ago, KNUDSON (1971) postulated that the rare childhood eye tumor retinoblastoma (Rb) is triggered by two successive genetic lesions, one inherited and the second one acquired. This hypothesis led more recently to the identification of the Rb gene. It encompasses 180 kb of DNA mapping to chromosome 13q14 encoding a 105-kDa nuclear phosphoprotein (pRb; FRIEND et al. 1986; MCGEE et al. 1989). The nuclear localization of pRb and its DNA-binding ability suggested a role in transcriptional regulation (BERNARDS et al. 1989). It was, however, not until it was discovered that this protein exists in the DNA tumor virus-transformed cell as a complex with the viral oncoproteins, the E1A protein (WHYTE et al. 1988), the SV40 large T antigen (DE CAPRIO et al. 1988), or the E7 protein from human papillomavirus (MUNGER et al. 1989) that the connection to tumor suppression became apparent. The DNA tumor viruses can obviously transform cells through their ability to eliminate a vital cellular growth-suppressing mechanism mediated through the Rb protein. All three viral oncoproteins share a small region of structural similarity that appears to complex with pRb. The latter has an oncoprotein-binding pocket that is the target for the viral proteins (HUANG et al. 1990). The E1A protein combines with a whole family of pRb-like proteins (which have different molecular weights and emanate from distinct genes) that all seem to contain a similar pocket for binding the E1A 243R protein (WHYTE et al. 1989). The E1A protein, therefore, traps the tumor-suppressive pRb and its relatives in an unphosphorylated form. Since phosphorylation of pRb is required to go through the normal growth cycle, the elimination of pRb in a complex with E1A leads to a constitutive growth potential.

Another aspect of this control mechanism was revealed by the demonstration that a normal cellular transcription factor, E2F, is required for the regulation of several genes involved in growth control. E2F was originally identified as a cellular DNA-binding protein that bound specific sequences in the adenovirus E2 promoter (KOVESDI et al. 1986). Among the E2F-responsive cellular genes, several are induced when $G_0$-arrested cells are forced to reenter the cell cycle. In fact, the sequence-specific DNA binding of E2F only results when two cellular proteins,

E2F1 and DP1, bind as heterodimers to E2F sites in various promoters (HELIN et al. 1992, 1993; HUBER et al. 1993). The E2F can also associate with the non-phosphorylated form of Rb protein, but does not bind to the phosphorylated type present at the $G_1$–S border. It appears that the binding of pRb to the E2F inhibits expression of the E2F-dependent genes, suggesting that E2F binding is a mechanism used to regulate cell cycle progression (HELIN et al. 1993). E2F can also associate with other members of the pRb family such as the p107, and the interaction between E2F and its associated proteins can be disrupted by adenovirus E1A and other DNA tumor virus oncoproteins that bind the Rb family proteins and block their ability to associate with E2F (FATTAEY et al. 1993). The end effect is, therefore, that the oncoproteins liberate the E2F from the Rb family proteins and in this way promote expression of E2F-dependent genes, leading to inappropriate cell cycle progression and thus to unrestricted growth.

The third element in the pathway of Rb control of cell growth was revealed when another group of transcription factors was characterized in detail. Since growth and differentiation are mutually exclusive phenomena, it was natural to examine whether the transition of growing cells to growth-arrested differentiated cells involved specific transcription factors. This was investigated by using myoblast differentiation as a model (OLSON 1990). The basic helix–loop–helix (bHLH) protein MyoD1 and its related family members were identified (WEINTRAUB et al. 1991), and it was demonstrated that each one of them alone can induce myotubular differentiation. Since muscle differentiation involves cell cycle arrest of the myoblasts, it was also analyzed whether the bHLH protein could arrest cells irrespective of the muscle differentiation program. An analysis of a set of MyoD1 mutants revealed that cell cycle arrest was due to a distinct moiety of MyoD1 separable from that required for muscle differentiation (SORRENTINO et al. 1990; CRESCENZI et al. 1990). However, the mechanism involved in MyoD regulation of the cell cycle became apparent only when it was found that MyoD directly and specifically binds to the unphosphorylated form of pRb and p107 (GU et al. 1993). The MyoD-binding pocket in pRb appears to be smaller than that required by E1A or E2F. Recently, it was established that other ubiquitously expressed bHLH transcription factors such as the E2 proteins (E12/E47) can also bind to the Rb family proteins (PEVERALI et al. 1994). Since these transcription factors can readily form heterodimers with MyoD1 or Myogenin, it is tempting to propose that the bHLH heterodimers bind pRb in an unphosphorylated form to control arrest in the growth cycle. When an oncogene such as the adenovirus E1A protein is overexpressed in normal cells, the E12–Rb complex dissociates and E1A irreversibly binds pRb, thereby making E2F available for unlimited growth. During growth induction with serum, the E2F heterodimers dissociate the bHLH from the pRb protein, leading to growth induction. Phosphorylation and dephosphorylation then probably mediate a control on access to E2F or to the bHLH at various stages of the growth cycle. This pathway of cell cycle control is schematically represented in Fig. 3, which demonstrates that the events first revealed by the binding of DNA tumor virus oncogenes to the Rb protein have expanded into a proposed pathway for growth control in a mammalian cell. It should be emphasized that this model

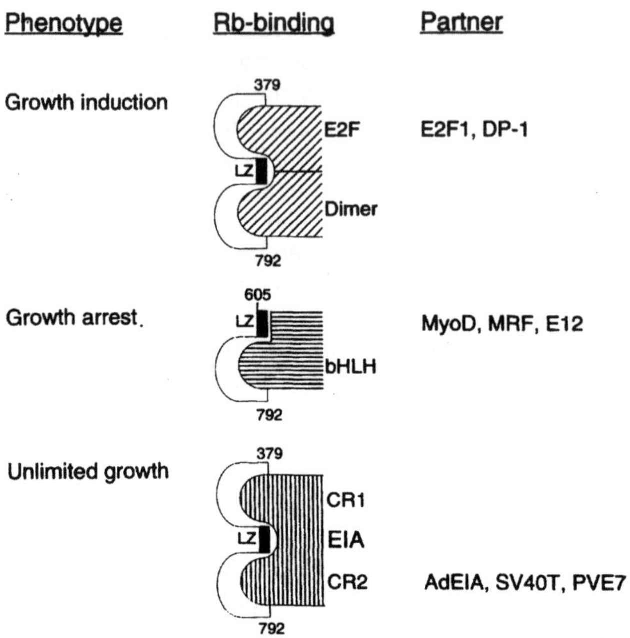


**Fig. 3.** An hypothesis explaining how retinoblastoma (Rb) protein interaction with different transcription factors or oncogenes may control growth. The three phenotypes—growth induction, growth arrest, and unlimited growth—are shown to the *left*, and the specific binding of various transcription factors to the retinoblastoma protein in the binding pocket between residues 379 and 792 are identified. The transcription factor partners are identified in the *right column*. Please note that, at growth arrest, only part of the retinoblastoma-binding pocket is occupied, making it easy to dissociate the factors during growth induction or unlimited growth. The viral oncogenes can obviously induce unlimited growth if introduced during both growth arrest and growth induction. (Reproduced from PEVERALI et al. 1994)

is still tentative, but the combined evidence allows at least a presentation of the hypothesis. It may serve as an experimental base for a simple control of the cell cycle in mammalian cell. Again, the adenovirus early gene E1A was instrumental in revealing a major control mechanism operating in mammalian cells.

## 5 E3 Region Is a Cassette for Host Cell Surveillance

The E3 region of adenovirus has been sequenced in several different serotypes and clearly contains nine overlapping mRNA generated by alternative splicing of a common transcript that initiates from one promoter but can terminate at two alternative polyA sites (WOLD and GOODING 1991). The E3 promoter is unique among the adenovirus promoters because it contains a binding site for the NFκB transcription factor (WILLIAMS et al. 1990), leading to independent transcription in lymphoid cells. This site may explain why adenoviruses persist in lymphoid

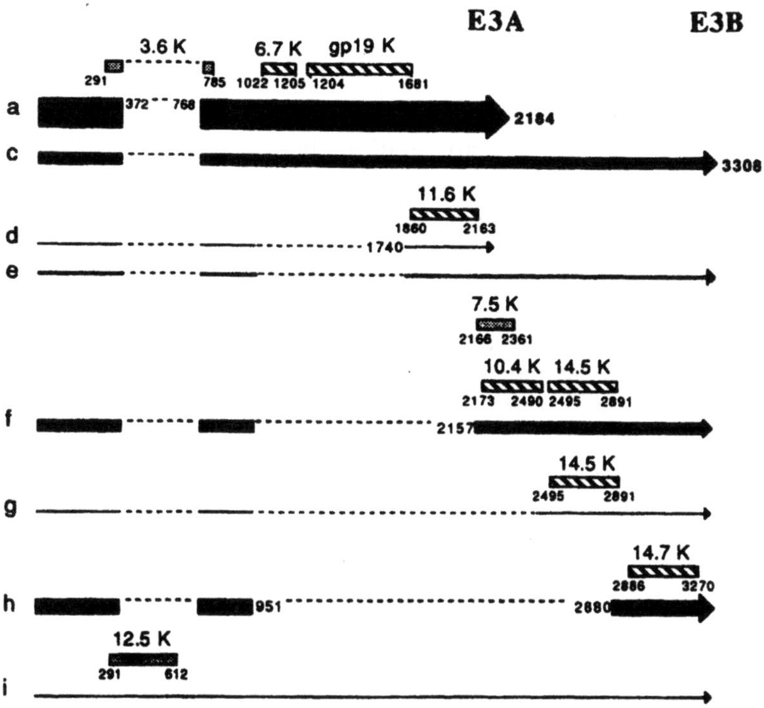


**Fig. 4.** Adenovirus E3 region. *Arrows* indicate the spliced structures of the mRNA, and the *thickness of the arrow* reflects the relative abundance of the mRNA. *Dashed lines* indicate introns. The two adenylation sites E3A and E3B are also identified. The *bars* above the arrows indicate the proteins encoded in the sequence. *Hatched bars* refer to proteins that have already been identified, and *stippled bars* are proteins that remain to be identified in vitro and in vivo. (Reproduced with permission from Wold and Gooding 1991)

tissue. Out of the nine predicted E3 protein sequences, six have been identified in infected cells by immunoprecipitation (Wold and Gooding 1991). None of the E3 proteins are required for adenovirus replication in cultured cells or for infection in several in vivo model systems (Ginsberg et al. 1989). In fact, E3 deletion mutants can replicate efficiently in human cell cultures. Nevertheless, since the E3 region is always maintained in natural isolates of adenovirus and shared by several serotypes, it may be important for in vivo infection of humans. Figure 4 shows a schematic illustration of the E3 region, indicating the abundance of the messenger RNA and showing the encoded proteins from this region of the adenovirus genome (Wold and Gooding 1991). The most abundant protein prevents lysis of adenovirus infected cells by cytotoxic T lymphocytes (CTL). This is a glycoprotein (GP19K) located in the membrane of the endoplasmic reticulum (ER) and containing a C-terminal transmembrane stretch followed by a 15-amino acid polar cytoplasmic domain. The first clue to the function of 19-kDa glycoprotein came when it was demonstrated that it can tightly bind to the class 1 antigens of the major histocompatibility complex (MHC) in adenovirus-transformed rat cells (KVIST et al. 1978). The binding is an intrinsic property of the protein. It is noncovalent and does not require glycosylation or other adenovirus proteins. It occurs after $\beta_2$-microglobulin has complexed with the heavy chains of the class 1 antigens of human, rat, and mouse origin. The GP19K binds in the peptide-docking site in the $\alpha_1$- and $\alpha_2$-domains of the heavy chain (BURGERT and KVIST 1987). The effect of the binding is to retain the class 1 antigens in the ER, thereby blocking transport to the cell surface and leading to cells that are less sensitive to cytotoxicity mediated by alloreactive CTL (PÄÄBO et al. 1989). The GP19K has different affinity for different class 1 antigens in the mouse. It interacts well with Kd, Ld, and Db, but less efficiently with other antigens (BURGERT and KVIST 1987). Hence, susceptibility to adenovirus infection in humans may be related to the affinity of individual class 1 antigens to GP19K. It is tempting to propose that GP19K blocks the peptide-binding site of the class 1 antigens (TOWNSEND et al. 1989), which seems to be a prerequisite for transport.

However, we must understand better the molecular details of the E3 region in the light of its role in model adenovirus infections. If the GP19K functions by preventing class 1 antigen expression, cells infected with the E3 deletion mutants should be destroyed in vivo and the mutated virus would therefore be less pathogenic. Deletion of the E3 region leads instead to an increased pathogenicity in model infections in cotton tail rats (GINSBERG et al. 1989) and to a dramatic increase in the lymphocyte and macrophage infiltration, suggesting that the symptoms are mainly caused by an inflammatory response. In the wild-type virus, class 1 molecules are absent at the plasma membrane of the infected cell, and no inflammatory response is observed. This disagreement between the in vivo and the in vitro results clearly emphasize our deficiency in understanding the processes involved in viral pathogenesis.

Although cytotoxic T cells are one way to eliminate virus-infected cells, tumor necrosis factor (TNF) may also be involved. TNF is a multifunctional cytokine that is secreted by activated macrophages and lymphocytes and that regulates several events in the inflammatory response (BEUTLER and CERAMI 1989). TNF is cytotoxic or cytostatic for certain tumor cells. C3HA mouse fibroblasts are lysed by TNF when infected by adenovirus mutants lacking the E3 region (GOODING et al. 1988), and stable E1A-transfected cell lines are susceptible to cytolysis by TNF (CHEN et al. 1987). These results led to the conclusion that to prevent TNF cytolysis, some E3 proteins are required, especially the E3 14.7K (GOODING et al. 1988). In fact 14 of 15 mouse cell lines infected with different adenovirus mutants were protected against TNF cytolysis by the 14.7K protein from the E3 region (GOODING et al. 1990). On the other hand, in human cells the E1B region 19K protein seems to be required for TNF resistance (GOODING et al. 1991). The 14.7K or the 10.4/14.5K proteins from the E3 region can, however, serve as a second defense line against TNF cytolysis for human cells. Thus, in human cells, four separate proteins might function independently to prevent TNF cytolysis, suggesting that TNF is an important antiviral defense mechanism of the host.

Furthermore, in addition to their role in preventing TNF cytolysis, two E3 proteins (10.4 and 14.5K) downregulate the epidermal growth factor (EGF) receptor in adenovirus-infected cells (CARLIN et al. 1989). Although both proteins are required, the 10.4K can act alone in mouse cells stably transfected with a human EGF receptor. Both these proteins are cytoplasmic, probably localized close to the plasma membrane to be able to interact with the EGF receptor from the inside. Whether this interaction is related to their ability to prevent TNF cytolysis has not been established.

All these results establish that the E3 region contains a cassette of genes that helps the virus to evade the immune system of the host by counteracting early defense mechanisms.

## 6 Adenovirus Provides Proof for the Discontinuous Gene

In a more general perspective, adenovirus probably gave its greatest contribution to molecular biology by providing the tool for the understanding of processing of pre-mRNA into mRNA in mammalian cells. The synthesis of mRNA in mammalian cells has been studied for a long time. In 1963 it was recognized that nonribosomal nuclear RNA and messenger RNA showed similar base composition (SCHERRER et al. 1963). That was the first primitive tool to study the pathway of mRNA processing. Due to the high GC content of ribosomal RNA in comparison to the low GC content of heterogenous nuclear RNA (hnRNA), the two entities could be identified (PENMAN et al. 1963). Previously, it had been demonstrated in mammalian cells that the 18S and 28S ribosomal RNA are derived from a large 45S preribosomal RNA precursor (SCHERRER and DARNELL 1962; SCHERRER et al. 1963). A crucial experiment establishing that mRNA was present both in hnRNA and in the mRNA associated with the polyribosomes was carried out using SV40 virus mRNA taking advantage of the newly developed DNA hybridization method (BENJAMIN 1966; LINDBERG and DARNELL 1970). However, to finally attack this problem, additional markers on both hnRNA and mRNA were required. The first of those markers was the 3' polyA tail of the mRNA, which was independently identified by several groups (KATES 1970; EDMONDS et al. 1971; LEE et al. 1971; DARNELL et al. 1971a). The connection, however, between hnRNA and mRNA polyadenylate (polyA) tracks was established in the adenovirus system through the detection of specific mRNA on the adenovirus genome (PHILIPSON et al. 1971; DARNELL et al. 1971b). The addition of polyA was obviously a post-transcriptional event, suggesting for the first time a precusor–product relationship between hnRNA and mRNA (DARNELL et al. 1971b). The next marker that became available was the capping of the 5' end of both hnRNA and mRNA, which could be detected with methyl labels (SHATKIN 1976; MOSS and KOCZOT 1976). It was first demonstrated in the adenovirus system that the nuclear polyA-containing RNA was, on average, five times longer than the polyadenylated mRNA from an internal region of the genome (WALL et al. 1972). In retrospect, knowing the mechanism for splicing, it should have been possible already at this time to foresee this mechanism (PERRY et al. 1976), since the conclusion from metabolic labeling studies clearly indicated that the 5' methyl groups in long hnRNA molecules (SALDITT-GEORGIEFF et al. 1976) and 3' polyA ends (DARNELL et al. 1971b) were preserved between hnRNA and mRNA. To explain these results, the prediction should have been that internal sequences had to be removed.

A specific and identifiable mRNA was, however, required to unequivocally establish the splicing principle. Two sets of experiments allowed a frontal attack on analysis of specific transcripts. One was the mapping of the promoter sites accomplished with ultraviolet (UV) mapping of the size of the transcripts (SAUERBIER 1976; GOLDBERG et al. 1977), and the other identified the size and the abundance of transcripts by nuclear run-on experiment (VENNSTRÖM and PHILIPSON 1977; WEBER et al. 1977; EVANS et al. 1977). The UV experiments correctly predicted the major late transcript to initiate at around 16.5 map units of the DNA. Together, these two sets of experiments established that the size of the major late transcription unit was around 23 kb and that at least 30 different mRNA were generated from this unit. Concurrently, it was established, although not yet interpretable, that all the late adenovirus mRNA contained a common oligonucleotide in the 5' end of the mRNA (GELINAS and ROBERTS 1977). Contributing to the dissection of the genome and the specific mRNA of adenovirus was, of course, the precise mapping of the regions expressed into mRNA, both early and late in the infectious cycle (PHILIPSON et al. 1974; SHARP et al. 1974). The final proof for the splicing of exons and removal of introns came from two laboratories based on the visualization of specific mRNA relative to the genome. Dr. Berget, a postdoctoral student in Dr. Phillip Sharp's laboratory at the Massachusetts Institute of Technology (MIT), and a group of young investigators at Cold Spring Harbor Laboratory provided the first evidence by hybridizing adeno mRNA to fragments and then to the adenovirus genome (BERGET et al. 1977; CHOW et al. 1977). For example, they were able to establish that the 3' poly(A) sequences in the hexon mRNA did not hybridize, but neither did the 5' end. When the whole DNA was used, specific loops were revealed, positioning the 5' and at three distinct positions far upstream of the sequences encoding the hexon protein. These positions corresponded to three exons close to the promoter forming the tripartite leader in all late mRNA derived from the MLP. This observation immediately led to the correct interpretation of the results with a common oligonucleotide in all late mRNA. Additional, more elaborate RNA–DNA hybrids of several different viral mRNA in Cold Spring Harbor (BROKER et al. 1977) finally established the discontinuous gene structure of the adenovirus genome. A reinterpretation of several results from other mammalian mRNA established that mammalian genes consist of exons and introns and that the intron sequence are deleted during post-transcriptional processing of the pre-mRNA into the final product presented to the translational machinery. This was a true creative revelation in the summer of 1977, leading everyone working in the field to the spontaneous comment "Why did I not think of that"? The adenovirus system definitely demonstrated the discontinuous gene concept, as summarized by PIERRE CHAMBON in the Cold Spring

Harbor Symposium of 1977. This led to the analysis of several different mammalian genes with regard to their DNA structure and the splicing pattern of their transcripts. It is fair to conclude that the adenovirus RNA splicing provided the key to how fragmented genes are arranged and mRNA generated. In retrospect, it is clear that a model system with well-characterized individual mRNA and genes was required to delineate the events.

This contribution was finally recognized in 1993 by the Nobel Committee in Stockholm giving the prize to Dr. Phillip Sharp and Dr. Richard Roberts for the discovery of RNA splicing. Subsequently, the adenovirus system has contributed to the understanding of the mechanism for splicing both by providing an in vitro splicing system as well as characterization of the small nuclear ribonucleoprotein particles (snRNP) involved in the splicing reaction. Fifteen years later, the major intranuclear particle involved in the splicing reaction, referred to as the spliceosome, is being dissected and analysed in all its details.

## 7 Preferential Translation and Transport of Viral mRNA

During the late phase of the adenovirus growth cycle, infected cells synthesize viral polypeptides exclusively. Although this could be due solely to a translational control mechanism, there are other ways by which viruses could mediate preferential protein synthesis. In pox and herpes simplex, virus-mediated degradation of host cell mRNA occurs (Rice and Roberts 1983; Nishioka and Silverstein 1977). Vesicular stomatitis virus generates a high concentration of viral mRNA, leading to preferential occupation of the sites available for the initiation of protein synthesis (Lodish and Porter 1980). In human immunodeficiency virus (HIV)-infected cells, inhibition of host cell transcription or host mRNA transport seems to be the major cause for preferential translation of viral messengers (McCune 1991). Finally, in influenza virus-infected cells, virus-mediated modification of cellular mRNA such as removal of the cap structure at the 5' end favors viral protein synthesis (Krug et al. 1979).

In the adenovirus system, cellular mRNA are not degraded during the late phase of productive infection. In fact, they are translationally competent, as revealed by extracting mRNA from late adenovirus-infected cells and using them in in vitro translation systems (Thimmappaya et al. 1982). Large amounts of both cellular and late viral polypeptides are synthesized in the in vitro system. The cellular mRNA are furthermore both capped and polyadenylated during the late phase of the infectious cycle (Thimmappaya et al. 1982). Thus, cellular mRNA are available at almost normal levels in the cytoplasm of late adenovirus-infected cells and are not modified in a manner that would prevent their translation, nor are they degraded. In fact, examination of some individual mRNA by cDNA cloning also showed the same cytoplasmic abundance of stable cellular mRNA in infected and uninfected cells (Babich et al. 1983). The conclusion, therefore, is that cellular mRNA are well represented in the cytoplasm at late times after infection. On the other hand, the majority (around 90%–95%) of the cytoplasmic mRNA associated with polysomes are derived from viral transcripts (LINDBERG and SUNDQUIST 1974). These mRNA correspond predominantly to late species transcribed from the viral MLP and, in addition, from two independent transcription units encoding proteins IX and IVa2. These observations could not be explained simply by an enormous amount of viral mRNA diluting the concentration of cellular mRNA. In fact, at late times after infection, adenovirus mRNA constitute only about 20% of the total cytoplasmic pool (PRICE and PENMAN 1972). Likewise, in vitro translation of cellular mRNA was efficient in extracts from late adenovirus-infected cells. Thus, adenovirus promotes the protein synthesis templated by its own late mRNA, while suppressing those of the host cell. Nevertheless, when the adenovirus infection enters the late phase, the majority of newly synthesized mRNA accumulating in the cytoplasm correspond to viral sequences. Around 90%–95% of the mRNA transported to the cytoplasm from the nucleus at late times emanate from the adenovirus genome (FLINT 1984). On the other hand, analysis of nuclear transcription rate for a variety of cellular genes showed that the levels of expression were essentially the same as that observed in uninfected cells (FLINT et al. 1983) These results suggest that there is no accelerated rate of nuclear degradation of host mRNA, but rather that the adenovirus genome provides a clear block in the transport of cellular mRNA during the late phase of the infectious cycle. This reduced transport of cellular mRNA appears to be caused by the adenovirus early E1B-55K protein, which is complexed with the E4-34K protein (BABISS and GINSBERG 1984; PILDER et al. 1986). Adenovirus mutants which fail to synthesize either of these two proteins are unable to transport viral mRNA selectively during late infection and continue to transport cellular mRNA at normal rates. Adenovirus thus possesses a mechanism for selective transport of its own mRNA, but that does not necessarily, by itself, explain the preferential translation of the viral over host mRNA. In fact, steady state levels of a number of short-lived cellular mRNA were reduced dramatically, as could be expected if preferential transport exists (FLINT et al. 1984). On the other hand, a large proportion of long-lived host mRNA are only slightly reduced in the late phase (BABICH et al. 1983). In addition to a transport block, there must, therefore, exist specific translational control mechanisms favoring translation of viral over cellular mRNA. In fact, some cellular mRNA such as β-tubulin escape the transport block and accumulate to near normal levels in the cytoplasm and are not translated (MOORE et al.1987). In addition, some drugs that are known to inhibit phosphoprotein phosphatases such as 2-aminopurine allow cellular protein synthesis without affecting the transport block of host mRNA during late infection (HUANG and SCHNEIDER 1990). The conclusion is that adenovirus, through a complex of two early proteins (E1B-55k and E4-34K), can establish a block for transport of newly synthesized cellular mRNA which is bypassed by viral mRNA. In addition, there must be specific translational control mechanisms operating to favor preferential accumulation of adenovirus mRNA in the polyribosomes of infected cells.

# 8 A Complex Mechanism for Virus-Mediated Translational Control

Three independent lines of evidence have provided clues for possible mechanisms of preferential translation of adenovirus mRNA late in infected cells. The results infer that more than one virus-encoded molecule is involved, suggesting that a complex mechanism involving both RNA and protein from the adenovirus genome in an integrated manner achieve the preferential translation of viral over cellular mRNA.

The first line involves a 100K protein originating in the L4 region. It is one of the first late polypeptides to be expressed and represents a very abundant viral protein (Oosterom-Dragon and Ginsberg 1980). It is associated with the hexon protein and required for folding of the hexon monomer into trimers in the cytoplasm (Cepko and Sharp 1983). It is also required for the production of viral particles, probably due to its fundamental role in hexon assembly. The 100K protein also appears to be associated with mRNA late in the infectious cycle (Lindberg and Sundquist 1974; Adam and Dreyfuss 1987). The dissociation constant approaches $10^{-8}$ and the bond is resistant to high salt concentrations. If the 100K protein is bound to cellular mRNA, as originally proposed (Lindberg and Sundquist 1974), it might reduce their translational efficiency or alternatively, if combined with viral mRNA as later indicated (Adam and Dreyfuss 1987), it may enhance viral mRNA translation. A temperature-sensitive (ts) mutant with a serine to proline substitution in the L4-100K protein (H5ts1; Hayes et al. 1990) shows a defect in late viral protein synthesis at the restrictive temperature. A characterization of the phenotype showed that the defect led to reduced translation of late viral mRNA, probably due a reduced efficiency of initiation. However, the efficient block in initiation of cellular mRNA was still present, suggesting that this protein is primarily involved in promoting translation of viral late mRNA and that other mechanisms must be involved in the inhibition of cellular mRNA translation. It has not yet been unequivocally established whether the L4-100K protein preferentially binds to viral messenger or cellular mRNA. It has been proposed that it might direct late viral mRNA to a specific translation compartment within the cell. The precise molecular mechanism mediated by the L4-100K protein during late adenovirus translation remains to be clarified; directly or indirectly, it appears, however, to be responsible for preferential adenovirus mRNA initiation.

Another viral gene that has been implicated in the viral translation control is the VA RNA I. This pol III product is synthesized in large amounts late in infection (Söderlund et al. 1976). Its involvement in adenovirus late translation was revealed by a mutant dl331, incapable of synthesizing VA RNA I (Thimmappaya et al. 1982). At late times in infection, this mutant showed inhibition in translation of viral mRNA which was not caused by defective viral DNA replication or mRNA accumulation. The translational defect affected initiation and was associated with initiation factor EIF2 (REICHEL et al. 1985; SCHNEIDER et al. 1985). Phosphorylation of EIF2 is mediated in animal cells by a protein kinase which is induced by interferon or viral infection and activated by low concentrations of double-stranded RNA. This kinase is called the dsRNA activated inhibitor (DAI) of protein synthesis because synthesis of an inactive form is induced by interferon, but activitated by dsRNA (HOVANESSIAN 1989). VA RNA I prevents the activation of DAI kinase by directly binding to the enzyme and blocking the entry of dsRNA (KATZE et al. 1987). This pathway blocks the antiviral response mediated by the DAI kinase during late adenovirus infection. An important body of evidence links this pathway to events in selective translation of late viral mRNAs. However, there is still a poor correlation between the phosphorylation state of EIF2 and the translation of viral and cellular mRNA, suggesting that these effects cannot fully account for all translational discrimination during late infection involving both preferential viral and inhibition of cellular translation (MATHEWS and SHENK 1991).

The third element implicated in adenovirus translational control are the three exons constituting the tripartite leader in the 5' noncoding region of all the mRNA originating from the MLP (LOGAN and SHENK 1984; BERKNER and SHARP 1985). In poliovirus-infected cells, protein synthesis from cellular mRNA is totally blocked. In contrast, adenovirus late viral mRNA translation is resistant to inhibition by superinfecting poliovirus (CASTRILLO and CARRASCO 1987). The basis for selective translation of poliovirus mRNA is the proteolytic degradation of a polypeptide that is a component of the initiation factor EIF4F, the cap-dependent RNA helicase (p220) which stimulates protein synthesis by unwinding the 5' end of the mRNA (SONENBERG 1987). The p220 is degraded during infection with polio and other picornaviruses. This degradation leads to preferential protein synthesis from picornavirus mRNA because they bind ribosomes internally and probably do not need the EIF4F to stimulate initiation (JANG et al. 1988). Translation in the absence of this factor is, therefore, unique to only a small number of mRNA such as the picornavirus and cellular mRNA that initiate at internal sites. It was, therefore, interesting when it was found that the adenovirus tripartite leader confers translation independence of the EIF4F activity in poliovirus-infected cells (DOLPH et al. 1988). Likewise, the tripartite leader directed efficient EIF4F-independent translation when attached to other mRNA expressed from transfected plasmids during poliovirus infection. The tripartite leader thus appears to confer translation independence for the initiation factor EIF4F.

Secondary structure analysis suggests that the tripartite leader probably adopts a confirmation which is not greatly influenced by the mRNA to which it is attached, suggesting that it may contain a structural element by itself which is E1F4F resistant. In accordance with a role for the tripartite leader, EIF4F activity is significantly reduced during late adenovirus infection, and 2-aminopurine, which prevents shutoff of cellular mRNA translation during late adenovirus infection, probably inhibits dephosphorylation of EIF4F without affecting adenovirus mRNA translation (HUANG and SCHNEIDER 1990, 1991). When the shutoff of host protein synthesis is prevented with 2-aminopurine, translation of cellular and several viral mRNA without a tripartite leader competes with those from the MLP.

In conclusion, adenovirus possesses several different mechanisms to achieve translational dominance and obviously three different entities within the adenovirus genome seem to affect the preferential synthesis of viral proteins at late times in the infectious cycle involving the L4-100K, the VA RNA I, and the tripartite leader in all viral mRNA from the MLP of the viral genome.

## 9 General Conclusion

The adenovirus system, which was introduced through the isolation of its first member in 1953 (Rowe et al. 1953; Hilleman and Werner 1954), has amply demonstrated its role as an extraordinarily useful model for gene expression in mammalian cells. By dissection of the pathways for specific viral gene expression, several cellular control mechanisms for transcription, RNA processing, mRNA transport, and translation have been elucidated.

During recent years, the adenovirus has become one of the preferred vectors for DNA transfer into mammalian cells, and a detailed characterization of these events might provide the tools required for gene therapy. The molecular approach in medicine today aims to correct many important genetic diseases by gene therapy focusing on introducing the right gene in the right cells. A viable alternative may be to achieve amplification of the product by providing the right DNA fragment in the right organ at the onset of the disease, which can easily be accomplished with the adenovirus gene delivery system. If this hypothesis proves correct, the adenovirus system may, in addition, become an archetype for gene delivery systems into several different human organs.

## References

Abmayr SM, Workman JL, Roeder RG (1988) The pseudorabies immediate early protein stimulates in vitro transcription by facilitating TFIID: protein interactions. Genes Dev 2: 542–553

Adam SA, Dreyfuss G (1987) Adenovirus proteins associated with mRNA and hnRNA in infected HeLa cells. J Virol 61: 3276–3283

Akusjärvi G, Philipson L, Pettersson U (1978) A protein kinase associated with adenovirus type 2. Virology 87: 276–286

Babich A, Feldman LT, Nevins JR, Darnell JE, Weinberger C (1983) Effect of adenovirus on metabolism of specific host mRNAs: transport control and specific translational discrimination. Mol Cell Biol 3: 1212–1221

Babiss LE, Ginsberg HS (1984) Adenovirus type 5 early region 1b gene product is required for efficient shutoff of host protein synthesis. J Virol 50: 202–212

Benjamin TL (1966) Virus-specific RNA in cells productively infected or transformed by polyoma virus. J Mol Biol 16: 359–373

Berget SM, Moore C, Sharp PA (1977) Spliced segments of the 5' terminus of adenovirus 2 late mRNA. Proc Natl Acad Sci USA 74: 3171–3175

Berk AJ (1986) Adenovirus promoters and E1A transactivation. Annu Rev Genet 20: 45–79

Berk AJ, Lee F, Harrison T, Williams J, Sharp PA (1979) A pre-early adenovirus 5 gene product regulates synthesis of early viral messenger RNAs. Cell 17: 935–944

Berkner KL, Sharp PA (1985) Effect of the tripartite leader on synthesis of a non-viral protein in an adenovirus 5 recombinant. Nucleic Acids Res 13: 841–857
Bernards R, Schackleford GM, Gerber MR, Horowitz JM, Friend SH, Schartl M, Bogenmann E, Rapaport JM, McGee T, Dryja TP et al. (1989) Structure and expression of the murine retinoblastoma gene and characterization of its encoded protein. Proc Natl Acad Sci USA 86: 6474–6478
Beutler B, Cerami A (1989) The biology of cachectin/TNF-a primary mediator of the host response. Annu Rev Immunol 7: 625–655
Blumenthal R, Seth P, Willingham MC, Pastan I (1986) pH-Dependent lysis of liposomes by adenovirus. Biochemistry 25: 2231–2237
Broker TR, Chow LT, Dunn AR, Gelinas RE, Hassell JA, Klessig DF, Lewis JB, Roberts RJ, Zain BS (1977) Adenovirus-2-messengers—an example of baroque molecular architecture. Cold Spring Harbor Symp Quant Biol 42: 531–553
Burgert HG, Kvist S (1987) The E3/19K protein of adenovirus type 2 binds to the domains of histocompatibility antigens required for CTL recognition. EMBO J 6: 2019–2026
Carlin CR, Tollefson AE, Brady HA, Hoffman BL, Wold WSM (1989) Epidermal growth factor receptor is downregulated by a 10,400 MW protein encoded by the E3 region of adenovirus. Cell 57: 135–144
Castrillo JL, Carrasco L (1987) Adenovirus late protein synthesis is resistant to the inhibition of translation induced by poliovirus. J Biol Chem 262: 7328–7334
Cepko CL, Sharp PA (1983) Analysis of Ad5 hexon and 100K ts mutants using conformation-specific monoclonal antibodies. Virology 129: 137–154
Chambon P (1977) Summary: the molecular biology of the eukaryotic genome is coming of age. Cold Spring Harbor Symp Quant Biol 42: 1209–1234
Chardonnet Y, Dales S (1970) Early events in the interaction of adenoviruses with HeLa cells I. Penetration of type 5 and intracellular release of the DNA genome. Virology 40: 462–477
Chen MJ, Holskin B, Strickler J, Gorniak J, Clark MA, Johnson PJ, Mitcho M, Shalloway D (1987) Induction by E1A oncogene expression of cellular susceptibility to lysis by TNF. Nature 330: 581–583
Chow LT, Gelinas RE, Broker TR, Roberts RJ (1977) An amazing sequence arrangement at the 5' ends of adenovirus 2 messenger RNA. Cell 12: 1–8
Cotten M, Wagner E, Zatloukal K, Birnstiel ML (1993) Chicken adenovirus (CELO virus) particles augment receptor-mediated DNA delivery to mammalian cells and yield exceptional levels of stable transformants. J Virol 67: 3777–3785
Crescenzi M, Fleming TP, Lassar AB, Weintraub H, Aaronson SA (1990) MyoD induces growth arrest independent of differentiation in normal and transformed cells. Proc Natl Acad Sci USA 87: 8442–8446
Darnell JE, Wall R, Tushinski RJ (1971a) An adenylic acid-rich sequence in messenger RNA of HeLa cells and its possible relationship to reiterated sites in DNA. Proc Natl Acad Sci USA 68: 1321–1325
Darnell JE, Philipson L, Wall R, Adesnik M (1971b) Processing of HnRNA to mRNA in HeLa cells: the posttranscriptional addition of polyadenylic acid sequences. Science 174: 507–510
De Caprio JA, Ludlow JW, Figge J, Shew JY, Huang CM, Lee WH, Marsilio E, Paucha E, Livingston DM (1988) SV40 large tumor antigen forms a specific complex with the product of the retinoblastoma susceptibility gene. Cell 54: 275–283
Defer C, Belin M-T, Caillet-Boudin M-L, Boulanger P (1990) Human adenovirus-host interactions: comparative study with members of subgroups B and C. J Virol 64 : 3661–3673
Dolph PJ, Racaniello V, Villamarin A, Palladino F, Schneider RJ (1988) The adenovirus tripartite leader may eliminate the requirement for cap-binding protein complex during translation initiation. J Virol 62: 2059–2066
Edmonds M, Vaughan MH, Nakazato H (1971) Polyadenylic acid sequences in the heterogenous nuclear RNA and rapidly-labeled polyribosomal RNA of HeLa cells: possible evidence for a precursor relationship. Proc Natl Acad Sci USA 68: 1336–1340
Evans RM, Fraser N, Ziff E, Weber J, Wilson M, Darnell JE (1977) The initiation sites for RNA transcription in Ad2 DNA. Cell 12: 733–739
Fattaey AR, Harlow E, Helin K (1993) Independent regions of adenovirus E1A are required for binding to and dissociation of E2F-protein complexes. Mol Cell Biol 13: 7267–7277
Flint SJ (1984) Adenovirus cytopathology Comp Virol 19: 297–358
Flint SJ, Shenk T (1989) Adenovirus E1A protein paradigm viral transactivator. Annu Rev Genet 23: 141–161
Flint SJ, Beltz GA, Linzer DIH (1983) Synthesis and processing of simian virus 40-specific RNA in adenovirus-infected simian virus 40-transformed human cell. J Mol Biol 167: 335–359
Flint SJ, Plumb MA, Yang UC, Stein S, Stein JL (1984) Effect of adenovirus infection on expression of human histone genes. Mol Cell Biol 4: 1363–1371

Fried M, Prives C (1986) The biology of simian virus 40 and polyomavirus. Cancer Cells 4: 1–16
Friend SH, Bernards R, Rogelj S, Weinberg RA, Rapaport JM, Albert DM, Dryja TP (1986) A human DNA segment with properties of the gene that predisposes to retinoblastoma and osteosarcoma. Nature 312: 643–646
Gaynor RB, Feldman LT, Berk AJ (1985) Viral immediate early proteins activate transcription of class III genes. Science 230: 447–450
Gelinas E, Roberts RJ (1977) Predominant 5'-undecanucleotide in adenovirus 2 late messenger RNAs. Cell 11: 533–544
Ginsberg HS (1984) A overview. In: Ginberg Hs (ed) The adenoviruses 1984. Plenum, New York pp 1–4
Ginsberg HS, Lundholm-Beauchamp U, Horstwood RL, Pernis B, Wold WSM, Chanock RM, Prince GA (1989) Role of early region 3 (E3) in pathogenesis of adenovirus disease. Proc Natl Acad Sci USA 86: 3823–3827
Goldberg S, Weber J, Darnell JE Jr (1977) The definition of a large viral transcription unit late in Ad2 infection of HeLa cells: mapping by effects of ultraviolet irradiation. Cell 10: 617–621
Gooding LR, Elmore LW, Tollefson AE, Brady HA, Wold WSM (1988) A 14,700 MW protein from the E3 region of adenovirus inhibits cytolysis by tumor necrosis factor. Cell 53: 341–346
Gooding LR, Sofola IO, Tollefson AE, Duerksen-Hughes P, Wold WSM (1990) The adenovirus E3-14.7K protein is a general inhibitor of tumor necrosis factor-mediated cytolysis. J Immunol 145: 3080–3086
Gooding LR, Aquino L, Duerksen-Hughes PJ, Day D, Horton TM, Yei S, Wold WSM (1991) The E1B 19,000-molecular-weight protein of group C adenoviruses prevents tumor necrosis factor cytolysis of human cells but not of mouse cells. J Virol 65: 3083–3094
Graham FL, Van der Eb AJ (1973) A new technique of the assay of infectivity of human adenovirus DNA. Virology 52: 456–467
Greber UF, Willets M, Webster P, Helenius A (1993) Stepwise dismantling of adenovirus 2 during entry into cells. Cell 75: 477–486
Green M (1970) Oncogenic viruses. Annu Rev Biochem 39: 701–756
Green M, Loewenstein PM, Pusztai R, Symington JS (1988) An adenovirus E1A protein domain activates transcription in vivo and in vitro in the absence of protein synthesis. Cell 53: 921–926
Gu W, Schneider JW, Condorelli G, Kaushal S, Mahdavi V, Nadal-Ginard B (1993) Interaction of myogenic factors and retinoblastoma protein mediates muscle cell commitment and differentiation. Cell 72: 309–324
Halbert DN, Cutt JR, Shenk T (1985) Adenovirus early region 4 encodes functions required for efficient DNA replication, late gene expression and host cell shutoff. J Virol 56: 250–257
Hardy S, Engel DA, Shenk T (1989) An adenovirus early region 4 gene product is required for induction of the infection-specific form of cellular E2F activity. Genes Dev 3: 1062–1074
Harris H (1988) The analysis of malignancy by cell fusion: the position in 1988. Cancer Res 48: 3302–3306
Hayes BW, Telling GC, Myat MM, Williams JF, Flint SJ (1990) The adenovirus L4 100-kilodalton protein is necessary for efficient translation of viral late mRNA species. J Virol 64: 2732–2742
Helin K, Lees JA, Vidal M, Dyson N, Harlow E, Fattaey A (1992) A cDNA encoding a pRB-binding protein with properties of the transcription factor E2F. Cell 70: 337–350
Helin K, Harlow E, Fattaey AR (1993) Inhibition of E2F-1 transactivation by direct binding of the retinoblastoma protein. Mol Cell Biol 13: 6501–6508
Hilleman MR, Werner JH (1954) Recovery of new agents from patients with acute respiratory illness. Proc Soc Exp Biol Med 85: 183–188
Hoeffler WK, Roeder RG (1985) Enhancement of RNA polymerase III transcription by the E1A gene product of adenovirus. Cell 41: 955–963
Horwitz MS (1990) Adenoviridae and their replication. In: Fields BN, Knipe DM (eds) Virology, Vol 1, 2nd edn. Raven, New York, pp 1679–1721
Hovanessian AG (1989) The double stranded RNA-activated protein kinase induced by interferon: dsRNA-PK. J Interferon Res 6: 641–647
Huang JT, Schneider RJ (1990) Adenovirus inhibition of cellular protein synthesis is prevented by the drug 2-aminopurine. Proc Natl Acad Sci USA 87: 7115–7119
Huang JT, Schneider RJ (1991) Adenovirus inhibition of cellular protein synthesis involves inactivation of cap-binding protein. Cell 65: 271–280
Huang S, Wand NP, Tseng BY, Lee WH, Lee EH (1990) Two distinct and frequently mutated regions of retinoblastoma protein are required for binding to SV40 T antigen EMBO J 9: 1815–1822
Huber HE, Edwards G, Goodhart PJ, Patrick DR, Huang PS, Ivey-Hoyle M, Barnett SF, Oliff A, Heimbrook DC (1993) Transcription factor E2F binds DNA as a heterodimer. Proc Natl Acad Sci USA 90: 3525–3529

Jang SK, Krausslich HG, Nicklin MJH, Duke GM, Palmenberg AC, Wimmer E (1988) A segment of the 5' nontranslated region of encephalomyocarditis virus RNA directs internal entry of ribosomes during in vitro translation. J Virol 62: 2636–2643

Jones N, Shenk T (1979) An adenovirus type 5 early gene function regulates expression of other early viral genes. Proc Natl Acad Sci USA 76: 3665–3669

Kates J (1970) Transcription of the vaccinia virus genome and the occurence of polyriboadenylic sequences in messenger RNA. Cold Spring Harbor Symp Quant Biol 35: 743–752

Katze MG, DeCorato D, Safer B, Galabru J, Hovanessian AG (1987) Adenovirus VAI RNA complexes with the 68 000 Mr protein kinase to regulate its autophosphorylation and activity. EMBO J 6: 689–697

Knudson AG (1971) Mutation and cancer: statistical study of retinoblastoma. Proc Natl Acad Sci USA 68: 820–823

Kovesdi I, Reichel R, Nevins JR (1986) Identification of a cellular transcription factor involved in E1A transactivation. Cell 45: 219–228

Krug RM, Broni BA, Bouloy M (1979) Are the 5' ends of influenza viral mRNAs synthesized in vivo donated by host mRNAs? Cell 18: 329–334

Kvist S, Östberg L, Persson H, Philipson L, Peterson PA (1978) Molecular association between transplantation antigens and a cell surface antigen in an adenovirus-transformed cell line. Proc Natl Acad Sci USA 75: 5674–5678

Lee SY, Mendecki J, Brawerman G (1971) A polynucleotide segment rich in adenylic acid in the rapidly-labeled polyribosomal RNA component of mouse sarcoma 180 ascites cells. Proc Natl Acad Sci USA 68: 1331–1335

Lindberg U, Darnell JE (1970) SV40-specific RNA in the nucleus and polyribosomes of transformed cells. Proc Natl Acad Sci USA 65: 1089–1096

Lindberg U, Sundquist B (1974) Isolation of messenger ribonucleoproteins from mammalian cells. J Mol Biol 86: 451–468

Lodish HF, Porter M (1980) Translational control of protein synthesis after infection by vesicular stomatitis virus. J Virol 36: 719–733

Logan J, Shenk T (1984) Adenovirus tripartite leader sequence enhances translation of mRNAs late after infection. Proc Natl Acad Sci USA 81: 3655–3659

Lonberg-Holm K, Philipson L (1969) Early events of virus-cell interaction in an adenovirus system. J Virol 4: 323–338

Longberg-Holm K, Crowell RL, Philipson L (1976) Unrelated animal viruses share receptors. Nature 259: 679681

Mathews MB, Shenk T (1991) Adenovirus virus-associated RNA and translation control. J Virol 65: 5657–5662

McCune JM (1991) HIV-1: the infective process in vivo. Cell 64: 351–363

McGee TL, Yandell DW, Dryja TP (1989) Structure and partial genomic sequence of the human retinoblastoma susceptibility gene. Gene 80 : 119–123

Meyers RE, Sharp PA (1993) TATA-binding protein and associated factors in polymerase II and polymerase III transcription. Mol Cell Biol 13: 7953–7960

Michael N, Spector D, Mavromara-Nazos P (1988) The DNA-binding properties of the major regulatory protein α4 of herpes simplex viruses. Science 239: 1531–1534

Miles BD, Luftig RB, Weatherbee JA, Weihing RR, Weber J (1980) Quantitation of the interaction between adenovirus types 2 and 5 and microtubules inside infected cells. Virology 105: 265–269

Moore M, Schaack J, Baim SB, Morimoto RI, Shenk T (1987) Induced heat shock mRNAs escape the nucleocytoplasmic transport block in adenovirus-infected HeLa cells. Mol Cell Biol 7: 4505–4512

Moran E, Mathews MB (1987) Multiple functional domains in the adenovirus E1A gene. Cell 48: 177–178

Moss B, Koczot F (1976) Sequence of methylated nucleotides at the 5'-terminus of adenovirus-specific RNA. J Virol 17: 385–392

Munger K, Werness BA, Dyson N, Phelps WC, Harlow E, Howley PM (1989) Complex formation of human papillomavirus E7 proteins with the retinoblastoma tumor suppressor gene produce. EMBO J 8: 4099

Nevins JR (1981) Mechanism of activation of early viral transcription by the adenovirus E1A gene product. Cell 26: 213–220

Nishioka Y, Silverstein S (1977) Degradation of cellular mRNA during infection by herpes simplex virus. Proc Natl Acad Sci USA 74: 2370–2374

Olson EN (1990) MyoD family: a paradigm for development? Genes Dev 4: 1454–1461

Oosterom-Dragon EA, Ginsberg HS (1980) Purification and preliminary immunological characterization of the type 5 adenovirus, nonstructural 100,000-dalton protein. J Virol 33: 1203–1207

Pääbo S, Severinsson L, Andersson M, Martens I, Nilsson T, Peterson PA (1989) Adenovirus proteins and MHC expression. Adv Cancer Res 52: 151–163
Penman S, Scherrer K, Becker Y, Darnell JE (1963) Polyribosomes in normal and poliovirus-infected HeLa cells and their relationship to messenger-RNA. Proc Natl Acad Sci USA 49: 654–662
Pereira HG, Huebner RJ, Ginsberg HS, Van der Veen J (1963) A short description of the adenovirus group. Virology 20: 613–620
Perry RP, Bard E, Hames BD, Kelley DE, Schibler U (1976) The relationship between hnRNA and mRNA. Prog Nucleic Acid Res Mol Biol 19: 275–291
Peverali FA, Ramquist T, Saffrich R, Pepperkok R, Barone MV, Philipson L (1994) Regulation of G1 progression by E2A and Id HLH-proteins. EMBO J 13: 4291–4301
Philipson L, Lonberg-Holm K, Pettersson U (1968) Virus-receptor interaction in an adenovirus system. J Virol 2: 1064–1075
Philipson L, Wall R, Glickman R, Darnell JE Jr (1971) Addition of polyadenylic acid sequences to virus specific RNA during adenovirus replication. Proc Natl Acad Sci USA 68: 2806–2809
Philipson L, Pettersson U, Lindberg U, Tibbetts C, Vennström B, Persson T (1974) RNA synthesis and processing in adenovirus infected cells. Cold Spring Harbor Symp Quant Biol 39: 447–456
Pilder S, Moore M, Logan J, Shenk T (1986) E1B-55K transforming polypeptide modulates transport or cytoplasmic stabilization of viral and host cell mRNAs. Mol Cell Biol 6: 470–476
Price R, Penman S (1972) Transcription of the adenovirus genome by an amanitine-sensitive ribonucleic acid polymerase in HeLa cells. J Virol 9: 621–626
Reichel PA, Merrick WC, Siekierka J, Mathews MB (1985) Regulation of a protein synthesis initiation factor by adenovirus virus-associated RNA. Nature 313: 196–200
Rice AP, Roberts BE (1983) Vaccinia virus induces cellular mRNA degradation. J Virol 47: 529–539
Richter JD, Slavicek JM, Schneider JF, Jones NC (1988) Heterogeneity of adenovirus type 5 E1A proteins: multiple serine-phosphorylations induce slow-migrating electrophoretic variants but do not affect E1A-induced transcriptional activation or transformation. J Virol 62: 1947–1955
Rosenfeld MA, Siegfried W, Yoshimura K, Yoneyama K, Fukayama M, Stier LE, Paakko PK, Gilardi P, Stradford-Perricaudet LD, Perricaudet M, Jallat S, Pavirani A, Lecocq JP, Crystal RG (1991) Adenovirus-mediated transfer of a recombinant $\alpha$1-antitrypsin gene to the lung epithelium in vivo. Science 252: 431–434
Rowe WP, Huebner RJ, Gillmore LK, Parrott RH, Ward TG (1953) Isolation of a cytogenic agent from human adenoids undergoing spontaneous degeneration in tissue cultures. Proc Soc Exp Biol Med 84: 570–573
Salditt-Georgieff M, Jelinek W, Darnell JE (1976) Methyl labeling of HeLa Cell hnRNA: a comparison with mRNA. Cell 7: 227–237
Sauerbier W (1976) UV damage at the transcriptional level. Adv Radiat Biol 6: 49–106
Scherrer K, Darnell JE (1962) Sedimentation characteristics of rapidly labelled RNA from HeLa cells. Biochem Biophys Res Commun 7: 486–490
Scherrer K, Latham H, Darnell JE (1963) Demonstration of an unstable RNA and of a precursor to ribosomal RNA in HeLa cells. Proc Natl Acad Sci USA 49 : 240
Schneider RJ, Safer B, Munemitsu S, Samuel CE, Shenk T (1985) Adenovirus VAI RNA prevents phosphorylation of the eukaryotic initiation factor 2 alpha subunit subsequent to infection. Proc Natl Acad Sci USA 82: 4321–4325
Sharp PA, Gallimore PH, Flint SJ (1974) Mapping of adenovirus 2 RNA sequences in lytically infected cells and transformed cell lines. Cold Spring Harbor Symp Quant Biol 39: 457–474
Shatkin AJ (1976) Capping of eukaryotic mRNAs. Cell 9: 645–654
Söderlund H, Pettersson U, Vennström B, Philipson L, Mathews MB (1976) A new species of virus-coded low molecular weight RNA from cells infected with adenovirus type 2. Cell 7: 585–593
Sonenberg N (1987) Regulation of translation by poliovirus. Adv Virus Res 33: 175–204
Sorrentino V, Pepperkok R, Davis RL, Ansorge W, Philipson L (1990) Cell proliferation inhibited by MyoD1 independently of myogenic differentiation. Nature 345: 813–815
Spalholz BA, Byrne JC, Howley PM (1988) Evidence for cooperativity between E2 binding sites in E2 transfiguration of bovine papillomavirus type 1. J Virol 62: 3143–3150
Tate VE, Philipson L (1979) Parental adenovirus DNA accumulates in nucleosome-like structure in infected cells. Nucleic Acids Res 6: 2769–2785
Thimmappaya B, Weinberger C, Schneider RJ, Shenk T (1982) Adenovirus VAI RNA is required for efficient translation of viral mRNAs at late times after infection. Cell 3: 543–551
Townsend A, Ohlen C, Bastin J, Ljunggren HG, Foster L, Karre K (1989) Association of class 1 major histocompatibility heavy and light chains induced by viral peptides. Nature 340: 443–448

Vaessen RTMJ, Houweling A, Israel A, Kourilsky P, Van der Eb A (1986) Adenovirus E1A-mediated replication of class I MHC expression. EMBO J 5: 335–341

Valentine RC, Pereira HG (1965) Antigens and structure of the adenovirus. J Mol Biol 13: 13–20

Varga MJ, Weibull C, Everitt E (1991) Infectious entry pathway of adenovirus type 2. J Virol 65: 6061–6070

Vennström B, Philipson L (1977) Fidelity of adenovirus RNA transcription in isolated HeLa cell nuclei. J Virol 22: 290–299

Wall R, Philipson L, Darnell JE (1972) Processing of adenovirus specific nuclear RNA during virus replication. Virology 50: 27–34

Weber J (1976) Genetic analysis of adenovirus type 2.III. Temperature sensitivity of processing of viral proteins. J Virol 17: 462–471

Weber J, Jelinek W, Darnell JE (1977) The definition of a large viral transcription unit late in Ad2 infection of HeLa cells: mapping of nascent RNA molecules in isolated nuclei. Cell 10: 611–616

Weinmann R, Brendler TG, Raskas JH, Roeder RG (1976) Low molecular weight viral RNAs transcribed by RNA polymerase III during adenovirus 2 infection. Cell 7: 557–566

Weintraub H, Davis R, Tapscott S, Thayer M, Krause M, Benezra R, Blackwell TK, Turner D, Rupp R, Hollenberg S, Zhuang Y, Lassar A (1991) The myoD gene family: nodal point during specification of the muscle cell lineage. Science 251: 761–766

Whyte P, Buchkovich KJ, Horowitz JM, Friend SH, Raybuck M, Weinberg RA, Harlow (1988) Association between an oncogene and an anti-oncogene: the adenovirus E1A proteins bind to the retinoblastoma gene product. Nature 334: 124–129

Whyte P, Williamson NM, Harlow E (1989) Cellular targets for transformation by the adenovirus E1A proteins. Cell 56: 67–75

Wickham TJ, Mathias P, Cherish DA, Nemerow GR (1933) Integrins $\alpha_v\beta3$ and $\alpha_v\beta5$ promote adenovirus internalization but not virus attachment. Cell 73: 309–319

Williams JL, Garcia J, Harrich D, Pearson L, Wu F, Gaynor R (1990) Lymphoid specific gene expression of the adenovirus early region 3 promoter is mediated by NF-κB binding motifs. EMBO J 9: 4435–4442

Wold WSM, Gooding LR (1991) Region E3 of adenovirus: a cassette of genes involved in host immunosurveillance and virus-cell interactions. Virology 184: 1–8

Ziff EB (1980) Transcription and RNA processing by the DNA tumour viruses. Nature 287: 491–499

# Adenovirus Structure by X-ray Crystallography and Electron Microscopy

P.L. Stewart[1,2] and R.M. Burnett[1]



## 1 Introduction

In the 40 years since the discovery of adenovirus, a wide range of biophysical and biochemical techniques have been applied to understanding the structure of the virion (Ginsberg 1979; Philipson 1983). Its size and complexity have presented a particular challenge, and this chapter describes a novel combination of two imaging techniques. Interpretation of the resultant images has relied on the wealth of earlier data. The aim has been to use X-ray crystallography to determine atomic structures of individual protein components and electron microscopy (EM) to determine the low-resolution structure of the intact complex. By accurately positioning the atomic structures within the reconstructed EM density and performing difference imaging, high-resolution structural information on the $150 \times 10^6$-Da complex has been obtained. In comparison with other mammalian viruses for which structures have been determined by X-ray crystallography alone, the adenovirus particle is much larger, roughly 18 times the mass of the picornaviruses (Rossmann et al. 1985; Hogle et al. 1985) and six times the mass of

[1] The Wistar Institute, 3601 Spruce Street, Philadelphia, PA 19104, USA
[2] Present address: Department of Molecular and Medical Pharmacology, UCLA School of Medicine, 10833 Le Conte Avenue, Los Angeles, CA 90024-6948, USA

the polyoma virus SV40 (LIDDINGTON et al. 1991). Thus far, the atomic structure of only one complete adenovirus capsid component, hexon, has been determined (ROBERTS et al. 1986; ATHAPPILLY et al. 1994). An early low-resolution structure of hexon (BURNETT et al. 1985) was used to develop a model for hexon packing in the adenovirus capsid (BURNETT 1985). In addition to hexon, there are six other viral proteins in the capsid and four viral proteins associated with the double-stranded DNA genome (Fig. 1). The three-dimensional structure of the entire particle has now been determined by cryoelectron microscopy together with three-dimensional image reconstruction (STEWART et al. 1991).

The combined crystallographic and EM approach was first applied to a viral capsid fragment called the group-of-nine hexons (GON), which are formed under mild dissociation conditions (FURCINITTI et al. 1989). Biochemical analysis showed the GON to be composed of nine hexons plus 12 copies of polypeptide IX, a 15-kDa component known to stabilize the hexon capsid. Scanning transmission electron microscope (STEM) images of the GON were averaged and enhanced with image processing. A similar two-dimensional array was built from a projection of the crystallographic hexon, and subtraction of the crystallographic array from the STEM image revealed the locations of polypeptide IX in two dimensions. This difference imaging technique has recently been extended to three dimensions in order to visualize all of the protein components in the adenovirus capsid (STEWART et al. 1993). As the distinctive triangular shape of hexon is clearly

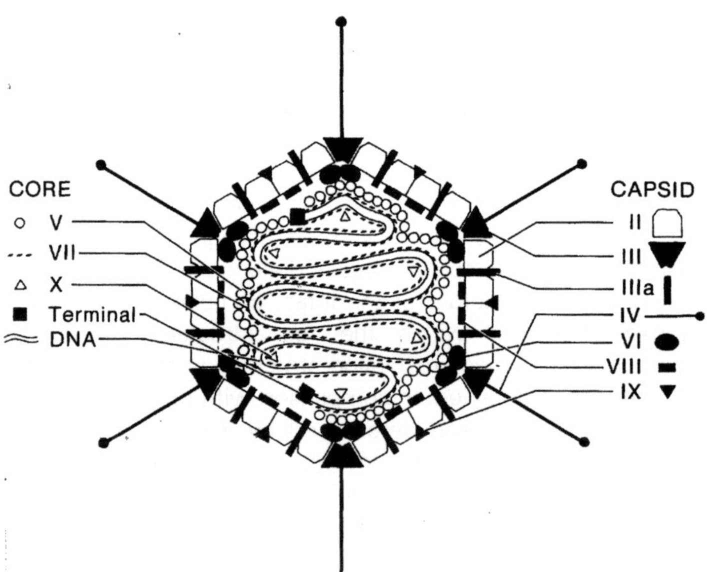


**Fig. 1.** A stylized section of the adenovirus particle summarizing current structural knowledge of the polypeptide components and the viral DNA. No real section of the icosahedral virion would contain all these components. This diagram is based on an earlier drawing (GINSBERG 1979) and reprinted from STEWART and BURNETT (1993)

visible on the surface of the EM image reconstruction (STEWART et al. 1991), the crystallographic hexon could be positioned accurately within the EM density. A complete capsid was then built from 240 crystallographic hexons and subtracted from the EM image reconstruction. The resulting three-dimensional difference map was analyzed using the known molecular weights and copy numbers of the various viral components (VAN OOSTRUM and BURNETT 1985).

## 2 X-Ray Crystallographic Structure of Hexon

The crystallographic structure of hexon from type 2 adenovirus (ROBERTS et al. 1986) has recently been refined to 2.5 Å resolution (ATHAPPILLY et al. 1994; R. MURALI and R.M. BURNETT, unpublished results). The hexon monomer has 967 amino acid residues that form two eight-stranded β-barrels and three extended loops. The loops from different subunits intertwine in the hexon trimer and are exposed on the surface of the virus. In the trimer, the six β-barrels, two from each monomer, sit at the corners of a hexagon to generate a pseudohexagonal base. This shape is convenient for close packing of the hexon capsomers on the surface of the virus. In the refined structure, three segments in surface loops and the N terminus are not defined, as they are too mobile to be observed by crystallography. The first residue of the N terminus that is visible lies on the part of the molecule that forms the inner capsid surface of the virion. Thus, the N-terminal stretch most likely interacts with the viral core DNA and protein. The hexon surface loops display the highest variability between adenovirus serotypes and contain most of the type-specific epitopes (TOOGOOD et al. 1989). Two neutralizing type-specific epitopes have been localized to mobile regions at the tops of two loops (TOOGOOD et al. 1992).

## 3 Cryoelectron Microscopy of the Intact Virion

Cryoelectron microscopy is an emerging structural technique (ADRIAN et al. 1984; DUBOCHET et al. 1985) that has been applied to numerous icosahedral viruses (OLSON et al. 1993; PRASAD et al. 1993; YEAGER et al. 1994). In this increasingly important technique, the sample is rapidly frozen in a cryogen and then kept at liquid nitrogen temperature during image collection. The rapid freezing prevents formation of crystalline ice and instead traps the water in a frozen amorphous state that preserves the biological specimen in conditions very similar to its normal aqueous environment. A low dose of electrons must be used to avoid vaporizing the frozen water. The main advantage of this technique is that the images directly reveal the density of any protein, DNA, or lipid material in the unstained sample. This contrasts with the conventional method, in which the presence of biological material must be inferred from a coating of an electron-dense metal stain. Combined with icosahedral reconstruction methods (CROWTHER 1971; FULLER 1987), cryoelectron microscopy has proven to be a powerful, albeit low-resolution, structural method for determining the structures of large viruses. Other macromolecular complexes have also been studied by this technique and the highest-resolution image reconstruction published so far is of the nicotinic acetylcholine receptor at 9 Å resolution (UNWIN 1993).

A typical cryoelectron micrograph of a field of type 2 adenovirus particles is shown in Fig. 2. After digitizing the film and determining the orientations of numerous viral particles, a three-dimensional image reconstruction was calculated and full icosahedral symmetry applied. The resulting image reconstruction of adenovirus from images of 29 different particles is shown in Fig. 3 (STEWART et al. 1991). Only portions of the structure which are truly icosahedral are faithfully reproduced in the reconstruction. Thus, the icosahedral protein capsid is well resolved in the reconstruction. The internal nucleoprotein core is less well

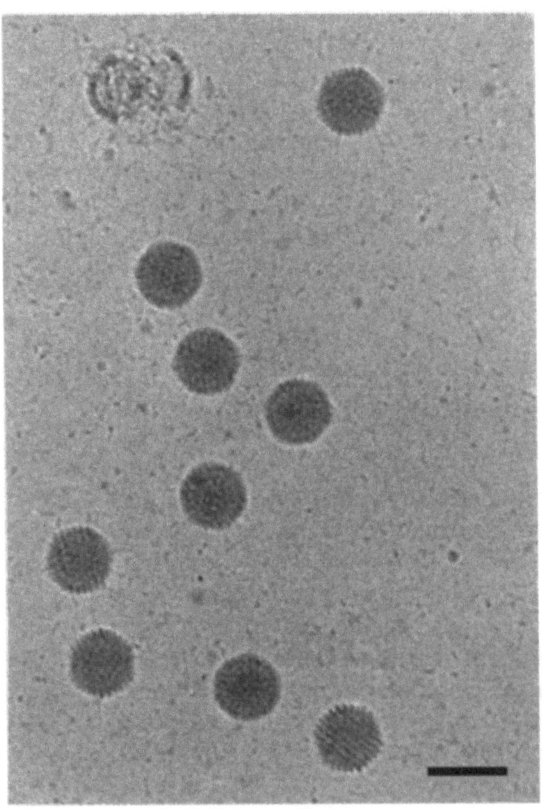

**Fig. 2.** A cryoelectron micrograph of human type 2 adenovirus. The virus particles are trapped in random orientations in a layer of amorphous ice stretched across a holey carbon grid. The image appears noisy because a low dose of electrons must be used to prevent damaging the unstained biological material. *Bar*, 1000 Å. Reprinted from STEWART et al. (1991)

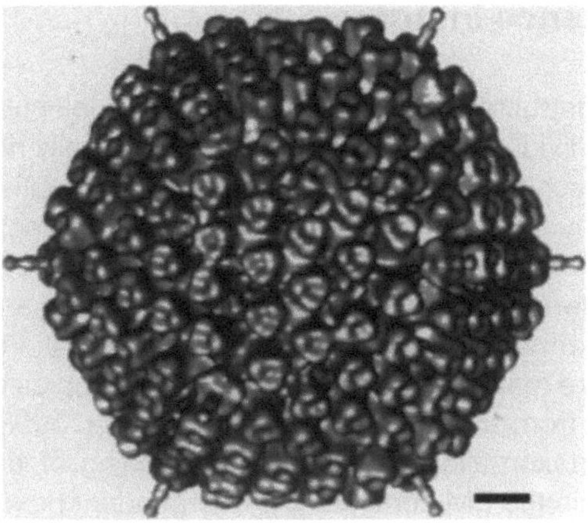

**Fig. 3.** The three-dimensional image reconstruction of the intact adenovirus particle viewed along an icosahedral threefold axis. Only a short portion of the protruding fiber was properly reconstructed because the fibers are bent and thus they do not completely conform to icosahedral symmetry. *Bar*, 100 Å. Reprinted from STEWART and BURNETT (1993)

ordered and so appears smeared in the reconstruction. The long, thin fibers protruding from the capsid, which are characteristic of adenovirus, are not perfectly straight and thus only a short portion is observed in the reconstruction. However, the entire length of the fiber is shown modeled in Fig. 4. During viral infection, the knob at the end of the fiber attaches to an unidentified cellular receptor and begins the infective process (PHILIPSON et al. 1968; DEFER et al. 1990). Internalization is mediated through a secondary integrin receptor that recognizes an Arg-Gly-Asp (RGD) peptide sequence on the penton base (WICKHAM et al. 1993; WHITE 1993). The crystal structure of the knob was recently determined and is described in the chapter by XIA et al. (this volume).

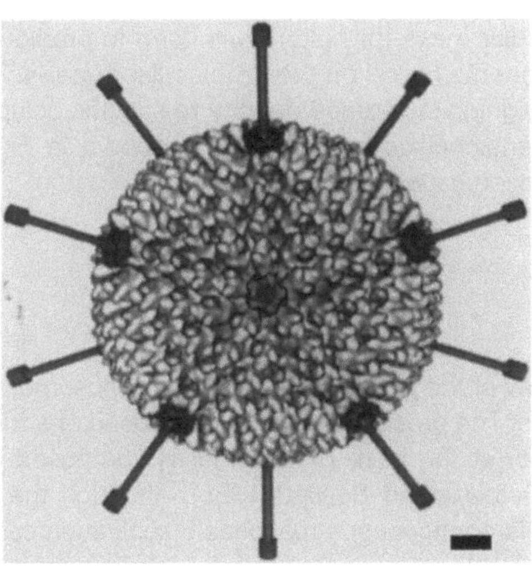

**Fig. 4.** The three-dimensional image reconstruction viewed along an icosahedral fivefold axis. The penton complex, consisting of the penton base in the capsid and the protruding fiber, is shown in *dark gray*. The shaft of the fiber (approximately 300 Å) and the knob at its end could not be reconstructed and so are modeled (WHITE 1993). *Bar*, 100 Å

## 4 Three-Dimensional Difference Imaging

It is often difficult to interpret reconstructed EM density of a macromolecular complex, as the boundaries of the individual protein components are usually not distinguishable. Several assignment schemes have been developed and applied to virus structures. One approach involves identifying specific protein components by labeling with Fab fragments from monoclonal antibodies (PRASAD et al. 1990). Another approach involves determining structures of subviral particles and subtracting them from the structure of the intact virion (YEAGER et al. 1994). For adenovirus, protein assignments were made after subtracting multiple copies of the crystallographic hexon structure from the EM density (STEWART et al. 1993). The distinctive triangular shape of hexon allowed accurate positioning of the crystallographic hexon structure within the EM density. Four independent hexon positions within the icosahedral asymmetric unit were refined using a cross-correlation coefficient between the X-ray and EM densities. Then 60-fold icosahedral symmetry was applied to generate coordinates for all 240 hexons in the viral capsid, corresponding to over 5 million nonhydrogen atoms (Fig. 5). As hexon is the major capsid protein, the resulting difference map showed, for the most part, density well separated from the other capsid proteins.

After adjusting the resolution of the crystallographic structure to the lower resolution of the EM reconstruction and then matching the mean and standard deviations of the calculated capsid and the EM density, a simple subtraction yielded a three-dimensional difference map (STEWART et al. 1993). Analysis of the difference map allowed assignment of density for six of the seven capsid components. In addition, EM density was observed for the mobile residues at the top of the hexon monomers. This observation illustrates the idea that the same amount of disorder that renders electron density invisible by X-ray crystallography will merely cause some density smearing in a lower-resolution EM reconstruction. All of the adenovirus structural proteins are listed in Table 1 along with their molecular masses and copy numbers (VAN OOSTRUM and BURNETT 1985). The observed ratio of volume to molecular mass for hexon was used to predict the volumes of the other capsid components based on their molecular masses. This information was quite useful in assigning difference density to specific polypeptides. A portion of the three-dimensional difference map is shown in Fig. 6, superimposed on the calculated hexon capsid.

## 5 Penton Base and Fiber

The difference map gave a clear view of the penton, the secondary capsomer that is located on each of the 12 vertices. The protruding fiber is composed of a trimer of polypeptide IV, while the density at the base of the fiber is composed of a pentamer of polypeptide III (VAN OOSTRUM and BURNETT 1985). Although the EM density from the two polypeptides is contiguous, a reasonable separation can be

$$TRI = \begin{matrix} 0.000 & 0.000 & 1.000 \\ 1.000 & 0.000 & 0.000 \\ 0.000 & 1.000 & 0.000 \end{matrix} \qquad \begin{matrix} M_2 = TRI \times M_1 \\ M_3 = TRI \times M_2 \end{matrix}$$

$$Hx = M_1 + M_2 + M_3$$

a

$$R_1 = \begin{matrix} 0.752 & -0.023 & -0.659 \\ 0.570 & 0.526 & 0.631 \\ 0.332 & -0.850 & 0.409 \end{matrix} \qquad R_2 = \begin{matrix} 0.721 & -0.046 & -0.691 \\ 0.598 & 0.544 & 0.588 \\ 0.349 & -0.838 & 0.420 \end{matrix}$$

$$T_1 = (-135.4\ \text{Å},\ -443.4\ \text{Å},\ 4.4\ \text{Å}) \qquad T_2 = (-42.5\ \text{Å},\ -445.8\ \text{Å},\ 8.4\ \text{Å})$$

$$R_3 = \begin{matrix} 0.713 & -0.096 & -0.695 \\ 0.681 & 0.333 & 0.653 \\ 0.168 & -0.938 & 0.302 \end{matrix} \qquad R_4 = \begin{matrix} 0.786 & -0.006 & -0.618 \\ 0.555 & 0.447 & 0.702 \\ 0.272 & -0.895 & 0.355 \end{matrix}$$

$$T_3 = (3.7\ \text{Å},\ -425.2\ \text{Å},\ 99.9\ \text{Å}) \qquad T_4 = (-95.3\ \text{Å},\ -426.8\ \text{Å},\ 91.3\ \text{Å})$$

$$\begin{aligned} Hx_1 &= R_1 \times Hx + T_1 \\ Hx_2 &= R_2 \times Hx + T_2 \\ Hx_3 &= R_3 \times Hx + T_3 \\ Hx_4 &= R_4 \times Hx + T_4 \end{aligned}$$

$$A1 = Hx_1 + Hx_2 + Hx_3 + Hx_4$$

b

$$C_1 = \begin{matrix} 0.500 & -0.309 & 0.809 \\ -0.309 & 0.809 & 0.500 \\ -0.809 & -0.500 & 0.309 \end{matrix} \qquad C_2 = \begin{matrix} -1.000 & 0.000 & 0.000 \\ 0.000 & -1.000 & 0.000 \\ 0.000 & 0.000 & 1.000 \end{matrix}$$

$$C_3 = \begin{matrix} -1.000 & 0.000 & 0.000 \\ 0.000 & 1.000 & 0.000 \\ 0.000 & 0.000 & -1.000 \end{matrix} \qquad C_4 = \begin{matrix} 0.000 & 0.000 & 1.000 \\ 1.000 & 0.000 & 0.000 \\ 0.000 & 1.000 & 0.000 \end{matrix}$$

$$\begin{aligned} A2 &= C_1 \times A1 \\ A3 &= C_1 \times A2 \\ A4 &= C_1 \times A3 \\ A5 &= C_1 \times A4 \end{aligned} \qquad \begin{aligned} A6 &= C_2 \times A1 \\ A7 &= C_2 \times A2 \\ A8 &= C_2 \times A3 \\ A9 &= C_2 \times A4 \\ A10 &= C_2 \times A5 \end{aligned}$$

$$\begin{matrix} A11 = C_3 \times A1 & A21 = C_4 \times A1 & A41 = C_4 \times A21 \\ \vdots & \vdots & \vdots \\ A20 = C_3 \times A10 & A40 = C_4 \times A20 & A60 = C_4 \times A40 \end{matrix}$$

c **Complete Icosahedral Capsid of 240 Hexons = A1 + A2 + . . . + A60**

**Fig. 5a–c.** Transformation matrices to generate a complete hexon capsid. **a** Hexon trimer (Hx) from crystallographic hexon monomer ($M_1$). The refined Cartesian crystallographic coordinates for a hexon monomer are available from the authors and will be deposited in the Brookhaven Protein Data Bank (ATHAPPILLY et al. 1994). **b** Capsid asymmetric unit (*A*1) from Hx. **c** Complete icosahedral capsid from *A*1

made on the basis of EM images of negatively stained isolated penton base and fiber (RUIGROK et al. 1990) and capsid fragments (VAN OOSTRUM et al. 1987a, b). These images indicate that fiber fills the approximately 30-Å-diameter central core of the penton base. While only a short portion of the fiber is observed in the difference map, the entire fiber is shown modeled in Fig. 7. The observed fiber density agrees with the left-handed triple-helical model proposed by STOUTEN et al. (1992).

**Table 1.** Adenovirus structural proteins (adapted from Stewart et al. 1993)

| Polypeptide | Molecular mass of monomer (daltons)[a] | Number of residues in monomer[a] | Biochemical copy number of monomer[a] | Copy number of monomer in current model |
|---|---|---|---|---|
| II (Hexon) | 109 077 | 967 | 720±7 | 720 |
| III (Penton base) | 63 296 | 571 | 56±1 | 60 |
| IIIa | | | | |
| External portion | 41 689 | 376[c] | | |
| Internal portion | 21 846 | 194[c] | | |
| Total | 63 535[b] | 570[b] | 68±2[b] | 60 |
| IV (fiber) | | | | |
| N terminus + two eighths of shaft | 13 912 | 133[c] | | |
| Six eighths of shaft + C terminus | 48 048 | 449[c] | | |
| Total | 61 960 | 582 | 35±1 | 36 |
| V (Core) | 41 631 | 368 | 157±1 | –[d] |
| Terminal[e] (core) | ~55 000 | ~500 | 2 | – |
| VI | | | | |
| Ordered portion | 9 773 | 91[c] | | |
| Disordered portion | 12 345 | 115[c] | | |
| Total | 12 118[b] | 206[b] | 342±4[b] | 360 |
| VII (Core) | 19 412 | 174 | 833±19 | – |
| VIII | 15 390[b] | 140[b] | 127±3[b] | – |
| IX | 14 339 | 139 | 247±2 | 240 |
| μ[f](Core) | ~4 000 | ~36 | ~104 | – |

[a] The molecular masses, residue numbers, and biochemically determined copy numbers for the monomers are from VAN OOSTRUM and BURNETT (1985) and updated with adenovirus protease cleavage sites from ANDERSON (1990)
[b] Includes the effect of updated adenovirus protease cleavage sites (ANDERSON 1990)
[c] Estimated division
[d] The dash indicates density not assigned in the three-dimensional difference map
[e] Estimates from REKOSH (1981)
[f] Estimates from HOSOKAWA and SUNG (1976)

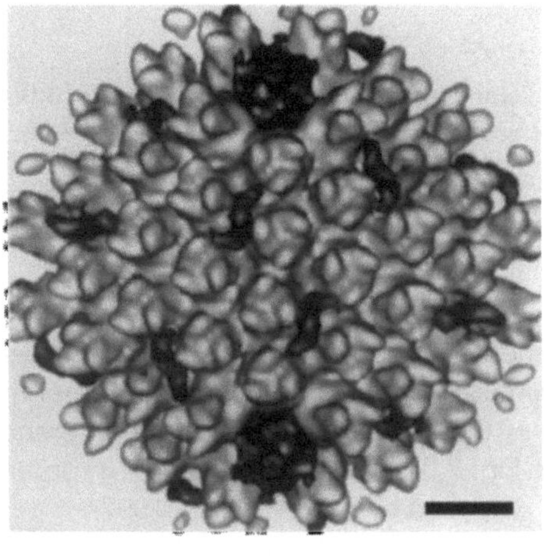

**Fig. 6.** A portion of the three-dimensional difference image (dark gray) superimposed on the hexon capsid density calculated from the crystallographic hexon structure (light gray). Density from the penton base and protruding fiber is seen at the two vertices. Elongated density from polypeptide IIIa is observed along the edges of the hexon capsid. *Bar*, 100 Å

**Fig. 7.** The penton complex is shown as a composite of the penton base from the three-dimensional image reconstruction and the fiber modeled using dimensions obtained from negatively stained electron microscopy (EM) images (RUIGROK et al. 1990). The distribution of beads of density along the fiber shaft in the EM images is in agreement with the three density beads (two buried within the penton base) that are observed in the EM difference density for the fiber. The first external bead visible in the figure is the EM density, and the remaining six beads are modeled. *Bar*, 50 Å

The penton base is surrounded by five hexons, known as peripentonal hexons as these are next to pentons. The interaction between the penton base and the surrounding peripentonal hexons can be examined in atomic detail, as shown in Fig. 8. One of the two hexon β-barrels fits nicely into a groove on the side of the penton base. A particular strand of this β-barrel is adjacent to the highly acidic region of 16 sequential aspartate and glutamate residues in the adenovirus type 2 hexon. A pH-mediated structural change of the viral capsid has been predicted to occur in the endosome during cell entry (WOHLFART 1988). A drop in pH could cause a shift in the position of the β-barrel attached to the acidic stretch and thus allow the penton base to dissociate from the virus. Several lines of evidence suggest that the penton base component pierces the endosomal membrane and thus allows the viral particle to escape into the cytoplasm

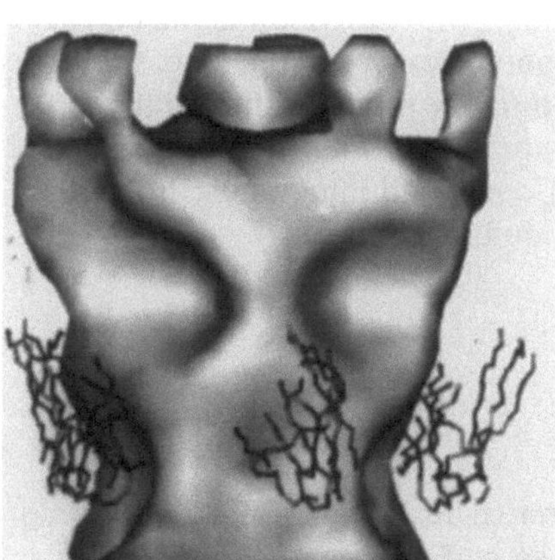

**Fig. 8.** The penton base and a short portion of the fiber are shown surrounded by β-barrels from five peripentonal hexons that nestle in grooves of the penton base. A pH-mediated shift in the positions of these β-barrels could result in the dissociation of the penton base from the virus particle. The structures are oriented such that the outside surface of the viral capsid is to the top and the inside surface to the bottom. *Bar*, 25 Å

(SETH et al. 1984; VARGA et al. 1991). This property of penton base is being used in gene therapy to enhance DNA transfection rates by permeabilizing cell membranes (ROSENFELD et al. 1992).

# 6 Polypeptide IIIa

Polypeptide IIIa is one of the three largest protein components in adenovirus (Table 1) and yet its position in the capsid was controversial for a long time. An estimate of 60 copies of polypeptide IIIa per virion (BOUDIN et al. 1980) and early biochemical studies, which suggested it was released with the peripentonal hexons (EVERITT et al. 1973) led to a model with five monomers of polypeptide IIIa in the vertex near the penton base. Adding to the confusion, apparently conflicting biochemical evidence placed this protein component on both the inner and outer surfaces of the protein capsid (EVERITT et al. 1975). A careful stoichiometric analysis by VAN OOSTRUM and BURNETT (1985), which showed that polypeptide IIIa is not released with the penton and peripentonal hexons, placed doubt on the original model.

The three-dimensional difference map (Fig. 6) revealed 60 copies of an elongated component that appears to span the hexon capsid (Fig. 9). EVERITT et al. (1975) first suggested that polypeptide IIIa might span the hexon capsid, and later CUILLEL et al. (1990) showed that recombinant IIIa in 1 *M* urea is long enough (200 Å) for this to be possible. The measured volume of the elongated component in the difference map is within 3% of the predicted volume for this 63.5-kDa protein (STEWART et al. 1993), giving strong support to the assignment of this density to polypeptide IIIa. In addition, there is no other density in the difference map that is unaccounted for and large enough to correspond to polypeptide IIIa. As the C-terminal end of the IIIa precursor is cleaved by the adenovirus protease (ANDERSON 1990), this end is most likely contained in the interior of the viral particle.

Just as the difference map allows the atomic interaction of hexons with the penton base to be analyzed, the interaction with polypeptide IIIa can also be evaluated. One striking interaction is observed between the long, thin arm of polypeptide IIIa and two β-strands in the tower region of hexon (Fig. 10). As this protein–protein binding site is characterized to within one residue, structure-based antiviral drug design using combined information from X-ray crystallography and EM is a real possibility in future research.

# 7 Polypeptides VI, VIII, and IX

The two-dimensional difference image of the GON, an adenoviral capsid fragment, revealed the locations of polypeptide IX, a 15-kDa component, within an

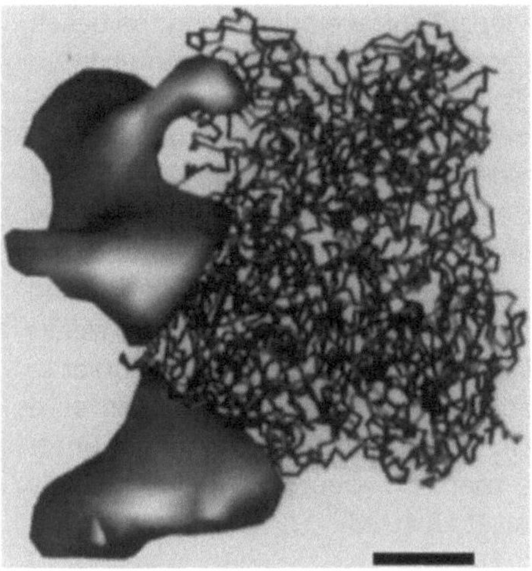

**Fig. 9.** Electron Microscopy (EM) density assigned to polypeptide IIIa is shown next to the Cα backbone of a complete crystallographic hexon trimer. The polypeptide IIIa density has a protruding narrow arm, approximately 15 Å in diameter, that contacts the tower region of one hexon. The structures are oriented such that the outside surface of the viral capsid is to the top and the inside surface to the bottom. *Bar*, 25 Å

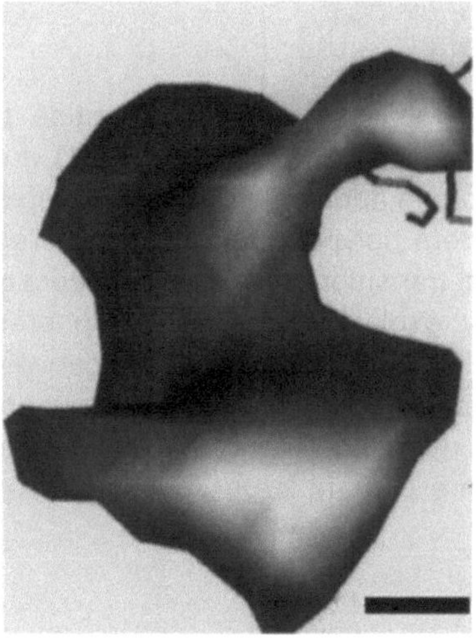

**Fig. 10.** An enlarged view of the external portion of polypeptide IIIa next to two β-strands from the same β-barrel of an adjacent hexon capsomer. Although only Cα backbone atoms are shown for simplicity, the side chain coordinates are known and so the polypeptide IIIa binding site is localized to within one residue on the crystallographic hexon structure. *Bar*, 12.5 Å

array of nine hexons (FURCINITTI et al. 1989). However, the two-dimensional projection image could not reveal the height of polypeptide IX with respect to the hexons. At the time, polypeptide IX was assumed to bind on the outer surface of the closely packed hexon bases, as mutant viruses lacking this protein can form intact virions (COLBY and SHENK 1981) but these are less stable than wild type. Polypeptide IX is produced late in infection and is not required for assembly, and so it presumably attaches to and stabilizes virions from the outside. The three-dimensional difference map confirmed the hypothesis that polypeptide IX binds on the outer surface of the capsid. The observed density for polypeptide IX closely matched the projected density obtained by two-dimensional difference imaging.

A ring of difference density on the inner capsid surface connecting the bases of the peripentonal hexons was assigned to polypeptide VI, a component known to bind DNA (RUSSELL and PRECIOUS 1982). This entirely internal density is consistent with the observation that both termini of polypeptide VI precursor are cleaved by the adenovirus protease (ANDERSON 1990). Interpretation of the EM difference density suggests that it is best associated with dimers of partially disordered polypeptide VI. Biochemical evidence suggested that this component forms dimers (EVERITT and PHILIPSON 1974), and a recent analysis of disulfide bond formation and protease cleavage also supports dimeric association (WEBSTER et al. 1993). Interestingly, the 11-amino acid carboxy terminus of the precursor to polypeptide VI, which is cleaved by the adenovirus protease, functions as a cofactor of the protease (MANGEL et al. 1993). The one capsid component unassigned in the three-dimensional difference map, polypeptide VIII, is probably disordered and on the inner surface (STEWART et al. 1993).

## 8 Core Components

At least four proteins, polypeptides V and VII (PHILIPSON 1983), polypeptide μ (HOSOKAWA and SUNG 1976), and a terminal DNA binding protein (REKOSH 1981), are complexed with the DNA in the core of the virion. In addition, volume estimates suggest that roughly half of the capsid component polypeptide VI is disordered and interacting with the internal DNA. A gradual transition from an icosahedrally ordered capsid to a disordered nucleus may explain the 12 large spheres observed in the core by ion etching (NEWCOMB et al. 1984) and by EM of negatively stained preparations of sarcosyl cores (BROWN et al. 1975). If the ring of polypeptide VI interacts with the internal viral DNA, it might tend to cause differential staining effects and loosely order the DNA in a slightly larger circular region matching the size of the observed spheres.

## 9 Conclusion

In conclusion, the combined approach of X-ray crystallography and EM has been very successful in elucidating the overall organization of this complex human virus. If the atomic structures of additional virion components are determined in the future, they can readily be interpreted in the context of the current model for the architecture of the entire virus particle.

*Acknowledgments.* We acknowledge the contributions of our many coworkers and collaborators during this ongoing project. In particular, we thank Dr. Stephen D. Fuller at the European Molecular

Biology Laboratory (EMBL) in Heidelberg, Germany. P.L.S. was a Helen Hay Whitney fellow during the period in which much of this research was carried out. R.M.B. is supported by grants from the National Science Foundation (MCB 92-07014), the National Institute of Allergy and Infectious Diseases (AI 17270), and by the Wistar Cancer Center (CA 10815).

## References


Adrian M, Dubochet J, Lepault J, McDowall AW (1984) Cryoelectron microscopy of viruses. Nature 308: 32–36

Anderson CW (1990) The proteinase polypeptide of adenovirus serotype 2 virions. Virology 177: 259–272

Athappilly FK, Murali R, Rux JJ, Cai Z, Burnett RM (1994) The refined crystal structure of hexon, the major coat protein of adenovirus type 2, at 2.9 Å resolution. J Mol Biol 242: 430–455

Boudin M-L, D'Halluin J-C, Cousin C, Boulanger P (1980) Human adenovirus type 2 protein IIIa. II. Maturation and encapsidation. Virology 101: 144–156

Brown DT, Westphal M, Burlingham BT, Winterhoff U, Doerfler W (1975) Structure and composition of the adenovirus type 2 core. J Virol 16: 366–387

Burnett RM (1985) The structure of the adenovirus capsid. II. The packing symmetry of hexon and its implications for viral architecture. J Mol Biol 185: 125–143

Burnett RM, Grütter MG, White JL (1985) The structure of the adenovirus capsid. I. An envelope model of hexon at 6 Å resolution. J Mol Biol 185: 105–123

Colby WW, Shenk T (1981) Adenovirus type 5 virions can be assembled in vivo in the absence of detectable polypeptide IX. J Virol 39: 977–980

Crowther RA (1971) Procedures for three-dimensional reconstruction of spherical viruses by Fourier synthesis from electron micrographs. Philos Trans R Soc Lond Biol 261: 221–230

Cuillel M, Cortolezzis B, Chroboczek J, Langowski J, Ruigrok RWH, Jacrot B (1990) Purification and characterization of wild-type and ts112 mutant protein IIIa of human adenovirus 2 expressed in Escherichia coli. Virology 175: 222–231

Defer C, Belin M-T, Caillet-Boudin M-L, Boulanger P (1990) Human adenovirus-host cell interactions: comparative study with members of subgroups B and C. J Virol 64: 3661–3673

Dubochet J, Adrian M, Lepault J, McDowall AW (1985) Cryoelectron microscopy of virtrified biological specimens. Trends Biochem Sci 10: 143–146

Everitt E, Philipson L (1974) Structural proteins of adenoviruses. XI. Purification of three low molecular weight virion proteins of adenovirus type 2 and their synthesis during productive infection. Virology 62: 253–269

Everitt E, Sundquist B, Pettersson U, Philipson L (1973) Structural proteins of adenoviruses. X. Isolation and topography of low molecular weight antigens from the virion of adenovirus type 2. Virology 52: 130–147

Everitt E, Lutter L, Philipson L (1975) Structural proteins of adenoviruses. XII. Location and neighbor relationship among proteins of adenovirion type 2 as revealed by enzymatic iodination, immunoprecipitation and chemical cross-linking. Virology 67: 197–208

Fuller SD (1987) The T = 4 envelope of Sindbis virus is organized by interactions with a complementary T = 3 capsid. Cell 48: 923–934

Furcinitti PS, van Oostrum J, Burnett RM (1989) Adenovirus polypeptide IX revealed as capsid cement by difference images from electron microscopy and crystallography. EMBO J 8: 3563–3570

Ginsberg HS (1979) Adenovirus structural proteins. In: Fraenkel-Conrat H, Wagner RR (eds) Comprehensive virology, vol 13. Plenum, New York, pp 409–457

Hogle JM, Chow M, Filman DJ (1985) Three-dimensional structure of poliovirus at 2.9 Å resolution. Science 229: 1358–1365

Hosokawa K, Sung MT (1976) Isolation and characterization of an extremely basic protein from adenovirus type 5. J Virol 17: 924–934

Liddington RC, Yan Y, Moulai J, Sahli R, Benjamin TL, Harrison SC (1991) Structure of simian virus 40 at 3.8-Å resolution. Nature 354: 278–284

Mangel WF, McGrath WJ, Toledo DL, Anderson CW (1993) Viral DNA and a viral peptide can act as cofactors of adenovirus virion proteinase activity. Nature 361: 274–275

Newcomb WW, Boring JW, Brown JC (1984) Ion etching of human adenovirus 2: structure of the core. J Virol 51: 52–56

Olson NH, Kolatkar PR, Oliveira MA, Cheng RH, Greve JM, McClelland A, Baker TS, Rossmann MG (1993) Structure of a human rhinovirus complexed with its receptor molecule. Proc Natl Acad Sci USA 90: 507–511

Philipson L (1983) Structure and assembly of adenoviruses. In: Doerfler W (ed) The molecular biology of adenoviruses. Curr Top Microbiol Immunol 109: 1–52

Philipson L., Lonberg-Holm K, Pettersson U (1968) Virus-receptor interaction in an adenovirus system. J Virol 2: 1064–1075

Prasad BVV, Burns JW, Marietta E, Estes MK, Chiu W (1990) Localization of VP4 neutralization sites in rotavirus by three-dimensional cryoelectron microscopy. Nature 343: 476–479

Prasad BVV, Prevelige PE, Marietta E, Chen RO, Thomas D, King J, Chiu W (1993) Three-dimensional transformation of capsids associated with genome packaging in a bacterial virus. J Mol Biol 231: 65–74

Rekosh D (1981) Analysis of the DNA-terminal protein from different serotypes of human adenovirus. J Virol 40: 329–333

Roberts MM, White JL, Grütter MG, Burnett RM (1986) Three-dimensional structure of the adenovirus major coat protein hexon. Science 232: 1148–1151

Rosenfeld MA, Yoshimura K, Trapnell BC et al. (1992) In vivo transfer of the human cystic fibrosis transmembrane conductance regulator gene to the airway epithelium. Cell 68: 143–155

Rossmann MG, Arnold E, Erickson JW et al. (1985) Structure of a human common cold virus and functional relationship to other picornaviruses. Nature 317: 145–153

Ruigrok RWH, Barge A, Albiges-Rizo C, Dayan S (1990) Structure of adenovirus fiber. II. Morphology of single fibres. J Mol Biol 215: 589–596

Russell WC, Precious B (1982) Nucleic acid-binding properties of adenovirus structural polypeptides. J Gen Virol 63: 69–79

Seth P, Fitzgerald D, Ginsberg H, Willingham M, Pastan I (1984) Evidence that the penton base of adenovirus is involved in potentiation of toxicity of Pseudomonas exotoxin conjugated to epidermal growth factor. Mol Cell Biol 4: 1528–1533

Stewart PL, Burnett RM (1993) Adenovirus structure as revealed by X-ray crystallography, electron microscopy, and difference imaging. Jpn J Appl Phys 32: 1342–1347

Stewart PL, Burnett RM, Cyrklaff M, Fuller SD (1991) Image reconstruction reveals the complex molecular organization of adenovirus. Cell 67: 145–154

Stewart PL, Fuller SD, Burnett RM (1993) Difference imaging of adenovirus: bridging the resolution gap between X-ray crystallography and electron microscopy. EMBO J 12: 2589–2599

Stouten PFW, Sander C, Ruigrok RWH, Cusack S (1992) New triple-helical model for the shaft of the adenovirus fibre. J Mol Biol 226: 1073–1084

Toogood CIA, Murali R, Burnett RM, Hay RT (1989) The adenovirus type 40 hexon: sequence, predicted structure and relationship to other adenovirus hexons. J Gen Virol 70: 3203–3214

Toogood CIA, Crompton J, Hay RT (1992) Antipeptide antisera define neutralizing epitopes on the adenovirus hexon. J Gen Virol 73: 1429–1435

Unwin N (1993) Nicotinic acetylcholine receptor at 9 Å resolution. J Mol Biol 229: 1101–1124

van Oostrum J, Burnett RM (1985) Molecular composition of the adenovirus type 2 virion. J Virol 56: 439–448

van Oostrum J, Smith PR, Mohraz M, Burnett RM (1987a) The vertex morphology of the adenovirus capsid. Ann NY Acad Sci 494: 423–426

van Oostrum J, Smith PR, Mohraz M, Burnett RM (1987b) The structure of the adenovirus capsid. III. Hexon packing determined from electron micrographs of capsid fragments. J Mol Biol 198: 73–89

Varga MJ, Weibull C, Everitt E (1991) Infectious entry pathway of adenovirus type 2. J Virol 65: 6061–6070

Webster A, Hay RT, Kemp G (1993) The adenovirus protease is activated by a virus-coded disulphide-linked peptide. Cell 72: 97–104

White JM (1993) Integrins as virus receptors. Curr Biol 3: 596–599

Wickham TJ, Mathias P, Cheresh DA, Nemerow GR (1993) Integrins $\alpha_v\beta_3$ and $\alpha_v\beta_5$ promote adenovirus internalization but not virus attachment. Cell 73: 309–319

Wohlfart C (1988) Neutralization of adenoviruses: kinetics, stoichiometry, and mechanisms. J Virol 62: 2321–2328

Yeager M, Berriman JA, Baker TS, Bellamy AR (1994) Three-dimensional structure of the rotavirus haemagglutinin VP4 by cryoelectron microscopy and difference map analysis. EMBO J 13: 1011–1018

# Structure of the Receptor Binding Domain of Adenovirus Type 5 Fiber Protein

D. XIA[1], L. HENRY[2], R.D. GERARD[2], and J. DEISENHOFER[1]



## 1 Introduction

In this chapter we briefly describe functional and structural properties of the recombinant knob domain of the adenovirus type 5 (Ad5) fiber protein. We were able to demonstrate that this domain forms a trimer in solution and can compete with intact virus for binding to cellular receptors (HENRY et al. 1994). We subsequently determined the crystal structure of knob at 1.7 Å resolution (XIA et al. 1994). The knob monomers are β-sandwich structures and are arranged as closely associated trimers in the crystal. We compared sequences of knob domains from different adenovirus serotypes and found patterns of highly conserved and highly variable sequences on the knob surface. We interpreted the conserved regions as probable sites of interaction of the knob ligand with the cellular receptors and the variable regions as antigenic sites which define the structural differences among serotypes.

[1] Howard Hughes Medical Institute, University of Texas Southwestern Medical Center, 5323 Harry Hines Boulevard, Dallas, TX 75235-9050, USA
[2] Departments of Biochemistry and Internal Medicine, University of Texas Southwestern Medical Center, 5323 Harry Hines Boulevard, Dallas, TX 75235-8573, USA

## 2 Cloning, Expression, Purification, and Functional Analysis of the Adenovirus Type 5 Knob Domain

The Ad5 knob domain, together with a single 15-residue repeat from the shaft domain (residues 386–501), has been cloned by polymerase chain reaction (PCR) amplification using Ad5 DNA (McGRORY et al. 1988) as a template. The entire polypeptide has 196 amino acid residues with a molecular weight of 21 279 Da. Subsequently, the cloned knob DNA was inserted into the bacterial expression vector pBEVpL. The expression of the recombinant knob protein in the *Escherichia coli* strain TG-1 was induced by growth in M9 media for 20–24 h and starvation for tryptophan in the late stages of growth (SHUBEITA et al. 1990). Yields of about 10 mg recombinant knob protein per liter *E. coli* cells can be routinely obtained. The expressed knob is a soluble protein and is purified using ammonium sulfate precipitation of the diethylaminoethyl (DEAE)-treated cell lysate, followed by Q-Sepharose chromatography.

The purified knob protein is able to bind to human HeLa or 293 cell membranes and to block subsequent infection by wild-type Ad5 in a manner which is dependent on the concentration of the recombinant knob. Scatchard analysis of knob binding to HeLa cells demonstrates binding to receptors with numbers and an affinity equivalent to those previously demonstrated for fiber binding. Antibodies raised against the knob protein and purified from a knob protein affinity column can neutralize Ad5 infectivity when premixed with the virus prior to infection of HeLa cell monolayers. Furthermore, antibodies to the knob domain are more efficient than those raised against the whole fiber in neutralizing intact virions (WADELL 1972). These results demonstrate that the expressed knob protein can specifically bind to the cell surface and is the ligand for the adenovirus receptor(s).

## 3 Structure Determination of the Adenovirus Type 5 Knob Domain

Knob crystals diffract X-rays to beyond 1.7 Å resolution; they have the symmetry of the cubic space group $P2_13$ with one monomer per crystallographic asymmetric unit and unit cell dimensions of $a=b=c=$ 86.4 Å (HENRY et al. 1994). The crystal structure of the knob protein was determined with the method of multiple isomorphous replacement at 3 Å resolution and subsequently refined at 1.7 Å resolution (XIA et al. 1994). The refined atomic model has a crystallographic $R$-factor of 0.162 for 17633 unique reflections in the resolution range between 12 Å and 1.7 Å ($R = \Sigma|F_{obs}-F_{calc}|/\Sigma F_{obs}$, where $F_{obs}$ and $F_{calc}$ are observed and calculated structure factor amplitudes, respectively); it includes 151 water molecules, which are mostly located at the protein surface.

## 4 Structure of the Knob Monomer

A Schematic drawing of the knob monomer structure (XIA et al. 1994) is shown in Fig. 1. It is a sandwich of two antiparallel β-sheets; the directions of the β-strands in the two β-sheets differ by an angle of about 30°. We call the β-sheet consisting of the strands J, C, B, and A the "V sheet" because its surface points towards the virion side in the trimer structure of the knob. The β-sheet containing strands D, I, H, and G is probably involved in the binding of cellular receptors; therefore we call it the "R sheet".

The β-strands of the knob include 65 residues, i.e., about one third of the polypeptide chain. The rest of the structure is made of loops and turns. There are six prominent loops with lengths between 8 and 55 residues; they are labeled AB, CD, DG, GH, HI and IJ after the β-strands they connect. These loops consist mostly of hydrophilic residues and contribute to the surface of the knob. With 55 residues the DG loop is the longest loop. It includes the two short β-strands, E and F, which extend the V sheet. The DG loop is the only loop that contributes to the hydrophobic core of the structure: the hydrophobic residues 494–505 fill a gap between the two β-sheets, giving the monomer a wedge-like shape (see Fig. 1). The two cysteine residues near the N terminus of the expressed knob are in their free thiol form and are readily accessible to heavy atom compounds.

The first seven residues (residues 386–392) of the recombinant knob protein are completely disordered in the crystal. Most of the remaining structure is well ordered, as is evident from the low *B* factors of the refined model (overall *B* factor

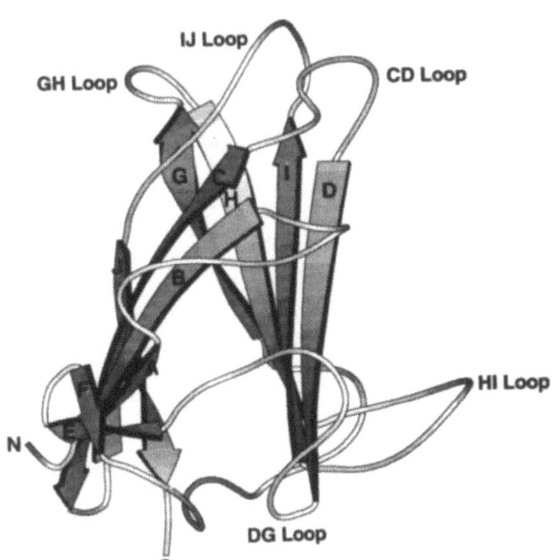


**Fig. 1.** Course of the polypeptide chain of the knob monomer. β-strands (drawn as *arrows*), loops, and chain termini are labeled. Figures 1, 2, and 4 were produced with the program MOLSCRIPT (KRAULIS 1991)

of 16.6 $\text{Å}^2$ for main chain atoms and 19.3 $\text{Å}^2$ for side chain atoms). Some amino acid residues in the loops have high *B* factors, especially residues in the HI loop, which have a mean *B* factor of 50 $\text{Å}^2$; this value indicates disorder or flexibility. Although the folding of the knob domain is similar to many other β-sandwich structures, the folding topology, i.e., the connectivity between β-strands, is different from any other known structure in this family, such as immunoglobulin domains (Deisenhofer 1981), Cu, Zn superoxide dismutase (Tainer et al. 1982), tumor necrosis factor (Sprang and Eck 1992), and viral coat proteins (Rossmann et al. 1985).

## 5 Knob Trimer

In the crystal, the Ad5 knob domains are arranged as trimers around the three-fold crystallographic symmetry axes (Xia et al. 1994). As shown in Fig. 2, the trimers visually resemble three-bladed propellers when viewed along this axis; the R sheets from the three monomers correspond to the surfaces of the three blades. The diameter of the knob trimer is 62 Å, and its height is 37 Å. Intermolecular contacts in the trimer cover about 1950 $\text{Å}^2$ of accessible surface area, a value typical for most oligomeric proteins (Janin et al. 1988). The monomer interfaces contain 30 water molecules which mediate charge interactions. The V sheet plays a major role in the

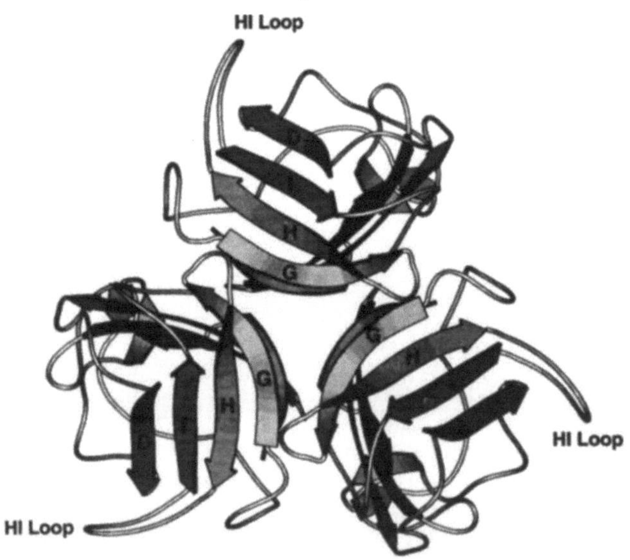


**Fig. 2.** The knob trimer, viewed along the three-fold symmetry axis. The R sheet with β-strands D, I, H, and G faces the cellular receptor

monomer–monomer interactions; in particular, the β-strands J and A are in close contact with their symmetry-related counterparts. The R sheets contribute to charge interactions across the threefold molecular symmetry axis with the β-strands G. Loops contribute very little to interactions in the knob trimer.

There is a deep depression around the threefold molecular symmetry axis on the surface of the trimer facing away from the virus surface. This depression is 24 Å in diameter at its rim and reaches as a narrowing channel 15 Å deep into the trimer. Its bottom is made of three symmetry-related cysteines (Cys 428) whose thiol groups are only 4.2 Å from each other. The residues lining the wall of the depression are mostly hydrophilic. Other prominent features on the surface of the knob trimer are valleys formed by β-strands G, H, I, and D, and the HI loops.

The amino termini of the monomers approach each other near the threefold molecular symmetry axis. The recombinant knob protein contains one sequence repeat of the shaft, but the absence of significant electron density indicates disorder in this region. The spontaneous trimerization of the recombinant knob monomers (HENRY et al. 1994) indicates that the shaft region is not needed for trimerization. Extensive interactions between monomers and large buried surface area upon monomer association suggest that the assembly of the adenovirus fiber protein may start with the folding of the knob monomer, followed by the trimerization of the knob domain; this in turn may trigger the trimerization of the shaft domain.

## 6 Sequence Conservation and the Putative Receptor Binding Sites

The aligned amino acid sequences of knob domains from four different adenovirus serotypes, Ad5, Ad2, Ad3, and Ad7 (SIGNAS et al. 1985; CHROBOCZEK and JACROT 1987), are shown in Fig. 3. Clearly, residues in the β-strands are more conserved than those in other parts of the structure, which makes us believe that the β-sandwich folding of Ad5 should be preserved in the other serotypes shown (XIA et al. 1994). The V sheet is more highly conserved than the R sheet. This is consistent with the view that the V sheet plays the major role in trimerization of knob domains through intermolecular interactions of conserved residues. The R sheet, on the other hand, is facing cellular receptors and antibodies and therefore would be expected to have a more variable surface.

Of the three most conserved regions in the sequence, the β-strand A near the N terminus may play an important role in defining the transition between the knob and shaft domains and ensuring correct folding of the shaft domain. The highly conserved J strands interact closely with each other in the trimer; knob truncated by deletion of the J strand does not trimerize (NOVELLI and BOULANGER 1991). The conserved sequence in the DG loop between residues 494 and 505 participates

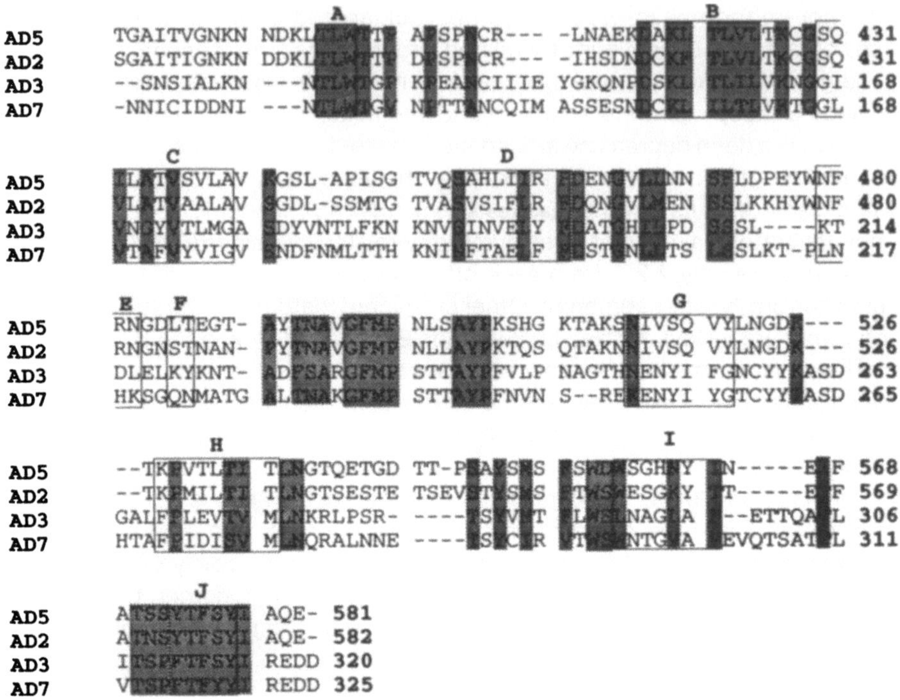

**Fig. 3.** Sequence alignment of the knob domains of the adenovirus fiber species adenovirus type 5 (Ad5), Ad2, Ad3, and Ad7. Ad2 and Ad5 belong to the same subgroup and share the same cellular receptor (PHILIPSON et al. 1968). Ad3 and Ad7 are also human adenoviruses, but bind to cellular receptor different from that for Ad2 and Ad5 (DEFER et al. 1990). The *gray, highlighted boxes* show the conserved residues among the knob sequences from different adenovirus species, and the *open boxes* indicate the β-strands in the structure of the type 5 adenovirus knob protein

in the formation of the hydrophobic core and helps to maintain the wedge-shaped monomer structure.

Ad5 and Ad2 belong to the adenovirus subgroup C and share the same cellular receptor (DEFER et al. 1990). Therefore, the amino acid residues around the receptor binding site should be conserved in these two knobs. Figure 4 is a Cα tracing of the trimeric knob. On top of the Cα tracing, amino acid residues of Ad5 that are different from those of Ad2 are shown as gray-shaded, space-filling models. These variations are distributed on the surface of the knob trimer, except for two regions in which residues from the two knobs are strictly conserved. One such region is located in the surface depression around the threefold molecular symmetry axis. The others are on the floors of three symmetry-related valleys.

Ad3 and Ad7 belong to adenovirus subgroup B and bind to the same cellular receptor, which is different from that for Ad5 and Ad2 (DEFER et al. 1990). The residues that are common between Ad5 and Ad2, but are different from the common residues between Ad3 and Ad7, are shown as unshaded, space-filling models in Fig. 4. These residues are concentrated in the surface depression and

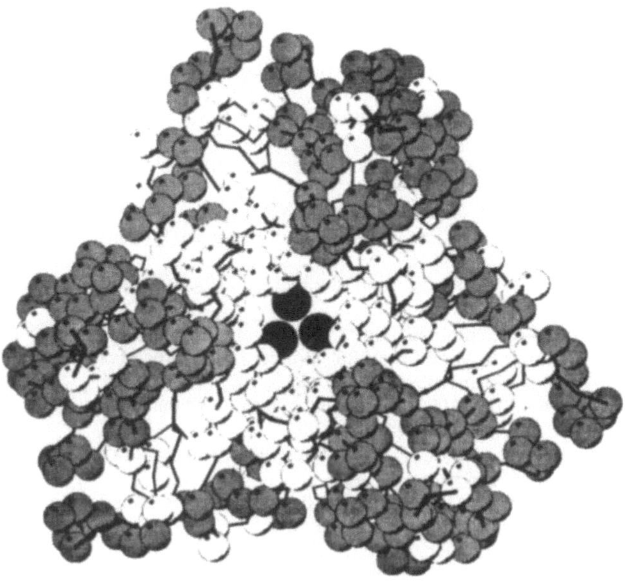

**Fig. 4.** The knob trimer, shown in the same orientation as in Fig. 2. The Cα trace is overlaid with gray-shaded, space-filling models of those amino acids which are different in adenovirus type 5 (Ad5) and Ad2. Amino acids which are identical in Ad5 and Ad2 and identical in Ad3 and Ad7, but different in the two pairs of serotypes, are shown as *white spheres*. The SH groups of the cysteine residues 428 near the threefold symmetry axis are shown as *dark-gray spheres*

on the floors of the valleys. Thus two receptor-binding modes can be postulated. One is the binding of the cellular receptor to the central surface depression with one receptor-binding site per knob trimer. This scenario would be similar to the binding of rhinovirus 14 to the intercellular adhesion molecule 1 (ICAM-1; Olson et al. 1993). Alternatively, if the receptor binds to the floor of the valley, each knob would have three receptor binding sites; this would resemble the binding of tumor necrosis factor to its cellular receptors (Sprang and Eck 1992). These alternative receptor binding modes can be discriminated by structure function analysis using site-specific mutagenesis.

## References

Chroboczek J, Jacrot B (1987) The sequence of adenovirus fiber: similarities and differences between serotypes 2 and 5. Virology 161: 549–554

Defer C, Belin M-T, Caillet-Boudin M-L, Boulanger P (1990) Human adenovirus-host cell interactions: comparative study with members of subgroups B and C. J Virol 64: 3661–3673

Deisenhofer J (1981) Crystallographic refinement and atomic models of a human Fc fragment and its complex with fragment B of protein A from Staphylococcus aureus at 2.9 and 2.8 Å resolution. Biochemistry 20: 2361–2370

Henry L, Xia D, Wilke ME, Deisenhofer J, Gerard RD (1994) Characterization of the knob domain of the adenovirus type 5 fiber protein expressed in *Escherichia coli*. J Virol 68: 5239–5246

Janin J, Miller S, Chothia C (1988) Surface, subunit interfaces and interior of oligomeric proteins. J Mol Biol 204: 155–164
Kraulis PJ (1991) MOLSCRIPT: a program to produce both detailed and schematic plots of protein structures. J Appl Crystallogr 24: 946–950
McGrory WJ, Bautista DS, Graham FL (1988) A simple technique for the rescue of early region 1 mutations into infectious human adenovirus type 5. Virology 163: 614–617
Novelli A, Boulanger PA (1991) Deletion analysis of functional domains in baculovirus-expressed adenovirus type 2 fiber. Virology 185: 365–376
Olson NH, Kolatkar PR, Oliveira MA, Cheng RH, Greve JM, McClelland A, Baker TS, Rossmann MG (1993) Structure of a human rhinovirus complexed with its receptor molecule. Proc Natl Acad Sci USA 90: 507–511
Philipson L, Lonberg-Holm K, Pettersson U (1968) Virus-receptor interaction in an adenovirus system. J Virol 2: 1064–1075
Rossmann MG, Arnold E, Erickson JW, Frankenberger EA, Griffith JP, Hecht HJ, Johnson JE, Kamer G, Lou M, Mosser AG, Rueckert RR, Sherry B, Vriend G (1985) Structure of a human common cold virus and functional relationship to other picornaviruses. Nature 317: 145–153
Shubeita HE, Gottey TL, Franke AE, Gerard RD (1990) Mutational and immunochemical analysis of plasminogen activator inhibitor-1. J Biol Chem 265: 18379–18385
Signas C, Akusjarvi G, Pettersson U (1985) Adenovirus 3 fiber polypeptide gene: implications for the structure of the fiber protein. J Virol 53: 672–678
Sprang SR, Eck MJ (1992) The 3-D structure of TNF. In: Beutler B (ed) Tumor necrosis factors: the molecules and their emerging role in medicine. Raven, New York, pp 11–31
Tainer JA, Getzoff ED, Beem KM, Richardson JS, Richardson DC (1982) Determination and analysis of the 2 Å structure of Cu, Zn superoxide dismutase. J Mol Biol 160: 181–217
Wadell G (1972) Sensitization and neutralization of adenovirus by specific sera against capsid subunits. J Immunol 108: 622–632
Xia D, Henry LJ, Gerard RD, Deisenhofer J (1994) Crystal structure of the receptor-binding domain of adenovirus type 5 fiber protein at 1.7 Å resolution. Structure 2: 1259–1270

# Virus Assembly

J.C. D'Halluin



## 1 Introduction

The basic structure of all virions is a nucleic acid molecule enclosed in a capsid. The capsid structures must be built of multiple copies of one or a few protein subunits, since viruses do not have enough genetic information to code for a great variety of proteins (Crick and Watson 1957). There are two ways in which identical asymmetrical subunits can assemble to build stable regular capsids: by forming closed shells or through helical assemblies. The assembly of animal viruses has benefited from bacteriophage studies. In bacteriophage systems, all virion proteins and other late proteins, such as scaffolding proteins, required for assembly are synthesized simultaneously and accumulate in a precursor pool. From the

INSERM U.124, 2. Place de Verdun, 59045 Lille Cedex, France

pool, they are withdrawn by specific interactions to form assembly intermediates that in turn are assembled into complete virions, a process that in many cases involves several enzymatic steps such as precursor cleavages. Some examples are provided by the T-even bacteriophags λ and p22; in vivo studies with conditional lethal mutants first showed the general pathway, but an in vitro assembly system of complementation was required to delineate the molecular details (EDGAR and WOOD 1966; EARNSHAW and CASJENS 1980; FULLER and KING 1980, 1982).

With eukaryotic viruses, it has been more difficult to: (i) identify all proteins involved in the assembly pathway; (ii) develop in vitro assembly systems. Moreover, in mammalian cells, many of the intermediate steps in virus assembly appear to involve interaction with specific cellular structures including internal membranes and/or nuclear matrix, and the construction of an in vitro system is extremely difficult. The development of new methods to characterize the proteins and nucleic acids associated with the assembly intermediates in association with temperature-sensitive (ts) and deletion mutants have, however, given some hope for the future. The knowledge of the pathway for assembly of complex DNA viruses has seen some development in the last 15 years, in parallel to the knowledge of the viral particle structure (STEWART et al. 1991). This chapter attempts to summarize the knowledge about the adenovirus assembly pathway. Reviews on this topic appeared 10 years ago or more (FLINT and BROKER 1981; PHILIPSON 1984).

## 2 Assembly of Capsomers

After the onset of viral DNA synthesis, the late phase of adenovirus-productive infection starts and viral mRNA encoding the structural proteins are preferentially translated. Most of the viral polypeptides are rapidly transported to the nucleus after being released from polyribosomes with a lag period of 3–6 min (HORWITZ et al. 1969; VELICER and GINSBERG 1970). The first step in adenovirus assembly is the formation of capsomers. The capsomers are the major structural units of the capsid (hexon, penton base, and fiber) assembled from monomeric polypeptide chains. This process may sometime require another polypeptide, a scaffolding protein, as is the case with the L4 100-kDa protein implicated in hexon trimerization. Where the oligomerization occurs in the cell is not yet clear, but it probably takes place shortly after translation.

### 2.1 Hexon

In vitro translation and a genetic approach suggest that assembly of hexons may require one additional viral polypeptide. In a heterologous translation system, the translation product of the hexon mRNA can be detected with antibodies against denatured hexons (STINSKI and GINSBERG 1974; ÖBERG et al. 1975). In contrast, when polysomes from infected cells are used, the trimer is formed (PERSSON et al. 1977). Using ts mutants (H5ts17 and H5ts20) from different complementation groups than the hexon gene, LEIBOWITZ and HORWITZ (1975) demonstrated that one additional gene product is required for hexon trimerization. By marker rescue, the two ts mutations have been mapped to the nonstructural L4 100-kDa protein gene (FROST and WILLIAMS 1978) and this localization was confirmed by DNA sequencing. These results were confirmed with other ts mutants located in the L4 100-kDa protein gene: H5ts115 and H5ts116 (OOSTEROM-DRAGON and GINSBERG 1981) and H2ts118 (D'HALLUIN et al. 1982). Moreover, a direct association between the 100-kDa protein and nascent hexon polypeptides was demonstrated using antibodies against both partners (OOSTEROM-DRAGON and GINSBERG 1981; CEPKO et al. 1981; CEPKO and SHARP 1982). Complexes of about 600 kDa were purified by gel filtration, suggesting a stoichiometric quantity of both partners (1/1; MORIN and BOULANGER 1984). The hexon assembly may therefore require the 100-kDa protein as a scaffolding protein. The protein domain of the two partners has not been identified yet, but may involve the reactive serine in the nascent hexon polypeptide that can interact with diisopropylfluorophosphate (DFP; DEVAUX and BOULANGER 1980). Once the hexon is formed, the 100-kDa protein dissociates from the trimeric unit. Next, after its trimerization, the hexon capsomers accumulate in the nuclei. The hexon transport requires another viral polypeptide which is associated with the hexon in the viral capsid: the precursor of the polypeptide VI (KAUFFMAN and GINSBERG 1976). At the end of the infectious cycle, the hexon protein represents approximately 10% of the total protein content of the cell.

## 2.2 Penton

The penton is made up of penton base and fiber. The genes encoding these polypeptides are located within the major late transcription units L2 and L5, respectively (MILLER et al. 1980). Several ts mutants have been isolated which map to the fiber region of the genome, but no ts mutants have yet been found in the penton base region. All defective fiber mutants have a mutation located in the fiber gene, suggesting that the fiber oligomerization does not require another viral function. This result has been confirmed by spontaneous trimerization of fiber mRNA translation products in a heterologous in vitro system, but in vitro trimerization occurs very slowly (NOVELLI and BOULANGER 1991).

The fiber is a trimer of identical subunits (SUNDQUIST et al. 1973a; VAN OOSTRUM and BURNETT 1985; VAN OOSTRUM et al. 1987) and shows a 15-amino acid pseudorepeat in the shaft region (GREEN et al. 1983). The N terminus of the fiber is located at the end of the shaft opposite the head, as predicted by the structural model proposed by GREEN et al. (1983), and it interacts in a noncovalent manner with the penton base (DEVAUX et al. 1987), probably by complementary peptide interaction (CAILLET-BOUDIN 1989). After synthesis, the subunits of the penton base and fiber accumulate in the nucleus of the infected cells (VELICER and GINSBERG 1970). During productive infection with wild-type (wt) virus, an excess of fiber is formed and the formation of penton base is probably rate limiting in the assembly process, since only minute amounts of free penton base can be found (BOUDIN et al. 1979). An efficacious in vitro assembly system for the penton has been described (BOUDIN and BOULANGER 1982; see below). The antigenicity of the fiber is changed during penton formation, possibly reflecting a three-dimensional structural change (BOUDIN and BOULANGER 1981). The formation of $ts^+$ recombinants in vivo establishes that a genus-specific area of the fiber moiety is recognized by the penton base (GINSBERG and YOUNG 1977; D'HALLUIN et al. 1982). It is not clear where the penton assembly occurs in the infected cells, though perhaps it is in the nucleus.

## 3 Assembly Intermediates

The pathway for adenovirus assembly was first analyzed by kinetic labeling experiments (HORWITZ et al. 1969; SUNDQUIST et al. 1973b; ISHIBASHI and MAIZEL 1974), but the large pool of structural proteins in infected cells and the fact that only a fraction of the capsomers are incorporated into virus particles (20% of hexon capsomers) hampered the interpretation of the results. The second approach was the use of defective ts mutants in capsid morphogenesis (WEBER 1976; EDVARDSSON et al. 1978; D'HALLUIN et al. 1978b), and the next step was the combination of ts mutants, pulse-chase experiments, and a mild technique of separation of assembly intermediates in the assembly pathway (EDVARDSSON et al. 1976; D'HALLUIN et al. 1978a,b).

The core proteins (V and pVII/ VII) appear in intact virions within 15 min after their synthesis, in sharp contrast to the lag period (60 min) required for the assembly of hexon polypeptides into virions (HORWITZ et al. 1969). This implies that core proteins become rapidly associated with DNA after their synthesis and that DNA and core protein(s) (complexes) are somehow inserted into previously assembled capsids. Thus, empty capsids, lacking viral DNA, would be expected to be precursors of mature virions, and considerable evidence in favor of this interpretation is available. From 13 h after infection, cells in culture infected by adenovirus type 2 (Ad2) or Ad3 contain both virions and empty capsids that can be readily separated by isopyknic centrifugation in CsCl gradients. These particles increase in concentration in parallel, and both are synthesized at maximum rates 24 h after infection (SUNDQUIST et al. 1973b). Although labeled amino acids are incorporated into incomplete particles without a lag, 60–80 min elapse before the label appears in mature virions. Pulse-chase experiments with Ad2-infected cells, which have a small pool of empty capsids, indicate a precursor–product relationship between incomplete and mature particles (SUNDQUIST et al. 1973b, ISHIBASHI and MAIZEL 1974). SUNDQUIST et al. (1973b) observed that labeled hexons do not appear in mature virions until at least 60 min after the addition of radioactive precursor, whereas there is no delay to their appearance in incomplete particles. On the other hand, labeled core proteins appear immediately in complete virions (SUNDQUIST et al. 1973b). These results are most consistent with the idea that mature virions form by the insertion of viral DNA and core proteins into preassembled capsids, and this process has been confirmed by studies using infected cells with ts mutants defective in virion assembly (see below).

Analysis of the protein composition of empty capsids from Ad2- and Ad3-infected cells also supports this contention. Capsid structures lacking DNA, purified from CsCl equilibrium gradients, contain no core protein V and VII (MAIZEL et al. 1968; PRAGE et al. 1972; ISHIBASHI and MAIZEL 1974), and proteins VI, VIII, and X also appear to be missing (SUNDQUIST et al. 1973b). However, several other polypeptides not present in mature virions were identified in incomplete particles (PRAGE et al. 1972; SUNDQUIST et al. 1973b; WADELL et al. 1973; ISHIBASHI and MAIZEL 1974; ROSENWIRTH et al. 1974; WINBERG and WADELL, 1977; EDVARDSSON et al. 1976). Some of these polypeptides have been identified as precursor polypeptides to virion proteins and designated pVI and pVIII (ANDERSON et al. 1973; ÖBERG et al. 1975). The small virion polypeptides, X–XII, might be products of these cleavage processes. Other polypeptides are found only in empty capsids and may be scaffolding proteins, such as the 50-kDa, the 39- to 40-kDa, and the 32- to 33-kDa proteins (EDVARDSSON et al. 1976; D'HALLUIN et al. 1978a). They are rapidly incorporated into empty capsids, but are not present as stable polypeptides in mature virions; some of them are proteolytically cleaved at a late stage in the assembly pathway.

Combining the results from studies on kinetics of labeling and protein composition, it appears that when empty capsids are assembled they contain precursor polypeptides and possibly scaffolding proteins; viral DNA and core proteins V and pVII are then incorporated either separately or together, and the final step involves proteolytic cleavage of the precursor polypeptides.

In the absence of an in vitro system, it is difficult to understand in detail the mechanism of formation of empty capsids. It seems that the self-assembly of capsomers is not sufficient and that the formation of assembly intermediates requires scaffolding proteins, as previously shown in prehead assembly in the bacteriophage system. The adenovirus assembly intermediates are fragile, and DNA in the nucleoprotein is easily sheared during the extraction procedure or/and isopyknic centrifugation in CsCl gradients (EDVARDSSON et al. 1976). After fixation by reversible cross-linkers, it is possible to isolate the "labile" intermediates in CsCl gradients and to investigate their polypeptide composition after reversion of the cross-linkage (D'HALLUIN et al. 1978a). By sucrose gradient centrifugation, two assembly intermediates sedimenting at 750S and 600S can be isolated from cells infected by Ad2. Following reversible cross-linking, the 750S species, which accumulates during chase, has the properties expected of complete virions. It has a density of 1.34 g/cm$^3$ in CsCl gradients and it contains full-length viral DNA. The polypeptide composition is similar to that identified in young virions. Both mature and young virions will sediment with the same velocity and equilibrate at the same density, and they can therefore be differentiated only by sodium dodecyl sulfate polyacrylamide electrophoresis (SDS-PAGE). The 600S intermediate peak, upon similar treatment, is resolved into two components with densities of 1.315 g/cm$^3$ and 1.37 g/cm$^3$. The lighter class contains precursor polypeptides pVI and pVIII, but no core polypeptides V or pVII/ VII, and it is associated with DNA that sediments at 11S in alkaline sucrose gradients. The heavier intermediates contain viral DNA with a sedimentation coefficient of 34S in alkaline sucrose gradients, but again no core polypeptides. Another difference between the two classes of intermediates is the presence in the lighter, but absence in the heavier of the polypeptides with apparent molecular weights of 50, 39, and 32–33 kDa. These proteins probably are scaffolding proteins necessary for forming the light intermediates (D'Halluin et al. 1978a). Using partial proteolysis, Persson et al. (1979) have shown that the 50-kDa protein is the IVa2 gene product, though the IVa2 protein is synthesized directly in its final form without maturation (54–56 kDa) and is present in a small amount in assembly intermediates (Winter and D'Halluin 1991). Weber and coworkers suggest that the 50-kDa protein is the core protein V (Weber and Khittoo 1983). Using specific antibodies, the 50-kDa protein has been shown to be different from core polypeptide V and protein IVa2 (Winter and D'Halluin 1991) and perhaps to be similar to the 52-kDa protein L1. The two other (39–40 and 32–33 kDa) presumptive scaffolding proteins have not yet been shown to be virally coded, but the 32- to 33-kDa protein is perhaps a gene product of the L4 region, as suggested by Morin and Boulanger (1984).

In kinetic labeling experiments, the light intermediates are detected before the heavy intermediates, and the label can be chased partially into the heavier form and finally into structures with a density of 1.34 g/cm$^3$, reminiscent of young virions (D'Halluin et al. 1978a). Although the capsid proteins accumulate in large excess, the formation of assembly intermediates requires continuous protein synthesis, perhaps for scaffolding protein(s) (D'Halluin et al. 1978b). On the basis of these observations, the scheme of viral assembly and maturation given in Fig. 1 can be proposed (D'Halluin et al. 1978a,b; D'Halluin 1980). The notion that the light intermediate is a true intermediate in virion assembly is supported by the results of experiments performed with the mutant H2ts112. At the nonpermissive temperature, cells infected by this mutant accumulate light intermediates. Following a temperature shiftdown, label incorporated into such light intermediates at the restrictive temperature can be chased into the heavy intermediates and then into the 750S form under conditions in which protein synthesis is inhibited by cycloheximide (D'Halluin et al. 1978b). The H2ts112 mutant was mapped in the IIIa gene, in which two amino acid changes were found (D'Halluin et al. 1982; Chroboczek et al. 1987). Similar results were obtained with the analyzed ts mutants in the IIIa region, H5ts 58 (Edvardsson et al. 1978) and H2ts 4 (Khittoo and Weber 1977); the pathway of virion assembly is blocked at the light intermediate stage, and a part of these particles can be matured after shift at the permissive temperature. It seems that the protein IIIa can directly or indirectly regulate encapsidation of the viral DNA.

Similarly, mutation in the L1 52- and 55-kDa proteins leads to accumulation of intermediate particles (Hasson et al. 1989). The nuclear phosphoproteins of 52 and 55 kDa have not been detected in mature virions (Lewis et al. 1985). In cells

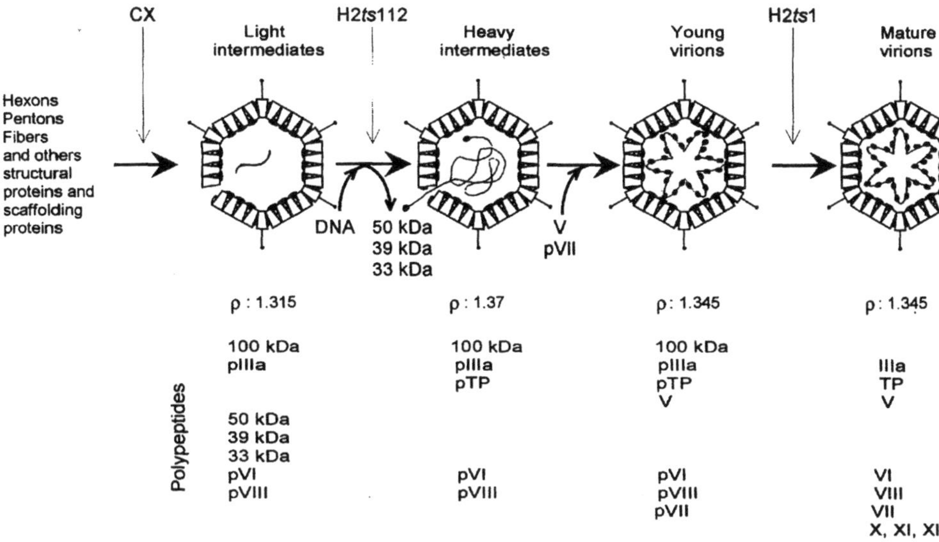


**Fig. 1.** Assembly pathway for adenoviruses. The structural proteins are synthesized and rapidly transported to the nucleus in the form of capsomers for hexon and penton. From subunits such as hexons or nanomers of hexons empty capsids are formed; this step requires scaffolding protein(s). The main assembly intermediates observed will contain some viral proteins absent in the mature virion. The light intermediates become heavy intermediates after the release of scaffolding protein(s) and association with viral DNA. The next step involves the young virions, where several polypeptides are present in a precursor form; this is followed by the final step involving proteolytic cleavage of at least five polypeptides in the virion structure. The different stages of virus assembly have been indicated at the *top* and the polypeptides only present in some structures are indicated *below*; prototype of defective mutants in the assembly pathway are indicated at the top of the *arrow* to the right side. CX, cycloheximide

infected with the mutant H5ts 369, the altered proteins (52 and 55 kDa) appear to be involved in the encapsidation step of virion assembly, but the proteins themselves are not components of mature virions. By immunoelectron microscopic studies, the 52- and 55-kDa proteins are found in close association with structures that appear to contain assembly virions (Hasson et al. 1989, 1992). These findings suggest that the 52- and 55-kDa proteins may be scaffolding proteins, while the 50-kDa protein may be considered an assembly intermediate. The products encoded by the L1 region (52- and 55-kDa and IIIa proteins) are involved in the encapsidation of viral DNA; the L1 region appears to be the DNA encapsidation operon.

## 4 DNA Encapsidation

In the case of bacteriophage λ, it has clearly been shown that the process by which phage DNA is selected from a pool of intracellular DNA for packaging involves the recognition of specific viral sequences, termed *cos* sites, by a phage-encoded protein, the terminase (Earnshaw and Casjens 1980; Feiss 1986). Binding of the terminase to its recognition sites results in the formation of a specific nucleoprotein complex, which is required for recognition and further interactions of the phage DNA with the empty proheads (EARNSHAW and CASJENS 1980; FEISS 1986). In adenovirus assembly, it is well established that the viral DNA is encapsided in empty capsids and viral sequences are required for its packaging. However, little information is available about the structure of viral DNA and its possible association with core proteins. Before discussing the mechanism of insertion of viral DNA into empty capsids and DNA sequences required, we should summarize the properties of the defective particles recovered from adenovirus-infected cells and the structure of the viral chromatin late in infection.

## 4.1 Defective Particles

Apart from mature virions on the one hand and structures lacking viral DNA on the other, various other defective particles can be isolated in CsCl gradients by virtue of their buoyant densities. The infectivity associated with these particles is orders of magnitude less than that of virions and is probably due to contamination with trace amount of infectious virions (PRAGE et al. 1972; WADELL et al. 1973; BURLINGHAM et al. 1974; NIIYAMA et al. 1975). Such defective particles have been isolated from cells infected by several adenoviral serotypes: Ad2 (RAINBOW and MAK 1970; BURLINGHAM et al. 1974; ROSENWIRTH et al. 1974; DANIELL 1976; TIJA et al. 1977), Ad3 (DANIELL 1976; WADELL et al. 1973), Ad7 (TIBBETTS 1977), Ad12 (MAK 1971; RAINBOW and MAK 1970; BURLINGHAM et al. 1974; MAK et al. 1979), canine adenovirus 1 (CAV1; SIRA et al. 1989), and equine adenovirus (ISHIYAMA et al. 1986). The viral factors that influence the production of defective particles are unknown, but the serial passage of stocks at high multiplicities of infection increased the yield of some types of particles (MAK 1971). The yields of incomplete particles vary considerably from one serotype to another. About 30% of the particles made in Ad3-infected cells are defective and form two characteristic bands in CsCl gradients (PRAGE et al. 1972; DANIELL 1976). In contrast, infections by Ad2 or Ad12 yield lower amounts of such particles (RAINBOW and MAK 1970; ROSENWIRTH et al. 1974; DANIELL 1976; TIJA et al. 1977). These defective particles contain a deletion of the viral genome that causes the change in buoyant density. These particles do not seem to be interfering (DANIELL 1976) and retain the ability to transform rodent cells (SCHALLER and YOHN 1974; IGARISHI et al. 1975), suggesting that the left-hand end of the genome is overrepresented.

Different populations of defective particles apparently contain cellular or viral DNA that varies both in quantity and nature. It has been claimed that they contain cellular DNA (TIJA et al. 1977; HAMMARSKJÖLD et al. 1977; KHITTOO and WEBER 1981), but precautions to remove unspecifically absorbed DNA by DNase digestion have not always been taken. In one report, using a two-step hybridization procedure cellular DNA was found to be covalently linked to viral DNA in defective particles (TIJA et al. 1977). It remains to be shown that such packaging is functionally significant, since cellular DNA can associate nonspecifically with empty capsids in vitro (TIBBETTS and GIAM 1979). The viral DNA contained in incomplete particles is very heterogeneous, ranging from about 10% of genome length to full size (BURLINGHAM et al. 1974; DANIELL 1976; TIBBETTS 1977). DNA of different densities isolated from defective particles of Ad2, Ad3, and Ad7 has been characterized by restriction endonuclease analysis, end-labeling, and heteroduplex mapping (DANIELL 1976; TIBBETTS 1977). Most molecules contain the normal left end of the adenoviral genome with a variable right terminus (TIBBETTS 1977). It seems clear that there is preferential packaging of DNA sequences starting at the left end of the genome, although in infected cells both ends of the genome were equally represented in the pool of subgenomic DNA molecules (DANIELL and MULLENBACH 1978). The presence of the recognition signal for packaging into empty capsids was first demonstrated by HAMMARSKJÖLD and WINBERG (1980): using plaque-purified Ad16 after multiple passages, they showed that duplication of the left-hand end at the right-hand end of the genome could encapside either end into defective particles. DNA contained in these particles from cells infected with several variants of multiple-passage virus was analyzed; 390 bp from the left-hand end are sufficient to allow packaging of DNA from the right-hand end and 290 bp appear to be insufficient. In conclusion, there is a sequence between 290 and 390 nucleotides from the left-hand end of the viral genome that is required for the packaging of Ad16 DNA (HAMMARSKJÖLD and WINBERG 1980). The left end of Ad3 DNA is also recognized as an encapsidation sequence (KOSTURKO et al. 1982). The packaging sequence is located on the right side of the left inverted terminal repeat (ITR).

## 4.2 Polarity and Sequences Required

A *cis*-acting packaging element has been localized in the left-hand end of the adenovirus chromosome during analysis of the E1A transcriptional control region (HEARING and SHENK 1983). The identified *cis*-acting packaging domain is located between nt 194 and 358 at the left end of the Ad5 genome (HEARING et al. 1987). This packaging domain is composed of at least five elements that are functionally redundant. Repeated sequences, termed the A repeat (consensus, A / TANA / T T T TG), were associated with packaging functions. The *cis*-acting components of the packaging signal appear to be subject to certain spatial constraints for function, possibly reflecting the coordinate binding of packaging proteins to these sites and from localization near the end of the viral DNA (GRÄBLE and HEARING 1990, 1992; for a review see P. HEARING, this volume).

## 4.3 Structure of Adenovirus Chromatin During the Encapsidation Process

Infectious adenovirus particles do not contain histones; the viral DNA is condensed into a core structure by viral basic protein V, VII, and μ (CORDEN et al. 1976;

D.T. Brown et al. 1975; Nermut 1978; Black and Center 1979; Sato and Hosokawa 1981; Vayda et al. 1983; Weber et al. 1983; Newcomb et al. 1984; Chatterjee et al. 1986a, b; Anderson et al. 1989; Wong and Hsu 1989). Electron microscopy reveals a nucleosome-like structure, but micrococcal nuclease digestion yields monomeric and heterogeneous DNA fragment lengths unlike the regular repeat pattern obtained with cellular chromatin (Corden et al. 1976; Mirza and Weber 1981, 1982; Vayda et al. 1983). During the lytic cycle, shortly after infection viral DNA acquires a nuclesomal structure similar to that of the host cell, suggesting a viral DNA–histone association (Daniell et al. 1981; Sergeant et al. 1979; Tate and Philipson 1979). Late in the productive cycle, cellular DNA and histone synthesis is inhibited and viral chromatin takes on a form similar to the viral core (M. Brown and Weber 1980; Daniell et al. 1981; Toth and Weber 1984). Viral nucleoprotein complexes have been isolated and shown to be similar, but not identical, to the core present in young virions (Weber and Philipson 1984). Kinetic experiments suggest that the DNA synthesized late in infection is condensed by core proteins and is destined for encapsidation (Dery et al. 1985). DNA–protein associations could dissociate during the procedure of purification of assembly intermediates using ammonium sulfate (D'Halluin et al. 1978a), although the core proteins dissociate from the viral DNA only at high salt concentration after disruption of young and mature virions. A possible explanation is that the viral DNA–core protein association is different before and after encapsidation and the heavy intermediate represents a very labile and quickly processed particle.

## 4.4 DNA Synthesis and Encapsidation

The first observation, by electron microscopy of an intermediate in adenovirus assembly that comprises a viral particle directly associated with replicating DNA, provided some evidence for the direct insertion of DNA into the empty capsid (Moncany et al. 1980). In some experiments, the inhibition of the viral replication by hydroxyurea did not affect the rate of virus assembly (D'Halluin et al. 1980b), but in others opposite results have been obtained (Weber et al. 1985).

The L1 52- and 55-kDa proteins, associated with assembling particles, were localized to regions within the infected cells nucleus that are distinct from viral DNA replication centers, indicating that replication and assembly of viral particles might occur in separate nuclear compartments (Hasson et al. 1992). The adenovirus assembly centers are found at the periphery of the nucleus (Puvion-Dutilleul et al. 1984; Hasson et al. 1989). Finally, the adenovirus assembly occurs in association with the nuclear matrix (NM; Khitto et al. 1986). Viral DNA appears also associated with the NM, and its association is mediated by the terminal protein (Schaack et al. 1990). Even though DNA replication and virus assembly seem to occur in separate nuclear compartments and the assembly pathway is not inhibited by inhibitors of DNA synthesis, it is unlikely that the capsid assembles as an independent structure, separate from the viral chromosome. If empty capsids were to assemble in the absence of viral DNA, then substantial quantities of such structures would accumulate in cells infected with H5*dl* 309-A5, a deletion mutant with a crippled *cis*-acting packaging element whose DNA is assembled into virions very inefficiently (HEARING and SHENK 1983). However, no empty capsids accumulate in H5*dl*309-A5-infected cells, suggesting that capsid formation occurs in the presence of DNA molecules carrying a functional *cis*-acting packaging element (HASSON et al. 1992). Analysis with several ts mutants and novobiocin suggests that factors implicated in the formation of the DNA synthesis initiation complex may play a role in viral DNA encapsidation (D'HALLUIN et al. 1980b). A rate-limiting step appears to occur after assembly of empty capsids, which results in accumulation of intermediate particles; this step is dependent on modifications in the conformation of the viral DNA.

## 5 Maturation of Young Virions

After DNA packaging, resulting young virions, harboring five precursor polypeptides, were processed into infectious viruses by proteolytic cleavages of precursors. The first evidence was provided by pulse-chase labeling experiments and analysis of particles in CsCl gradients (ISHIBASHI and MAIZEL 1974; SUNDQUIST et al. 1973b). Strong evidence indicating the necessity of proteolytic cleavage for virion maturation has come from a study of the mutant H2ts1 (WEBER et al. 1975; WEBER 1976). Cells infected with H2ts1 and maintained at the restrictive temperature accumulate virion structures, but none of the precursor polypeptides pVI, pVII and pVIII is processed. Thus, the lesion in H2ts1 blocks the appearance of six posttranslational products, polypeptides VI, VII, VIII, X, XI, and XII, all of which are normally associated with mature virions. Two other viral proteins are processed during the final step in maturation: the terminal protein covalently linked to the DNA, which is synthesized as an 87-kDa species, is cleaved into the 55-kDa species during virion maturation (CHALLBERG et al. 1980; LEWIS and MATHEWS 1980; STILLMAN et al. 1981); the polypeptide IIIa synthesized as a precursor form of the 67-kDa protein is processed to the 66-kDa species (BOUDIN et al. 1980; CUILLEL et al. 1987). The proteolytic cleavages necessary to generate mature virion proteins are inhibited by temperature shift up, but occur after temperature shift down of cells infected with H2ts1 (WEBER 1976). The young virions accumulated at the nonpermissive temperature are not infectious (MIRZA and WEBER 1979; HANNAN et al. 1983).

The H2ts1 mutation has been localized by restriction endonuclease analysis of intertypic ts$^+$ recombinants (HASSELL and WEBER 1978) in the L3 region. The gene identified by sequencing encodes a 23-kDa protein (KRUIJER et al. 1980; AKUSJÄRVI et al. 1981). The endopeptidase activity has been detected in nuclei from infected cells and inside mature virions, but not in the H2ts1 young virions accumulated at the nonpermissive temperature (BHATTI and WEBER 1979). The endopeptidase is a phosphoprotein enzyme that could be matured by autocatalysis (CHATTERJEE and FLINT 1987). Comparisons of the sequences of proteins VI and VII with the translated gene sequences of their precursors suggest the sequences cleaved in these proteins (Sung et al. 1977, 1983; Akusjärvi and Persson 1981). The similarities of amino acids in the vicinity of the cleavage sites in pVI and pVII led Tremblay et al. (1983) to propose that the proteinase has a specificity for Gly—Ala bonds. Sequence analysis of the other Ad2 proteinase substrates led these authors to suggest that Ala—Gly bonds may also be cleaved. Based on an assumed preference for bonds containing glycine and alanine, Hannan et al. (1983) predicted possible cleavage sites in pVIII. A similar sequence in the pTP (MTGG V) could be a target for the proteinase activity (Smart and Stillman 1982). Finally, the cleavage specificity has been determined using synthetic peptides. All peptides cleaved are of the form M(L)XGX G or M(L)XGG X (Webster et al. 1989). The adenoviral proteinase recognizes a specific secondary structure formed by a sequence of at least five amino acids, the main determinants of specificity lying two and four residues to the N-terminal side of the bond cleaved (Webster et al. 1989). Cleavage sites are conserved in the precursor pVI of Ad3 and Ad12 (Heysen et al. 1991; Freimuth and Anderson 1993). Similar proteinase cleavage sites occur in some protamine precursors (Wouters-Tyrou et al. 1991).

A purified recombinant endoprotease protein shows no activity (Anderson 1990), but protease activity is restored by addition of H2*ts*1 young virions, suggesting the presence of cofactors. Two cofactors are required for proteinase activity: viral DNA and a viral peptide present at the carboxy terminus of pVI (11 amino acids; Mangel et al. 1993). The viral DNA and C-terminal peptide of pVI are required for proteinase activity to ensure that virion precursor proteins are processed only after virion assembly. The use of DNA as cofactor for a proteinase activity is unprecedented (Mangel et al. 1993).

Antigenic analysis of the intermediates during virus assembly suggests that several proteins in the vertex region (penton base, fiber, and IIIa) are not recognized by antibodies to mature structural units until the final maturation involving proteolytic cleavage has occurred. These results suggest that the conformation of the vertex protein changes during the maturation process (D'Halluin et al. 1980a).

When young virions accumulated at the nonpermissive temperature in cells infected with H2*ts*1 and H2*ts*104 were dissociated with a mild detergent such as sodium deoxycholate (DOC), the core structure obtained with mature virions by the same treatment was destroyed (Mirza and Weber 1977; D'Halluin et al. 1980a). Thereby the low affinity of precursor pVII for viral DNA at low ionic strength (Everitt et al. 1977; Mirza and Weber 1977) was confirmed, though the interaction of protein VII and its precursor with viral DNA appears to be similar in in vitro studies (Chatterjee et al. 1986b).

## 6 Genetics of Assembly

Knowledge of the pathway of adenovirus assembly has benefited from studies with conditional mutants. Almost all ts mutants that affect proteins in the L2, L3, and L4 regions of the late transcription unit fail to accumulate empty capsids and assembly intermediates (EDVARDSSON et al. 1978; MARTIN et al. 1978), except H2*ts*1, mutated in the endoprotease gene (L3 region), which accumulates young virions at the nonpermissive temperature (WEBER 1976). The proteins encoded by regions L2, L3, and L4 are all structural proteins required for assembly intermediates, except proteins V and pVII. No ts mutant has been described in gene-encoding viral basic proteins. Ts mutants defective in the formation of the fiber encoded by the L5 region form a small amount of empty capsids and a very small quantity of mature virions (EDVARDSSON et al. 1978; D'HALLUIN et al. 1980a; CHEE-SHEUNG and GINSBERG 1982); similar results are obtained with fiber-deletion mutants (FALGOUT and KETNER 1987; 1988).

The L1 region encodes the 52- and 55-kDa proteins and the IIIa protein; cells infected with ts mutants in this region, H5*ts* 369 (52 and 55 kDa) and H2*ts* 112 (IIIa), accumulate light assembly intermediates at the nonpermissive temperature (HASSON et al. 1989; D'HALLUIN et al. 1978b). The L1 gene products are involved in the encapsidation of the viral DNA.

In the capsid, the polypeptide IX appears to associate with a group-of-nine hexon unit, and it has been suggested that this polypeptide plays a cementing role in the capsid structure (EVERITT et al. 1975; BOULANGER et al. 1979). This function has been confirmed by the study of the deletion mutant H5*dl* 313, which lacks the polypeptide IX gene; the mutant virions appear to be more heat labile than wt particles. Therefore, the polypeptide IX is not essential for assembly of Ad5 virions (COLBY and SHENK 1981).

It is possible to divide the adenovirus mutants defective in virus assembly into six groups:

1. Mutants defective in the formation of capsomers with a mutation in the gene whose product is absent in the mature virion (e.g., L4 100-kDa protein, involved in the trimerization of the hexon polypeptides).
2. Mutants defective in the formation of capsomers with mutations in genes encoding structural proteins, thereby failing to form capsids.
3. Mutants blocked in the stage of light assembly intermediates; the mutations are located in the L1 region.
4. Mutants with reduced yields of all viral particles such as the fiber mutants.
5. Mutants in the endoprotease gene (L3); cells infected with this mutant produced only young virions at the nonpermissive temperature.
6. Mutants affected in the stability of the viral particles, e.g., when the polypeptide IX gene is deleted.

## 7 In Vitro Assembly

In order to understand in detail the mechanism of assembly of virions, the availability of an in vitro system would be necessary. The complexity of the adenovirus capsid and the number of different capsomers do not permit the elaboration of an in vitro system, but a number of attempts to generate capsid-like structures with different virion substructures have been successful. Hexon capsomers, purified from infected cells, may be used to assemble nanomers (group-of-nine) of hexons, associated with the faces of the icosahedron (PETTERSSON 1971; BOULANGER 1975). The nanomers of hexons, purified from virions disrupted with sodium deoxycholate at 56°C, can reaggregate at low pH values into pairs and rings of five nanomers. In some cases, they can form icosahedral shells containing 20 nanomers; electron microscopic and ultracentrifugation analyses indicate that they resemble adenoviral capsids that lack the 12 vertices, each of which comprises a penton and five peripentonal hexons (PEREIRA and WRIGLEY 1974). These results suggest that the nanomers of hexons, isolated from disrupted virions, may self-assemble to form a cage-like structure. However, the possibility that the hexon-associated proteins VIII (pVIII) and IX have a role in the assembly process in vivo is not demonstrated. The precursor of polypeptide VIII as well as polypeptide IX is involved in the stability of the capsid (LIU et al. 1985). The precursor of polypeptide VI is involved in the transport of the hexon capsomers to the nucleus (KAUFFMAN and GINSBERG 1976). Polypeptide IX appears to be nonessential for the assembly of the viral shell (COLBY and SHENK 1981).

An in vitro assembly system for pentons has been described (BOUDIN and BOULANGER 1982) using purified fiber from Ad2 wt infection and penton base accumulated in cells infected with H2*ts* 125, a fiber-defective ts mutant (BOUDIN et al. 1983). The assembly occurs over a broad pH range (5.5–9) and at ionic strength from 0.05 to 1.0. The association between the penton base and the fiber is a reversible reaction with a dissociation constant of $2.10^{-7}$ (BOUDIN and BOULANGER 1982).

The penton units and protein IIIa are also associated with the empty capsids; this particle cannot be reconstructed in vitro, suggesting that several additional steps are involved in vivo. It therefore appears unlikely that formation of empty capsids can be explained by a self-assembly mechanism. Scaffolding proteins are required in the first step of intermediate assembly (EDVARDSSON et al. 1976; D'HALLUIN et al. 1978a). Formation of an empty capsid appears to be a prerequisite, however, for the insertion of the viral DNA into the virion.

In contrast to the λ system, we do not have an in vitro adenovirus packaging system is not at our disposal, although *cis*-acting sequences required for DNA encapsidation were identified in vivo. Cellular DNA readily associates with empty capsids, but fails to encapside into adenovirus empty capsids (TIBBETTS and GIAM 1979). The development of an in vitro packaging system will facilitate the understanding of the viral DNA insertion in the empty capsid and allow the use of adenovirus vectors in gene therapy by increasing the size of foreign genes in adenovirus vector. This development of an in vitro packaging system requires identification of protein(s) which control direct elimination of scaffolding proteins and recognize the *cis*-acting sequence involved in viral DNA packaging.

## 8 Conclusion and Perspectives

The assembly of adenovirus particles is a complex process involving first the assembly of major capsomers by oligomerization of individual polypeptide chains. From the capsomers, the capsid structure is then generated in a process involving four additional structural polypeptides and three nonstructural polypeptides, probably as scaffolding proteins. At least five virion polypeptides are present as precursor proteins, i.e., pVI, pVII, pVIII, pIIIa, and pTP, and several classes of particles appear to be assembly intermediates on the pathway to infectious virions. Figure 1 show a tentative diagram of the assembly pathway. The light intermediates appear to contain all the virion proteins except for the core proteins. In addition, three polypeptides (50, 39–40, and 32–33 kDa), are associated with these particles and may be scaffolding proteins. The DNA is then inserted into light intermediates via a specific recognition sequence located at the left-hand end of the genome. The recognized sequence is made up of six repeats. The DNA encapsidation requires the removal of scaffolding proteins of assembly intermediates, perhaps by proteolysis. The insertion of core proteins which leads to formation of heavy intermediates is the next step. These young virions contain the precursor polypeptides in an uncleaved form. After activation of the endoprotease activity by viral DNA and a peptide, the precursor polypeptides are cleaved at specific sites. The virion is now infectious and the viral cycle is complete.

The development of an in vitro assembly system would permit the identification, in detail, of the mechanisms involved in the assembly pathway and the development of a packaging system of adenoviral vectors for use in gene therapy.

*Acknowledgments.* Some of results described here were obtained in collaboration with my colleagues, in particular P.A. Boulanger, C. Cousin, G.R. Martin, M. Milleville, and N. Winter. I thank Dr. Chroboczek and Dr. Caillet-Boudin for reading the manuscript and for fruitful discussion. This work was supported by the Institut National de la Santé et de la Recherche Médicale, the Université de Lille II, the Fondation pour la Recherche Médicale, and the ARC (Association pour la Recherche contre le Cancer, grant no. 415-89).

## References

Akusjärvi G, Persson H (1981) Gene and mRNA for precursor polypeptide VI from adenovirus type 2. J Virol 38: 469–482

Akusjärvi G, Zabielski J, Perricaudet M, Pettersson U (1981) The sequence of the 3' non-coding region of the hexon mRNA discloses a novel adenovirus gene. Nucleic Acids Res 9 : 1–17

Anderson CW (1990) The proteinase polypeptide of adenovirus serotype 2 virions. Virology 177: 259–272

Anderson CW, Baum PR, Gesteland RF (1973) Processing of adenovirus type 2 proteins: hexon, fiber, component IX, early protein 1B–15K. Virology 104: 27–41

Anderson CW, Young ME, Flint SJ (1989) Characterization of the adenovirus 2 virion protein, mu. Virology 172: 506–512

Bhatti AR, Weber J (1979) Protease of adenovirus type 2. Subcellular localization. J Biol Chem 254: 12265–12268
Black BC, Center MS (1979) DNA-binding properties of the major core protein of adenovirus 2. Nucleic Acids Res 6: 2329–2353
Boudin ML, Boulanger PA (1981) Antibody-triggered dissociation of adenovirus penton capsomer. Virology 113: 781–786
Boudin ML, Boulanger PA (1982) Assembly of adenovirus penton base and fiber. Virology 116: 589–604
Boudin ML, Moncany M, D'Halluin JC, Boulanger PA (1979) Isolation and characterization of adenovirus type 2 vertex capsomer (penton base). Virology 92: 125–128
Boudin ML, D'Halluin JC, Cousin C, Boulanger PA (1980) Human adenovirus type 2 protein IIIa. Maturation and encapsidation. Virology 101: 144–156
Boudin ML, Rigolet M, Lemay P, Galibert F, Boulanger P (1983) Biochemical and genetical characterization of a fiber-defective temperature-sensitive mutant of type 2 adenovirus. EMBO J 2: 1921–1927
Boulanger PA (1975) Adenovirus assembly: self-assembly of partially digested hexons. J Virol 16: 1678–1682
Boulanger PA, Lemay P, Blair GE, Russell WC (1979) Characterization of adenovirus protein IX. J Gen Virol 44: 783–800
Brown DT, Westphal M, Burlingham BT, Winterhoff U, Doerfler W (1975) Structure and composition of the adenovirus type 2 core. J Virol 16: 366–387
Brown M, Weber J (1980) Virion core-like organisation of intranuclear adenovirus chromatin late in infection. Virology 107: 306–310
Burlingham BT, Brown DT, Doerfler W (1974) Incomplete particles of adenovirus. I. Characteristics of the DNA associated with incomplete adenovirions of types 2 and 12. Virology 60: 419–430
Caillet-Boudin ML (1989) Complementary peptide sequences in partner proteins of the adenovirus capsid. J Mol Biol 208: 195–198
Cepko CL, Sharp PA (1982) Assembly of adenovirus major capsid protein is mediated by a non-virion protein. Cell 31: 407–415
Cepko CL, Changelian PA, Sharp PA (1981) Immunoprecipitation with two-dimensional pools as a hybridoma screening technique: production and characterization of monoclonal antibodies against adenovirus 2 proteins. Virology 110: 385–401
Challberg MD, Desiderio SV, Kelly TJ Jr (1980) Adenovirus DNA replication in vitro: characterization of a protein covalently linked to nascent DNA strands. Proc Natl Acad Sci USA 77: 5105–5109
Chatterjee PK, Flint SJ (1987) Adenovirus type 2 endopeptidase: an unusual phosphoprotein enzyme matured by autocatalysis. Proc Natl Acad Sci USA 84: 714–718
Chatterjee PK, Vayda ME, Flint SJ (1986a) Identification of proteins and protein domains that contact DNA within adenovirus nucleoprotein cores by ultraviolet light crosslinking of oligonucleotides 32P-labelled in vivo. J Mol Biol 188: 23–37
Chatterjee PK, Yang UC, Flint SJ (1986b) Comparison of the interactions of the adenovirus type 2 major core protein and its precursor with DNA. Nucleic Acids Res 14: 2721–2735
Chee-Sheung CC, Ginsberg HS (1982) Characterization of a temperature-sensitive fiber mutant of type 5 adenovirus and effect of the mutation on virion assembly. J Virol 42: 932–950
Chroboczek J, Viard F, D'Halluin JC (1987) Human adenovirus temperature-sensitive mutant 112 contains three mutations in the protein IIIa gene. Gene 49: 157–160
Colby WW, Shenk T (1981) Adenovirus type 5 virions can be assembled in vivo in the absence of detectable polypeptide IX. J Virol 39: 977–980
Corden J, Engelking M, Pearson GD (1976) Chromatin like organization of the adenovirus chromosome. Proc Natl Acad Sci USA 73: 401–404
Crick FHC, Watson JD (1957) Structure of small viruses. Nature 177: 473–476
Cuillel M, Milleville M, D'Halluin JC (1987) Expression of the human Ad2 IIIa protein in E coli. Gene 55: 295–301
Daniell E (1976) Genome structure of incomplete particles of adenovirus. J Virol 19: 685–708
Daniell E, Mullenbach T (1978) Synthesis of defective viral DNA in HeLa cells infected with adenovirus type 3. J Virol 26: 61–70
Daniell E, Groff DE, Fedor MJ (1981) Adenovirus chromatin structure at different stages of infection. Mol Cell Biol: 1094–1105
Dery CV, Toth M, Brown M, Horvath J, Allaire S, Weber JM (1985) The structure of adenovirus chromatin in infected cells. J Gen Virol 66: 2671–2684
Devaux C, Boulanger P (1980) Reactive serine in human adenovirus hexon polypeptide. Virology 102: 94–106

Devaux C, Caillet-Boudin ML, Jacrot B, Boulanger P (1987) Crystallization, enzymatic cleavage, and the polarity of the adenovirus type 2 fiber. Virology 161: 121–128
D'Halluin JC (1980) Génétique de l'adénovirus. Bull Pasteur Paris 78: 347–405
D'Halluin JC, Martin GR, Torpier G, Boulanger PA (1978a) Adenovirus type 2 assembly analyzed by reversible cross-linking of labile intermediates. J Virol 26: 357–363
D'Halluin JC, Milleville M, Boulanger PA, Martin GR (1978b) Temperature-sensitive mutant of adenovirus type 2 blocked in virion assembly: accumulation of light intermediate particles. J Virol 26: 344–356
D'Halluin JC, Milleville M, Martin GR, Boulanger PA (1980a) Morphogenesis of human adenovirus 2. Study with fiber⁻ and fiber–penton base⁻ defective temperature-sensitive mutants. J Virol 33: 88–99
D'Halluin JC, Milleville M, Boulanger PA (1980b) Effects of novobiocin on adenovirus DNA synthesis and encapsidation. Nucleic Acids Res 8: 88–99
D'Halluin JC, Cousin C, Boulanger PA (1982) Physical mapping of adenovirus type 2 ts mutations by restriction nuclease analysis of interserotypic recombinants. J Virol 41: 410–413
Earnshaw WC, Casjens SR (1980) DNA packaging by the double-stranded DNA bacteriophages. Cell 21: 319–331
Edgar RS, Wood WB (1966) Morphogenesis of bacteriophage T4 in extracts of mutant-infected cells. Proc Natl Acad Sci USA 55: 498–502
Edvardsson B, Everitt E, Jörnvall H, Prage L, Philipson L (1976) Intermediates in adenovirus assembly. J Virol 19: 533–547
Edvardsson B, Ustacelebi S, Williams J, Philipson L (1978) Assembly intermediates among adenovirus type 5 temperature-sensitive mutants. J Virol 25: 641–651
Everitt E, Lutter L, Philipson L (1975) Structural proteins of adenoviruses. XII. Location and neighbor relationship among proteins of adenovirion type 2 as revealed by enzymatic iodination, immunoprecipitation and chemical cross-linking. Virology 67: 197–208
Everitt ES, Meador A, Levine AS (1977) Synthesis and processing of the precursor to the major core protein of adenovirus type 2. J Virol 21: 199–214
Falgout B, Ketner G (1987) Adenovirus early region 4 is required for efficient virus particle assembly. J Virol 61: 3759–3768
Falgout B, Ketner G (1988) Characterization of adenovirus particles made by deletion mutants lacking the fiber gene. J Virol 62: 622–625
Feiss M (1986) Terminase and the recognition, cutting and packaging of lambda chromosomes. Trends Genet 2: 100–104
Flint SJ, Broker TR (1981) Lytic infection by adenoviruses. In: Tooze J (ed) The molecular biology of tumor viruses, DNA tumor viruses, 2nd rev edn. Cold Spring Harbor Press, Cold Spring Harbor NY, pp 443–546
Freimuth P, Anderson CW (1993) Human adenovirus serotype 12 virion precursors pMu and pVI are cleaved at amino-terminal and carboxy-terminal sites that conform to the adenovirus 2 endoproteinase cleavage consensus sequence. Virology 193: 348–355
Frost E, Williams J (1978) Mapping temperature-sensitive and host-range mutations of adenovirus type 5 by marker rescue. Virology 91: 39–50
Fuller MT, King J (1980) Regulation of coat protein polymerisation by the scaffolding protein of bacteriophage P22. Biophys J 32: 381–401
Fuller MT, King J (1982) Assembly in vitro of bacteriophage P22 procapsids from purified coat and scaffolding subunits. J Mol Biol 156: 633–665
Ginsberg HS, Young CHS (1977) Genetics of adenoviruses. Compr Virol 9: 27–88.
Gräble M, Hearing P (1990) Adenovirus type 5 packaging domain is composed of a repeated element that is functionally redundant. J Virol 64: 2047–2056
Gräble M, Hearing P (1992) cis and trans requirements for the selective packaging of adenovirus type 5 DNA. J Virol 66: 723–731
Green NM, Wrigley NG, Russell WC, Martin SR, McLahlan AD (1983) Evidence for a repeating cross β-sheet structure in the adenovirus fibre. EMBO J 2: 1357–1366
Hammarskjöld ML, Winberg G (1980) Encapsidation of adenovirus 16 DNA is directed by a small DNA sequences at the left end of the genome. Cell 20: 787–795
Hammarskjöld ML, Winberg G, Norrby E, Wadell G (1977) Isolation of incomplete adenovirus 16 particles containing viral and host cell DNA. Virology 82: 449–461
Hannan C, Raptis LH, Dery CV, Weber J (1983) Biological and structural studies with an adenovirus type 2 temperature-sensitive mutant defective for uncoating. Intervirology 19: 213–223
Hassell JA, Weber J (1978) Genetic analysis of adenovirus type 2 VIII. Physical locations of temperature-sensitive mutations. J Virol 28: 671–678

Hasson TB, Soloway PD, Ornelles DA, Doerfler W, Shenk T (1989) Adenovirus L1 52- and 55-kilodation proteins are required for assembly of virions. J Virol 63: 3612–3621
Hasson TB, Ornelles DA, Shenk T (1992) Adenovirus L1 52- and 55-kilodalton proteins are present within assembling virions and colocalize with nuclear structures distinct from replication centers. J Virol 66: 6133–6142
Hearing P, Shenk T (1983) The adenovirus type 5 E1A transcriptional control region contains a duplicated enhancer element. Cell 3: 695–703
Hearing P, Samulski RJ, Wishart WL, Shenk T (1987) Identification of a repeated sequence element required for efficient encapsidation of the adenovirus type 5 chromosome. J Virol 61: 2555–2558
Heysen A, Verwaerde P, D'Halluin JC (1991) Nucleotide sequence and regulation of the adenovirus type 3 E2A early promoter. Virology 181: 241–250
Horwitz MS, Scharff MD, Maizel JV (1969) Synthesis and assembly of adenovirus 2. I. Polypeptide synthesis, assembly of capsomers and morphogenesis of the virion. Virology 39: 682–694
Igarishi K, Niiyama Y, Tsukamoto K, Kurokawa T, Sugino Y (1975) Biochemical studies on bovine adenovirus type 3. II. Incomplete virus. J Virol 16: 634–641
Ishibashi M, Maizel JV Jr (1974) The polypeptides of adenovirus. V. Young virions, structural intermediate between top components and aged virions. Virology 57: 409–424
Ishiyama T, Shinagawa M, Sato G, Fujinaga K, Padmanabhan R (1986) Generation of packaging-defective DNA molecules of equine adenovirus. Virology 151: 66–76
Kauffman RS, Ginsberg HS (1976) Characterization of a ts, hexon transport mutant of type 5 adenovirus. J Virol 19: 643–658
Khittoo G, Weber J (1977) Genetic analysis of adenovirus type 2. VI. A temperature-sensitive mutant defective for DNA encapsidation. Virology 81: 126–137
Khitoo G, Weber J (1981) The nature of the DNA associated with incomplete particles of adenovirus type 2. J Gen Virol 54: 343–355
Khitoo G, Delorme L, Dery CV, Tremblay ML, Weber JM, Bibor-Hardy V, Simard R (1986) Role of the nuclear matrix in adenovirus maturation. Virus Res 4: 391–403
Kosturko LD, Sharnick SV, Tibbetts C (1982) Polar encapsidation of adenovirus DNA: cloning and DNA sequence of the left end of adenovirus type 3. J Virol 43: 1132–1137
Kruijer W, van Schaik FMA, Sussenbach JS (1980) Nucleotide sequence analysis of a region of adenovirus 5 DNA encoding a hitherto unidentified gene. Nucleic Acids Res 8: 6033–6042
Leibowitz J, Horwitz MS (1975) Synthesis and assembly of adenovirus polypeptides. III. Reversible inhibition of hexon assembly in adenovirus type 5 temperature-sensitive mutants. Virology 66: 10–24
Lewis JB, Mathews M (1980) Control of adenovirus early gene expression: a class of immediate early products. Cell 21: 303–313
Lewis JB, Fahnestock ML, Hardy MM, Anderson C (1985) Presence in infected cells of nonvirion proteins encoded by adenovirus messenger RNAs of the major late transcription regions L0 and L1. Virology 143: 452–466
Liu GQ, Babiss LE, Volkert FC, Young CS, Ginsberg HS (1985) A thermolabile mutant of adenovirus 5 resulting from a substitution mutation in the protein VIII gene. J Virol 53: 920–925
Maizel JV Jr, White DO, Sharff MD (1968) The polypeptides of adenovirus. II. Soluble proteins, cores, top components and the structure of the virion. Virology 36: 126–136
Mak S (1971) Defective virions in human adenovirus type 12. J Virol 7: 426–433
Mak I, Ezoe H, Mak S (1979) Structure and function of adenovirus type 12 defective virions. J Virol 32: 240–250
Mangell WF, McGrath WJ, Toledo DL, Anderson CW (1993) Viral DNA and a viral peptide can act as cofactors of adenovirus virion proteinase activity. Nature 361: 274–275
Martin GR, Waroquier R, Cousin C, D'Halluin JC, Boulanger PA (1978) Isolation and phenotypic characterization of adenovirus type 2 temperature sensitive mutants. J Gen Virol 41: 303–314
Miller JS, Ricciardi RP, Roberts BE, Patterson BM, Mathews MB (1980) Arrangement of messenger RNAs and protein coding sequences in the major late transcription unit of adenovirus 2. J Mol Biol 142: 455–488
Mirza MA, Weber J (1977) Genetic analysis of adenovirus type 2. VII. Cleavage modified affinity for DNA of internal virion proteins. Virology 80: 83–97
Mirza MA, Weber J (1979) Uncoating of adenovirus type 2. J Virol 30: 462–471
Mirza MA, Weber JM (1981) Structure of adenovirus chromatin as probed with restriction endonucleases. Virology 108: 351–360
Mirza MA, Weber J (1982) Structure of adenovirus chromatin. Biochim Biophys Acts 696: 76–86
Moncany ML, Revet B, Girard M (1980) Characterization of a new adenovirus type 5 assembly intermediate. J Gen Virol 50: 33–37

Morin N, Boulanger P (1984) Morphogenesis of human adenovirus type 2: sequence of entry of proteins into previral and viral particles. Virology 136: 153–167
Nermut MV (1978) Structural elements in adenovirus cores. Studies by means of freeze-fracturing and ultrathin sectioning. Arch Virol 57: 323–337
Newcomb WW, Boring JW, Brown JC (1984) Ion etching of human adenovirus 2: structure of the core. J Virol 51: 52–56
Niiyama Y, Igarishi K, Tsukamoto K, Kurokawa T, Sugino Y (1975) Biochemical studies on bovine adenovirus type 3: I. Purification and properties. J Virol 28: 236–248
Novelli A, Boulanger PA (1991) Assembly of adenovirus type 2 fiber synthesized in cell-free translation system. J Biol Chem 266: 9299–9303
Öberg B, Saborio J, Persson T, Everitt E, Philipson L (1975) Identification of the in vitro translation products of adenovirus mRNA by immunoprecipitation. J Virol 15: 199–207
Oosterom-Dragon EA, Ginsberg HS (1981) Characterization of two temperature-sensitive mutants of type 5 adenovirus with mutations in the 100,000-dalton protein gene. J Virol 40: 491–500
Pereira HG, Wrigley NG (1974) In vitro reconstitution, hexon bonding and handedness of incomplete adenovirus capsid. J Mol Biol 85: 617–631
Persson H, Öberg B, Philipson L (1977) In vitro translation with adenovirus polyribosomes. J Virol 21: 187–198
Persson H, Mathisen B, Philipson L, Pettersson U (1979) A maturation protein in adenovirus morphogenesis. Virology 93: 198–208
Pettersson U (1971) Structural proteins of adenoviruses. VI. On the antigenic determinants of the hexon. Virology 93: 198–208
Philipson L (1984) Adenovirus assembly. In: Ginsberg HS (ed) The adenoviruses Plenum, New York, pp 309–337
Prage L, Höglund S, Philipson L (1972) Structural proteins of adenoviruses. III. Characterization of incomplete particles of adenovirus type 3. Virology 49: 745–757
Puvion-Dutilleul F, Pédron J, Cajean-Feroldi C (1984) Identification of intranuclear structures containing the 72K DNA-binding protein of human adenovirus type 5. Eur J Cell Biol 34: 313–322
Rainbow AJ, Mak S (1970) Functional heterogeneity of virions of human adenovirus types 2 and 12. J Virol 5: 188–197
Rosenwith B, Tija S, Westphal M, Doerfler W (1974) Incomplete particles of adenovirus. II. Kinetics of formation and polypeptide composition of adenovirus type 2. Virology 60: 431–437
Sato K, Hosokawa K (1981) The structure of adenovirion chromatin revealed by ultraviolet light-induced cross-linking. Biochem Biophys Res Commun 101: 1318–1323
Schaack J, Ho WY-W, Freimuth P, Shenk T (1990) Adenovirus terminal protein mediates both nuclear matrix association and efficient transcription of adenovirus DNA. Gene Dev 4: 1197–1208
Schaller JP, Yohn DS (1974) Transformation potentials of the non-infectious (defective) component in pools of adenoviruses type 12 and simian adenovirus 7. J Virol 14: 392–401
Sergeant A, Tigges MA, Raskas HJ (1979) Nucleosome-like structural subunits of intranuclear parental adenovirus type 2 DNA. J Virol 29: 888–898
Sira S, Dobbs M, Abouhaidar MG, Campbell JB (1989) Incomplete particles of canine adenovirus type 1: occurrence and structure. Virology 173: 188–195
Smart JE, Stillman BW (1982) Adenovirus terminal protein precursor: partial amino acid sequence and the site of covalent linkage to DNA. J Biol Chem 257: 13499–13506
Stewart PL, Burnett RM, Cyrklaff M, Fuller SD (1991) Image reconstruction reveals the complex molecular organization of adenovirus. Cell 67: 145–154
Stillman BW, Lewis JB, Chow LT, Mathews MB, Smart JE (1981) Identification of the gene and mRNA for the adenovirus terminal protein precursor. Cell 23: 497–508
Stinski MF, Ginsberg HS (1974) Antibody to the type 5 adenovirus hexon polypeptide: detection of nascent polypeptides in cytoplasm of infected KB cells. Intervirology 4: 226–236
Sundquist B, Pettersson U, Thelander L, Philipson L (1973a) Structural proteins of adenoviruses. IX. Molecular weight and subunit composition of adenovirus type 2 fiber. Virology 51: 252–256
Sundquist B, Everitt E, Philipson L, Höglund S (1973b) Assembly of adenoviruses. J Virol 11: 449–459
Sung MT, Lischwe MA, Richards JC, Hosokawa K (1977) Adenovirus chromatin I. Isolation and characterization of the major core protein VII and precursor pro-VII. J Biol Chem 252: 4981–4987
Sung MT, Cao TM, Coleman RT, Budelier KA (1983) Gene and protein sequences of adenovirus protein VII, a hybrid basic chromosomal protein. Proc Natl Acad Sci USA 80: 2902–2906
Tate V, Philipson L (1979) Parental adenovirus DNA accumulates in nuclesome-like structures in infected cells. Nucleic Acids Res 6: 2769–2785
Tibbetts C (1977) Viral DNA sequences from incomplete particles of human adenovirus type 7. Cell 12: 243–249

Tibbetts C, Giam CZ (1979) In vitro association of empty adenovirus capsids with double-stranded DNA. J Virol 32: 995–1005
Tija S, Fanning E, Schick J, Doerfler W (1977) Incomplete particles of adenovirus type 2. III. Viral and cellular DNA sequences in incomplete particles. Virology 76: 365–379
Toth M, Weber J (1984) Nuclease sensitivity of adenovirus type 2 chromatin in lytic infection. J Gen Virol 65: 1611–1615
Tremblay ML, Dery CV, Talbot BG, Weber J (1983) In vitro cleavage specificity of the adenovirus type 2 proteinase. Biochim Biophys Acta 743: 239–245
Van Oostrum J, Burnett RM (1985) Molecular composition of the adenovirus type 2 virion. J Virol 56: 439–448
Van Oostrum J, Smith PR, Mohraz M, Burnett RM (1987) The structure of the adenovirus capsid. III. Hexon packing determined from electron micrografs of capsid fragments. J Mol Biol 198: 73–89
Vayda ME, Rogers AE, Flint SJ (1983) The structure of nucleoprotein cores released from adenovirions. Nucleic Acids Res 11: 441–460
Velicer LF, Ginsberg HS (1970) Synthesis, transport and morphogenesis of type 5 adenovirus capsid proteins. J Virol 5: 338–352
Wadell G, Hammarskjöld ML, Varsanyi T (1973) Incomplete particles of adenovirus type 16. J Gen Virol 20: 287–302
Weber J (1976) Genetic analysis of adenovirus type 2. III. Temperature-sensitivity of processing of viral proteins. J Virol 17: 462–471
Weber JM, Khittoo G (1983) The role of phosphorylation and core protein V in adenovirus assembly. J Gen Virol 64: 2063–2068
Weber JM, Philipson L (1984) Protein composition of adenovirus nucleoprotein complexes extracted from infected cells. Virology 135: 321–327.18
Weber JM, Begin M, Khittoo G (1975) Genetic analysis of adenovirus type 2. II. Preliminary phenotypic characterization of temperature sensitive mutants. J Virol 15: 1049–1056
Weber JM, Khittoo G, Bhatti R (1983) Adenovirus core proteins. Can J Microbiol 29: 235–241
Weber JM, Dery CV, Mirza MA, Horvath J (1985) Adenovirus DNA synthesis is coupled to virus assembly. Virology 140: 351–359
Webster A, Russell S, Talbot P, Russell WC, Kemp GD (1989) Characterization of the adenovirus proteinase: substrate specificity. J Gen Virol 70: 3225–3234
Winberg G, Wadell G (1977) Structural polypeptides of adenovirus type 16 incomplete particles. J Virol 22: 389–401
Winter N, D'Halluin JC (1991) Regulation of the biosynthesis of subgroup C adenovirus protein IVa2. J Virol 65: 5250–5259
Wong ML, Hsu M (1989) Linear adenovirus DNA is organized into supercoiled domains in virus particles. Nucleic Acids Res 17: 3535–3550
Wouters-Tyrou D, Chartier-Harlin MC, Martin-Panthieu A, Boutillon C, van Dorsselare A, Sautière P (1991) Cuttlefish spermatid-specific protein T. Molecular characterization of two variants T1 and T2, putative precursors of protamine variants Sp1 and Sp2. J Biol Chem 266: 17388–17395

# Selective Encapsidation of Adenovirus DNA

S.I. Schmid and P. Hearing



## 1 Introduction

The selective packaging of adenovirus DNA into a capsid raises a number of interesting questions. How is the viral DNA specifically selected from the pool of viral and cellular DNA for encapsidation? What protein–protein and protein–DNA interactions are involved in this process? Are cellular proteins involved in adenovirus packaging? Is the packaging process coordinated with viral DNA replication? Does the virus share a packaging mechanism with one or more of the well-characterized prokaryotic phages? The focus of this review will be on the specific *cis*-acting elements that direct the selective encapsidation of adenovirus DNA into virions at late times after infection and the potential viral proteins that may be involved in this process.

The nonenveloped adenovirus particle contains a protein shell consisting of multiple proteins and a DNA–protein core. The polypeptide composition of the adenovirus particle has been carefully evaluated (Van Oostrom and Burnett 1987) and contains minimally 12 distinct virus-encoded proteins (the structural proteins hexon, penton, fiber, IIIa, VI, VIII, and IX; the core proteins V, VII, and $\mu$; and the nonstructural proteins proteinase and terminal protein). The virus capsid is assembled by the association of two different capsomeres: the hexon capsomere, composed of multiple (nine to 12) hexon molecules, and the penton

---

Department of Molecular Genetics and Microbiology, Health Sciences Center, State University of New York, Stony Brook, NY 11794, USA

capsomere, composed of five penton molecules and three fiber proteins. The hexon capsomeres assemble to form the 20 sides of the virion particle, while the penton capsomeres comprise the 12 vertices of the virion. There is some evidence that the hexon capsomeres can assemble into virus-like particles in the absence of the full composition of penton capsomeres (PEREIRA and WRIGLEY 1974), perhaps in analogy to phage prohead particles that contain a portal vertex used for the insertion of viral DNA during the packaging process. The viral core is composed of the linear double-stranded DNA genome, with covalently linked terminal protein at each 5'terminus, in association with three viral core proteins; V, VII, and $\mu$ (BROWN et al. 1975; VAYDA et al. 1983). The highly basic polypeptide VII is closely associated with the viral DNA, and protein V associates with this DNA–protein complex. The exact structure of the adenovirus core within the assembled virus particle is somewhat controversial, but it appears that the core DNA–protein complex is positioned in 12 globular domains with the outside loop of each domain in juxtaposition with each penton vertex (BROWN et al. 1975; NEWCOMB et al. 1984).

The assembly of virus particles (reviewed in detail by J.-C. D'HALLUIN, this volume) proceeds through an ordered series of assembly events. The assembly of the virus particle has been probed using viral temperature-conditional mutants blocked at different stages of assembly at the restrictive temperature and by pulse-chase kinetic analyses (D'HALLUIN, et al. 1978a, b, 1980a; EDVARDSSON et al. 1976, 1978; SUNDQUIST et al. 1973). Briefly, the first recognizable viral assembly intermediate is a light intermediate particle (buoyant density of 1.315 g/cc in a CsCl equilibrium gradient). These particles contain the capsid structural components and no or very little viral DNA and associated core proteins. Additionally, light intermediate particles contain several proteins that exit the particle during maturation (the IVa2 50-kDa polypeptide and an unknown 39-kDa protein; D'HALLUIN, et al. 1978a; EDVARDSSON et al. 1976; PERSSON et al. 1979) that may represent the adenoviral equivalent of phage scaffolding proteins. The light intermediate particles mature into heavy intermediate particles (1.37g/cc buoyant density) with the insertion of viral DNA. The 50-kDa and 39-kDa polypeptides are released from the particle during this maturation step. The heavy intermediate particles appear to lack core proteins which enter the particle during the next maturation step with the formation of young virus particles (1.34 g/cc buoyant density). As the final step in maturation, the virus-encoded and encapsidated proteinase performs numerous cleavages of multiple viral proteins to generate the mature, infectious virion (ANDERSON 1990; ANDERSON et al. 1973; WEBER 1976; WEBER et al. 1977). Four minor virus-encoded proteins appear to either enhance the assembly of subviral components and/or stabilize viral protein–protein interactions, and hence particle integrity, once formed (PHILIPSON 1983; STEWART et al. 1991). Protein IIIa is found within the virus particle at the vertex regions and may link the capsid to the core structure. Proteins VI and VIII are both found at capsid edges within the particle, where they appear to bridge, and perhaps promote assembly of and/or stabilize, hexons on adjacent facets.

Protein IX is found between hexons within a facet and has been shown to be important for the stability of mature virus particles (COLBY and SHENK 1981; FURCINITTI et al. 1989). The salient conclusion from these analyses is that adenovirus virion assembly likely follows an ordered series of maturation events with a capsid prohead assembled as the initial target for the DNA encapsidation process. In this way, the assembly of infectious adenovirus particles may follow the paradigm of prokaryotic phage assembly.

## 2 Incomplete Particles of Adenovirus

Infection with human adenoviruses, including types 2, 3, 12, and 16, in tissue culture systems has shown that a given serotype yields several classes of viral particles. These different classes can be distinguished from each other and individually purified on CsCl equilibrium gradients due to their distinct buoyant densities (BURLINGHAM et al. 1974; DANIELL, 1976; ROSENWIRTH et al. 1974; SUNDQUIST et al. 1973; WADELL et al. 1973). Only one of these different types of particles constitutes complete infectious virus. The particles isolated from the remaining bands, when analyzed by electron microscopy, have a morphology that resembles that of complete adenovirus, and yet they are weakly infectious or noninfectious. Therefore, they have been classified as incomplete virions. The number of bands representing discrete, incomplete virion particles as well as their predominance in an infected cell lysate relative to complete virions is characteristic for each adenovirus serotype, but independent of the cell line or culture conditions. For example, for adenovirus type 2 (Ad2), five different classes of incomplete particles were isolated and purified. It has been ruled out that these are breakdown products of complete adenovirus particles, since complete and incomplete virions are synthesized in parallel during infection (ROSENWIRTH et al. 1974). Pulse-chase experiments with incomplete and complete particles initially suggested a precursor–product relationship: radioactivity is first detected in incomplete virions, peaking at about 2 h after the pulse, and then decreases continuously. This decrease coincides with a linear rise of radioactivity in complete virions (SUNDQUIST et al. 1973). Analysis of the labeling kinetics of individual proteins further corroborated the precursor–product hypothesis: labeled hexon polypeptides, appeared in incomplete particles immediately after the pulse, but with a lag phase of at least 60 min into mature virions. In contrast, labeled core polypeptides, which are found in association with the viral genome in intact virions, are incorporated into complete virions without a lag phase. Additionally, the formation of incomplete particles is more sensitive to inhibition of protein synthesis than the formation of mature virions. These results strongly suggest that viral DNA and core proteins are inserted into preformed, empty capsids to yield infectious virions.

## 3 Polar Encapsidation of Adenovirus DNA

Incomplete particles of lower density than mature virions were isolated from Ad2, Ad3, and Ad7 to study the structure of the encapsidated genome (DANIELL 1976; TIBBETTS 1977). The lightest incomplete particles seem to be completely devoid of viral DNA, and it was concluded that they represent empty capsids as the earliest precursors of complete virions. Among the other defective virions, a linear relationship between the length of the encapsidated viral DNA and the density of the corresponding group of incomplete particles could be established. The size of the genome ranged from 15% of the length of the complete genome to full size. It is not known why specific size classes are predominant leading to the formation of distinct virion bands on a CsCl gradient. Contradictory reports exist on the question of whether the formation of incomplete genomes naturally occurs during replication or whether it is artificially created during the preparation by the shearing of incompletely packaged DNA. Both scenarios, however, are not mutually exclusive. Restriction endonuclease analysis of the packaged subgenomes revealed that sequences derived from the left end are strikingly overrepresented, which initially suggested that DNA packaging occurs in a polar fashion from left to right (DANIELL 1976; TIBBETTS 1977). In support of this model, both ends of the genome were equally represented in the pool of subgenomic adenovirus DNA in infected cells (DANIELL and MULLENBACH 1978). The first indication for the presence of *cis*-acting sequences in the adenovirus genome that direct selective DNA encapsidation from the left end came from studies with naturally occurring evolutionary variants of Ad16 (HAMMARSKJÖLD and WINBERG 1980). A duplication of the left end 390, but not 290 base pairs (bp), at the right end of the viral genome allowed DNA packaging to initiate from both ends of the genome. This was demonstrated by the recovery of equal amounts of left- and right-end fragments from the subgenomes of incomplete particles of the variant, in contrast to the predominance of left-end fragments in the parental wild-type virus. Consequently, it was suggested that the region between nucleotides (nt) 290 and 390 harbors an essential signal for polar viral DNA packaging. The fact that this sequence element acts in *cis* was demonstrated by the inability of the 390-bp duplication mutant to support bipolar DNA encapsidation of variants with smaller duplications at their right-end termini in mixed viral infections. A number of Ad3 variant viruses were selected after repeated high-multiplicity passage in HeLa cells (Robinson and Tibbetts 1984). These variants carry various mutations in their left-end 750 bp, but retain full capacity for growth and polar packaging of their genomes. Mutations of sequences within nt 136–318, the region immediately adjacent to the inverted terminal repeat (ITR), did not affect viral growth, whereas the maintenance of nt 319–390 in all the mutants suggested that this region is indispensable. This interval coincides with the above-mentioned region between nt 290 and 390 in Ad16 that is important for DNA encapsidation. Sequence comparisons of Ad3 with Ad5 and Ad12, representatives of adenovirus subgroups A, B, and C, revealed that the interval between nt 237 and 491 is highly conserved, which might similarly reflect the importance of a packaging signal located within this region (KOSTURKO et al. 1982). This may indicate that all adenoviruses may depend on a *cis*-acting DNA encapsidation signal and consequently employ a similar mechanism for selective DNA encapsidation from the left end. A noticeable feature within this region is a series of adenine/thymidine (AT)-rich stretches (see below).

## 4 *Cis*-Acting Sequences Involved in Packaging Specificity

The corresponding *cis*-acting packaging domain of Ad5 was localized to the left end of the genome during deletion analysis of the E1A transcriptional control region. The packaging domain overlaps with two distinct enhancer elements (Fig.1). Enhancer element I is repeated and specifically stimulates transcription of E1A upon binding of a nuclear HeLa cell factor, EF-1A (BRUDER and HEARING 1989). Deletion of the repeated elements decreases E1A expression approximately five-hold (HEARING and SHENK 1986). Efficient complementation in trans of element I viral mutants is achieved by propagation of the virus in 293 cells, a human embryonic kidney cell line that constitutively expresses the Ad5 E1A and E1B genes (GRAHAM et al. 1977). Enhancer element II augments transcription in *cis* from all the early transcription units by an unknown mechanism (HEARING and SHENK 1986). Element II mutations result in a three-to seven fold decrease in early

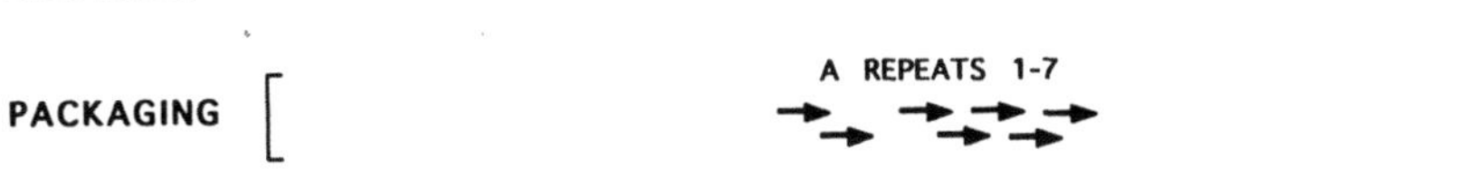


**Fig. 1.** The adenovirus packaging domain. Shown at the *top* is a schematic view of the left end of adenovirus type 5 including the inverted terminal repeat *(ITR)* and core region of replication *(CORE ORI)*, the packaging domain and E1A enhancer region *(PACKAGING/ENHANCER)*, and TATA box *(T/A)* for E1A transcription (direction of transcription indicated by an *arrow*). Nucleotide numbers *(Nt)* are relative to the left terminus. The E1A enhancer elements *(I and II)* and the packaging repeats *(A REPEATS 1-7)* described in the text are indicated

gene expression and, since some early gene products are required for DNA replication, element II mutations result in a corresponding reduction in virus yield. This *cis*-acting defect can be complemented in *trans* by coinfection of an element II mutant with a wild-type virus that provides all the early gene products (HEARING and SHENK 1986). Replication defects due to element II mutations will result in the production of a reduced pool of viral DNA available for packaging and can therefore not be distinguished from a packaging defect in single infections of 293 cells. For that reason, in addition to determination of viral yields in single infections, packaging efficiency can be measured directly by a comparison of the pool of viral nuclear DNA available for packaging to the amount of viral DNA that was actually packaged in a coinfection with wild-type virus (HEARING et al. 1987).

An Ad5 mutant virus lacking the packaging domain within the interval of nt 194–358 at the left end was nonviable in 293 cells (HEARING and SHENK 1983). Reconstruction of this same deletion into a virus that had the right-end ITR substituted by the left-end 353 bp restored the viability of the resulting virus (HEARING and SHENK 1983). Optimal growth was presumably obtained by transferring the left-end packaging signal to the right end of the genome. The fact that the Ad5 packaging domain indeed represents a functional unit was demonstrated by the construction of a series of mutant viruses that contained the packaging domain deleted and reinserted at a unit at different locations (HEARING et al. 1987). Viral growth was minimally impaired when the packaging domain was moved almost 300 bp to the right of its original location. Placing it another 350 bp to the right, however, abolished virus viability. The packaging domain could be inverted without any effect on virus yield, but inversion impaired its positional flexibility to some extent such that it could not be moved more than 100 bp pairs inward from the left end without rendering the resulting virus nonviable. These results also suggest strongly that in order to be functional, the adenoviral encapsidation signal has to be positioned near either terminus of the viral chromosome. Due to an absolute requirement of sequences in the ITR for initiation of DNA replication (Fig.1, core origin), it has not been possible to determine yet whether this region also functions in *cis* to support viral DNA packaging.

By the analysis of a series of deletion, insertion, and linker scanning mutations within the packaging domain (GRÄBLE and HEARING 1990, 1992), seven AT-rich repeats were identified as functional packaging elements (Fig. 2). They share a loosely defined consensus motif, 5'–GTN$_{3-4}$TTTG–3'. Sequential deletion of these repeats from either side of the packaging domain revealed that they are functionally redundant. Five of the A repeats (AI–AV) are located within the region between nt 194 and 358. Their deletion, as mentioned above, is sufficient to render the mutant virus nonviable. Viral growth can be restored to a certain extent, however, by the insertion of multiple copies of an oligonucleotide containing the AT-rich motif of one of the packaging elements (AII) into this same deletion background, which confirmed the respective element as a functional

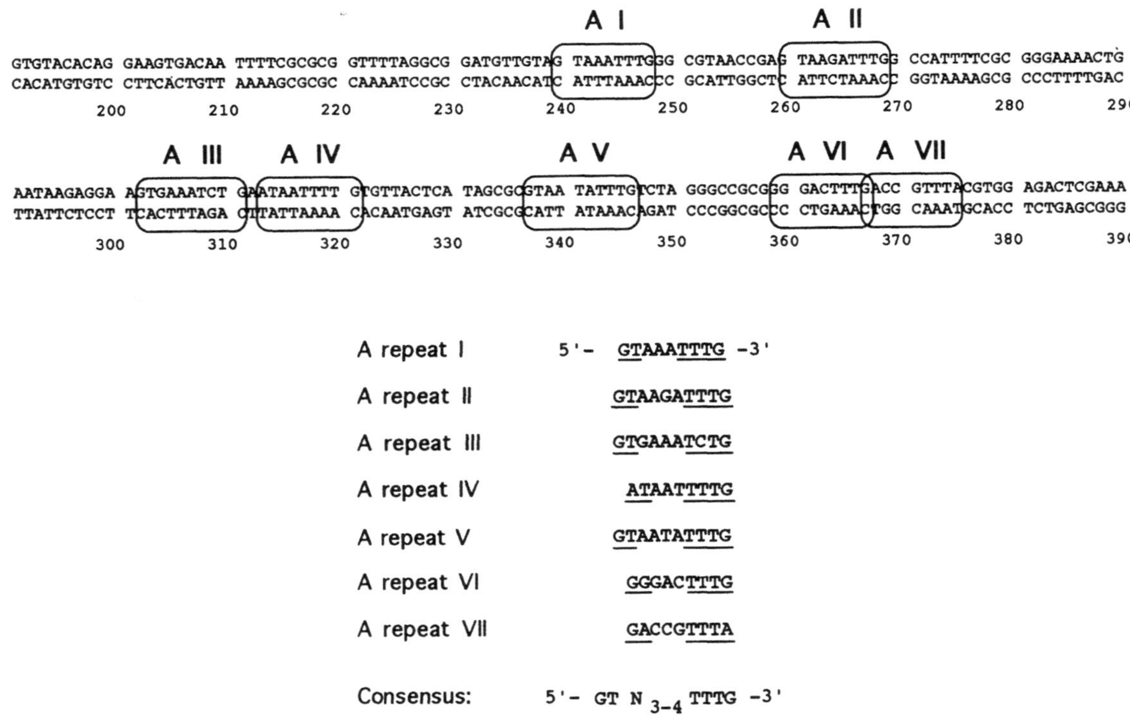


**Fig. 2.** Nucleotide (nt) sequence of the adenovirus type 5 (Ad5) packaging region. The nt sequence of the Ad5 genome between nt 191 and 390 is shown. *Numbers* below the sequence correspond to nt relative to the left terminus. A repeats I—VII are *encircled* sequence of the individual A repeats is shown at the bottom along with a consensus sequence drawn from a comparison of these sites

packaging unit (GRÄBLE and HEARING 1990). Two additional packaging elements are located to the right of nucleotide 358 (Fig.2, A repeats VI and VII; GRÄBLE and HEARING 1992).

Figure 3 shows an alignment of the nt sequences from the left-end termini of adenovirus types 5, 3, 4, 7, 12, 16, and 40, including the packaging domain with the seven AT-rich elements. Among all of the viruses, a high degree of sequence conservation can be observed within the ITR, since it contains the viral origin of replication with binding sites for cellular factors. Striking conservation is also observed for the E1A transcriptional start site and promoter. In between those two areas, the most highly conserved nt are either part of binding sites for transcription factors such as EF-1A and E2F, or they are contained in one of the packaging elements described above. The seven packaging elements, except for A repeat II, which is deleted in Ad12 and Ad40, are relatively conserved, which emphasizes their importance in an evolutionary sense.

Linker scanning mutagenesis of individual repeats in the context of a minimal packaging domain demonstrated that inactivation of individual repeats impairs packaging efficiencies of the resulting mutant viruses to different extents (GRÄBLE and HEARING 1992). These results indicate that in spite of their functional redundancy, the different packaging elements are not functionally equivalent to each other. A comparison of the degree of sequence conservation of the different packaging elements (Fig. 3) supports this idea in an interesting manner. A repeats III and IV, which functionally seem to represent the least important members of the A repeat family, are also the least conserved elements. They seem to have maintained a general AT richness rather than the consensus motif of the packaging elements mentioned above. The striking maintenance of several nt of A repeat III is probably the result of a conservation of the EF-1A binding sites within a region of overlap with A repeat III. Two pairs of A repeats, repeats I and II and repeats V and VI, which represent functionally more significant members of the A repeat family (GRÄBLE and HEARING 1990, 1992), are seperated from each other within each pair by exactly 21 bp or two helical turns of the DNA. Putative packaging factors bound to these repeats would be positioned on the same face of the DNA helix, in order to possibly interact with each other and/or with additional factors bound outside the packaging domain.

**Fig. 3.** Comparison of the nucleotide sequences of the left 450 bp of adenovirus (Ad) type 5, 3, 4, 7, 12, 16, and 40. The DNA sequences of the left ends of various Ad serotypes were aligned using SeqApp sequence editor and analysis program. *Dashes* indicate gaps introduced in specific sequences for alignment. *Shaded regions* indicate sequences where identical nucleotides were found in five out of seven positions. The sequences corresponding to the inverted terminal repeat *(ITR)*, *EF-1A* and *E2F* transcription factor binding sites, and A repeat packaging elements *(AI–AVII)* are indicated above the sequences

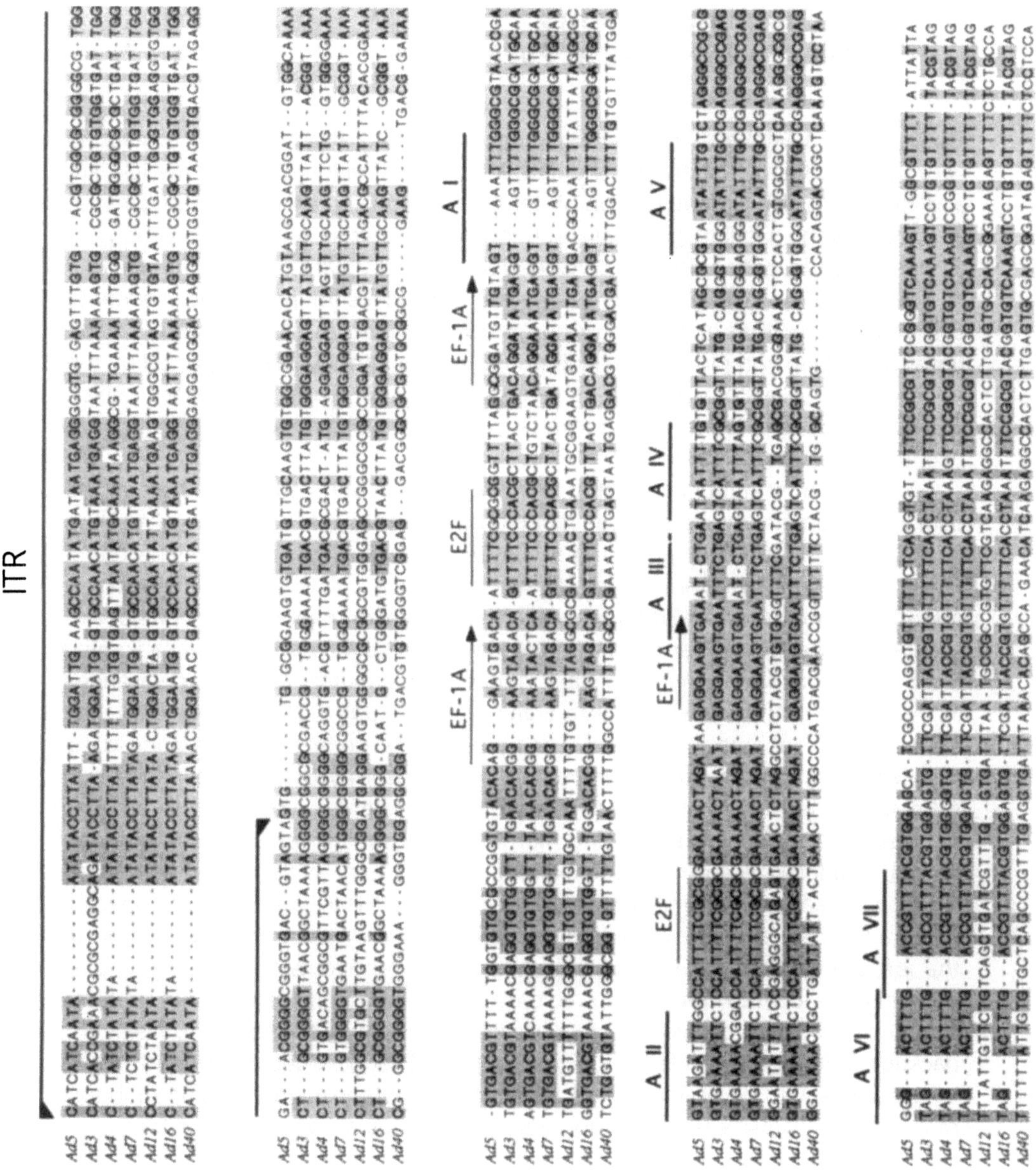
ITR
EF-1A
E2F
EF-1A
A I
A II
E2F
EF-1A
A III
A IV
A V
A VI
A VII
Ad5
Ad3
Ad4
Ad7
Ad12
Ad16
Ad40

## 5 *Trans*-acting Components That May Be Involved in Packaging

That in fact a limiting *trans*-acting component interacts with the *cis*-acting packaging domain was strongly indicated in cotransfection experiments (GRÄBLE and HEARING 1992). Cotransfection of cells with Ad5 DNA and a plasmid carrying an excess of the packaging domain resulted in a substantial decrease in virus yield as compared to cotransfection of wild-type genomes with nonspecific plasmid sequences. Presumably, the presence of an excess of packaging elements competed for a *trans*-acting factor and thus prevented the formation of a functional packaging complex on the Ad5 genome. This notion was supported by the fact that total levels of viral DNA as well as late mRNA were not affected, which indicates that, consistent with a packaging defect, the observed defect must have occurred very late in infection.

Analysis of mutant viruses that carry alterations in the spacing between the packaging domain and sequences located either to the left or to the right of the packaging domain also support the notion that *trans*-acting components interact with the packaging elements and adjacent sequences. Viruses carrying insertion mutations between the wild-type packaging domain, or truncated versions of this region, and the ITR first emphasized the likely interactions between a packaging factor or factors bound to the packaging domain and factors bound to sequences in the ITR or at the terminus (GRÄBLE and HEARING 1990, 1992). Specifically, the deletion of A repeats III, IV, and V resulted in the generation of a packaging-deficient virus. The growth of this mutant virus was improved by the insertion of 4 bp to the left of the truncated packaging domain (i.e., altering the spacing between the remaining packaging elements and the terminus of the viral genome). In addition, moving the packaging domain 89 nt to the left of its original location increased virus yield above wild-type levels (HEARING et al. 1987). Spacing constraints between the *cis*-acting packaging domain and elements located to the left are consistent with the aforementioned requirement of the packaging domain to be located near a genomic terminus and support the notion of protein–protein interactions between putative packaging factors bound to the *cis*-acting packaging elements and factors bound to the ITR sequences. In support of this idea, studies with several temperature-sensitive mutants using specific inhibitors of initiation of DNA replication suggested that proteins involved in the formation of the DNA synthesis initiation complex may also play a role in DNA encapsidation (D'HALLUIN et al. 1980b).

Similar results have been obtained with insertion or deletion mutations to the right of the packaging domain. An insertion of six nt to the right of a minimal packaging domain (containing A repeats V, VI and VII) resulted in a decrease in virus yield and packaging efficiency, which originally suggested spacing constraints between the packaging domain and elements located to the right of it. That in fact one or more additional packaging elements exist to the right of A repeat VII that are indispensable for efficient virus growth, at least in the context of a minimal packaging domain, was recently confirmed by the nonviability of a mutant adenovirus (containing A repeats V and VI) that additionally carries a large deletion initiating to the right of AVI and extending into the open reading frame of E1A (S. SCHMID, unpublished results). Propagation of this virus was attempted in 293 cells to complement the E1A-negative phenotype. The observed lethal defect of this additional deletion is most likely due to inefficient encapsidation of the viral genome, especially in view of the fact that another adenovirus E1A-deletion mutant, *dl* 312, which carries a similar deletion but contains a wild-type packaging domain, has wild-type growth properties in 293 cells (JONES and SHENK 1979). This implicates factors bound to the right of the defined packaging domain to play a role in viral packaging in addition to factors bound to ITR sequences and proteins bound to the packaging domain itself.

The search for such proteins will be complicated by the possibility that multiple proteins might be involved and by the fact that no information is currently available about the identity of viral or cellular gene products that may be directly involved in the packaging process. Considering adenovirus-encoded proteins, the IVa2 protein, which is present in light intermediate particles but not in heavy intermediates or mature particles (D'HALLUIN et al. 1978a; EDVARDSSON et al. 1976; PERSSON et al. 1979), represents a possible candidate for a scaffolding or packaging substrate recognition product. Other viral proteins that may have a direct involvement in the packaging process are suggested through the analysis of temperature-conditional mutants or missense mutants and include the single-stranded DNA-binding protein (DBP), the covalently linked terminal protein (TP), and the L1 52- and 55-kDa proteins. The possible role of DBP in packaging is suggested by the analysis of a temperature-sensitive DBP mutant (*ts* 19; ROOVERS et al. 1990) that accumulates light intermediate particles at the restrictive temperature (EDVARDSSON et al. 1978). The known DNA-binding properties of DBP are attractive with respect to the possible recognition of specific packaging sequences although this has yet to be tested. Similarly, the possible role of the L1 52- and 55-kDa proteins in packaging was suggested by the analysis of a temperature-sensitive mutant (*ts* 369) which accumulates incomplete particles at the restrictive temperature that carry only small segments of the left end of the viral genome (HASSON et al. 1989). What role these proteins play in the packaging process is unknown. Lastly, the role of TP in viral packaging is suggested by a codon insertion mutant of TP which maintains full replicative properties in vitro yet is nonviable in vivo (FREDMAN et al. 1991; FREIMUTH and GINSBERG 1986). This phenotype is consistent with a defect in DNA packaging, as are the positive or negative effects on packaging efficiency described above of different spacing mutations between the packaging domain and the viral genomic terminus. The possible role of TP in the packaging process is also suggested by analogy to bacteriophage $\phi$29 (reviewed in FUJISAWA and HEARING 1994). Bacteriophage $\phi$29 is a double-stranded bacteriophage with a number of similarities to adenovirus; it has a linear, double-stranded DNA genome that is replicated by strand displacement using a protein (gp3) that is covalently attached to the 5'terminus as a primer (functionally equivalent to adenovirus TP). The genome of ϕ29 is packaged in a polar fashion from left to right and requires a packaging protein (gp16) and a single-stranded RNA transcribed at the left end of the ϕ29 genome. The ϕ29 terminal protein gp3 is intimately involved in the packaging process. Whether adenovirus encodes an ATPase that is equivalent to gp16 has not been determined, nor has a role for RNA been tested in the adenovirus packaging process. A test of the latter possibility must await the development of an in vitro packaging system or the genetic identification of such an RNA component.

## 6 Perspectives

We support the following model to account for the selectivity and polarity of the process of adenovirus DNA encapsidation as well as to interpret the results of extensive genetic analyses of the *cis*-acting packaging domain. We propose, in analogy to bacteriophage lambda, that *cis*-acting packaging sequences in the adenovirus genome constitute binding sites for trans-acting packaging factors. The fact that those binding sites are functionally redundant might be the evolutionary result of a competition for the binding of one or more limiting *trans*-acting components, whereby the presence of multiple sites would increase the likelihood of packaging factor binding. Based on the analysis of spacing mutations in and around the packaging domain, we would anticipate that protein–protein interactions of one or distinct packaging factors are required for functional activity of this region. The AT-rich repeats might represent an array of interspersed binding sites that form a specific nucleoprotein complex at the left end of the adenovirus chromosome upon binding of packaging factors, which would mark the respective genome as a bona fide packaging substrate. Proteins bound to sequence elements to the right as well as to the left of the packaging domain additionally seem to be involved in the formation or stabilization of this complex. This suggests that the formation of a higher order nucleoprotein structure via multiple DNA–protein and protein–protein interactions is a prerequisite for DNA encapsidation. The possibility that DNA bending, as a consequence of the AT-rich character of the packaging elements (Eckdahl and Anderson 1988; Ohyama and Hashimoto 1989), is an additional requirement for packaging adds a further level of complexity. Formation of a specific nucleoprotein complex at the left-end terminus of the adenoviral chromosome would then allow for physical interaction with the viral prohead and subsequent selective encapsidation from left to right.

## References

Anderson CW (1990) The proteinase polypeptide of adenovirus serotype 2 virions. Virology 177: 259–272

Anderson CW, Baum PR, Gesteland RF (1973) Processing of adenovirus 2 induced proteins. J Virol 12: 241–252

Brown DT, Westphal M, Burlingham BT, Winterhoff U, Doerfler W (1975) Structure and composition of the adenovirus type 2 core. J Virol 16: 366–387
Bruder JT, Hearing P (1989) Nuclear factor EF-1A binds for the adenovirus E1A core enhancer element and to other transcriptional control regions. Mol Cell Biol 9: 5143–5153
Burlingham BT, Brown DT, Doerfler W (1974) Incomplete particles of adenovirus. I. Characteristics of the DNA association with incomplete adenovirions of types 2 and 12. Virology 60: 419–430
Colby WW, Shenk T (1981) Adenovirus type 5 virions can be assembled in vivo in the absence of detectable polypeptide IX. J Virol 39: 977–980
Daniell E (1976). Genome structure of incomplete particles of adenovirus. J Virol 19: 685–708
Daniell E, Mullenbach T (1978) Synthesis of defective viral DNA in HeLa cells infected with adenovirus type 3. J Virol 26: 61–70
D'Halluin JC, Martin GR, Torpier G, Boulanger PA(1978a) Adenovirus type 2 assembly analysed by reversible cross-linking of labile intermediates. J Virol 26: 357–363
D'Halluin J-C, Milleville M, Boulanger PA, Martin GR (1978b) Temperature sensitive mutant of adenovirus type 2 blocked in virion assembly: accumulation of light intermediate particles. J Virol 26: 344–356
D'Halluin J-C, Milleville M, Martin GR, Boulanger P (1980a) Morphogenesis of human adenovirus type 2 studied with fiber and penton base-defective temperature-sensitive mutants. J Virol 33: 88–99
D'Halluin J-C, Milleville M, Boulanger PA (1980b) Effects of novobiocin on adenovirus DNA synthesis and encapsidation Nucleic Acids Res. 8: 1625–1641
Eckdahl TT, Anderson JN (1988) Bent DNA is a conserved structure in an adenovirus control region. Nucleic Acids Res. 16: 2346
Edvardsson B, Everitt E, Jörnvall E, Prage L, Philipson L (1976) Intermediates in adenovirus assembly. J Virol. 19: 533–547
Edvardsson B, Ustacelebi S, Williams J, Philipson L (1978) Assembly intermediates among adenovirus type 5 temperature-sensitive mutants. J Virol 25: 641–651
Fredman JN, Pettit SC, Horwitz MS, Engler JA (1991) Linker insertion mutations in the adenovirus preterminal protein that affect DNA replication activity in vivo and in vitro. J Virol 63: 4591–4597
Freimuth PI, Ginsberg HS (1986) Codon insertion mutants of the adenovirus terminal protein. Proc Natl Acad Sci USA 83: 7816–7820
Fujisawa H, Hearings P (1994) Selective packaging of double-stranded DNA viruses. Seminl 5: 5–13
Furcinitti PS, van Oostrum J, Burnett RM (1989) Adenovirus polypeptide IX revealed as capsid cement by difference images from electron microscopy and crystallography. EMBO J 8: 3563–3570
Gräble M, Hearing P (1990) Adenovirus type 5 packaging domain is composed of a repeated element that is functionally redundant. J Virol 64: 2047–2056
Gräble M, Hearing P (1992) Cis and trans requirements for the selective packaging of adenovirus DNA. J Virol 66: 723–731
Graham FL, Smiley J, Russell WC, Nairu R (1977) Characteristics of a human cell line transformed by DNA from human adenovirus type 5. J Gen Virol 36: 59–72
Hammarskjöld ML, Winberg G (1980) Encapsidation of adenovirus 16 DNA is directed by a small DNA sequence at the left end of the genome. Cell 20: 787–795
Hasson TB, Soloway PD, Ornelles DA, Doerfler W, Shenk T (1989) Adenovirus L1 52- and 55-kilodalton proteins are required for assembly of virions. J Virol 63: 3612–3621
Hearing P, Shenk T (1983) The adenovirus type 5 E1A transcriptional control region contains a duplicated enhancer element. Cell 33: 695–703
Hearing P, Shenk T (1986) The adenovirus E1A enhancer contains two functionally distinct domains: one is specific for E1A and the other modulates all early units in cis. Cell 45: 229–236
Hearing P, Samulski RJ, Wishart WL, Shenk T (1987) Identification of a repeated sequence element required for efficient encapsidation of the adenovirus type 5 chromosome. J Virol 61: 2555–2558
Jones N, Shenk T (1979) Isolation of adenovirus type 5 host range deletion mutants defective for transformation of rat embryo cells. Cell 17: 683–689
Kosturko LD, Sharnick SV, Tibbetts C (1982) Polar encapsidation of adenovirus DNA: cloning and DNA sequence of the left end of adenovirus type 3. J Virol 43: 1132–1137
Newcomb WW, Boring JW, Brown JC (1984) Ion etching of human adenovirus 2: structure of the core. J Virol 51: 52–56
Ohyama T, Hashimoto S (1989) Upstream half of adenovirus type 2 enhancer adopts a curved DNA conformation. Nucleic Acids Res 17: 3845–3853
Pereira HG, Wrigley NG (1974) In vitro reconstitution, hexon-bonding and handedness of incomplete adenovirus capsids. J Mol Biol 85: 617–633

Persson H, Mathiesen B, Philipson L, Pettersson U (1979) A maturation protein in adenovirus morphogenesis. Virology 93: 198–208
Philipson L (1983) Structure and assembly of adenoviruses. Curr Top Microbiol Immunol 109: 1–52
Robinson CC, Tibbetts C (1984) Polar encapsidation of adenovirus DNA: evolutionary variants reveal dispensable sequences near the left ends of Ad3 genoms. Virology 137: 276–286
Roovers DJ, Young CSH, Vos HL, Sussenbach JS (1990) Physical mapping of two temperature-sensitive adenovirus mutants affected in the DNA polymerase and DNA binding protein. Virus Genes 4: 53–61
Rosenwirth B, Tjia S, Westphal M, Doerfler W (1974) Incomplete particles of adenovirus. II. Kinetics of formation and polypeptide composition of adenovirus type 2. Virology 60: 431–437
Stewart PL, Burnett RM, Cyrklaff M, Fuller SD, (1991) Image reconstruction reveals the complex molecular organization of adenovirus. Cell 67: 145–154
Sundquist B, Everitt E, Philipson L, Hoglund S (1973) Assembly of adenoviruses. J Virol 11: 449–459
Tibbetts C (1977) Viral DNA sequences from incomplete particles of human adenovirus type 7. Cell 12: 243–249
Van Oostrom J, Burnett RM (1987) Molecular composition of the adenovirus type 2 virion. J Virol 56: 439–448
Vayda ME, Rogers AE, Flint SJ (1983) The structure of nucleoprotein cores released from adenoviruses. Nucleic Acids Res 11: 441–460
Wadell G, Hammerskjold M, Varsanyi T (1973) Incomplete particle of adenovirus type 16. J Gen Virol 20: 287–302
Weber J (1976) Genetic analysis of adenovirus type 2. III. Temperature-sensitivity of processing of viral proteins. J Virol 17: 462–471
Weber J, Begin M, Carstens EB (1977) Genetic analysis of adenovirus type 2. IV. Coordinate regulation of polypeptides 80k, IIIa and V. Virology 76: 709–724

# Role of Adenovirus Structural Components in the Regulation of Adenovirus Infection

W.C. Russell and G.D. Kemp



## 1 Prologue

Adenoviruses contain at least 15–16 proteins in a complex assembly with the virus double-stranded genomic DNA. The disposition of the major proteins in the structure has been assigned primarily on the basis of scrutiny of the morphological features of the virus as demonstrated by electron microscopy (Valentine and Pereira 1965), by cross-linking studies (Everitt et al. 1975; Chatterjee et al. 1985) and more recently by the technique of difference imaging, which utilises the results obtained by high-resolution X-ray crystallography of the major capsid protein, the hexon, to reveal the probable location of the minor capsid proteins (Stewart et al. 1991, 1993). However, detailed knowledge of the arrangement of proteins within the capsid and their relationship to the virus genome is still rather rudimentary.

The morphology of the virion is described in detail in the chapter by Stewart and Burnett (this volume) and a summary of the structural proteins, nomenclature and some of their characteristics are given in Table 1.

As well as providing a protective shell for the virus genome, the structural components play a key role in targeting the infecting particle to the appropriate cellular compartments and most likely in regulating the significant initial events in

School of Biological and Medical Sciences, Division of Cell and Molecular Biology, University of St. Andrews, Irvine Building, North Street, St. Andrews, Fife, KY16 9AL, Scotland, UK

**Table 1.** Structural proteins of adenovirus (Human type 2)

| Name | Molecular Mass (to the nearest KDa)[a] | Characteristics |
|---|---|---|
| Hexon (ii) | 109 | Major capsomer with type, sub-type and group-specific determinant |
| Penton base (III) | 63 | Apical capsomer recognising cellular integrins |
| IIIA | 63 | Phosphoprotein cleaved by protease from a precursor (65 KDa)[b] |
| Fibre (IV) | 62 | Knobbed fibre attached to penton base; responsible for initial attachment to cell receptors |
| IVα2 | 51 | Binds to virus DNA |
| Core protein 1 (V) | 42 | Loosely bound to virus DNA in a core complex with VII |
| Terminal protein (TP) | 38 | Derived from precursor pre-terminal protein (pTp) of 75 kDa by virus protease; covalently linked to viral DNA; essential for nuclear matrix localization and for viral DNA replication[b] |
| Protease | 23 | Essential for maturation of virion |
| VI | 22 | Precursor pVI (27 kDa) cleaved by viral protease; associated with hexon at apex |
| Core Protein 2 (VII) | 19 | Precursor (22 kDa) cleaved by viral protease; tightly bound to virus DNA[b] |
| VIII | 15 | Precursor (25 kDa) cleaved by viral protease[b] |
| IX | 14 | Bound to hexon to form group-of-nine hexons (GON) |
| XI and XII and others | <10 | Derived from protease processing of other viral polypeptides[b] |
| μ | 3 | Cleaved by viral protease from precursor X (11 kDa) to give a very basic peptide tightly associated with viral DNA[b] |

[a] The molecular mass derived by sodium dodecyl sulphate (polyacrylamide gel electrophoresis) (SDS-PAGE) may be significantly different.

[b] Peptide fragments from protease cleavage may or may not be bound to the larger fragment in the virion.

transcription and replication. At the later stages of infection, both the cessation of macromolecular synthesis and assembly of virons are probably governed by the characteristics of the structural proteins and their interaction with each other.

In this survey of these structural proteins we shall consider their role at various stages of host cell infection:

1. Adsorption and penetration
2. "Uncoating" and transport to the nucleus
3. Transcription and replication
4. Virion maturation

## 2 Adsorption and Penetration

It has been recognised for some time that the penton complex plays a key role in the interaction of adenoviruses with the host cell and that the fibre component is pertinent to the initial recognition of the plasma membrane (PHILIPSON et al. 1968). The penton base component has also been implicated in membrane interaction since it can cause cell rounding and appears to be the instigator of the "early" cytopathic events in infection (PEREIRA 1958; RUSSELL et al. 1967).

The fibre protein has a trimeric structure (STOUTEN et al. 1992) with different lengths according to the virus serotype (SIGNAS et al. 1985; KIDD et al. 1993). This latter feature is directly related to the size of the fibre polypeptide. Different serotypes have polypeptides with structurally similar N and C termini, but the middle stem region demonstrates a characteristic 15-amino acid repeat motif (GREEN et al. 1983) with the number of motifs varying, e.g. six and 22 giving fibre lengths of 10 and 30 nm, respectively. The C termini of the three polypeptides have been shown to form the distal knob of the protein, whereas the N termini form a tail which interacts in a non-covalent manner with the pentameric penton base.

An interesting observation was made by KIDD et al. (1993), who demonstrated that adenovirus type 40 (Ad40) contained two fibre genes which were both expressed, providing distinctly different lengths of fibre (19 and 31 nm). Only one fibre was associated with each penton base in the virus particle, but these apparently could be of either length—a distinct difference from the case of some of the *Aviadenoviridae*, where the virus particles were demonstrated to have two fibres associated with the apices (GELDERBLOM and MAICHLE-LAUPPE 1982). It has been speculated that the fibre length may be related to the surface density of the cell receptors (KIDD et al. 1993) and that greater length may facilitate attachment if binding to more than one receptor is necessary. This supposition is consistent with the indication that there may well be a conserved flexible kink in the fibre shaft structure (RUIGROK et al. 1990). The number of cell receptors available for virus binding has been determined for a range of cells (PHILIPSON et al. 1968; DEFER et al. 1990; BELIN and BOULANGER 1993) and varies from 3 to $6 \times 10^3$. Adsorption appears to be followed by an energy-dependent clustering of virus particles at the cell surface (PATTERSON and RUSSELL 1983). Internalisation of the virus particles then occurs via clathrin-coated vesicles into endosomes (CHARDONNET and DALES 1970; SETH et al. 1986). The finding that cytochalasin B, an inhibitor of microfilament function, appeared to block this internalisation step (PATTERSON and RUSSELL 1983) implicated the involvement of the cytoskeleton, and changes noted in intracytoplasmic cables on infection (BELIN and BOULANGER 1993) supported this contention.

A number of studies have attempted to characterise the cell receptors involved in the binding process and one to five major polypeptide species have been noted (HUGHES and MAUTNER 1973; HENNACHE and BOULANGER 1977; SVENSSON et al. 1981; BELIN and BOULANGER 1993). The latter investigation has also shown that different receptors may be required for different adenovirus subgenera.

It has been assumed that the fibre knob structure is primarily responsible for the initial recognition events and this supposition has been strengthened by the finding that fibre monoclonal antibodies (Mab) which neutralise the virus and inhibit haemagglutination recognise epitopes on the sequences relating to the knob structure (WATSON et al. 1988), although a few of them also recognise the β-bend region on the shaft proximal to the presumed knob (W.C. RUSSELL unpublished data).

More recent investigations, however, have clarified these initial events considerably in that they demonstrate that there are at least two separate events leading to adsorption and internalisation of the virus particle. The first of these involves recognition by the fibre and then a second step, via the penton base, demonstrates the importance of integrins (WICKHAM et al. 1993) and divalent ions (see SVENSSON and PERSSON 1984). These findings are consistent with the observation of a conserved RGD motif in the penton base polypeptide and the ability of RGD and other related peptides to complete for virus adsorption (BELIN and BOULANGER 1993). Mutations altering the RGD sequence also affect the cell-rounding characteristics of the virus (BAI et al. 1993). Integrins are αβ-heterodimers and since there are at least 14 known α–subunits and 8 β-subunits (HYNES 1992), the array of integrins expressed in cells is quite complex and will vary between cell types and with cellular environments. In the study noted above (WICKHAM et al. 1993), the integrins involved appeared to be of the $\alpha_v$-group with ligands of vitronectin and fibronectin, but recognition of other integrins by the penton base cannot be excluded. Since integrins are important in cell signalling, the reaction with the penton base may alter intracellular phosphorylation patterns (see RUSSELL and BLAIR 1977) as well as cytoskeletal interactions leading to permeabilisation. This may explain the ability of adenovirus to stimulate the uptake of other macromolecules—a property which has been utilised to promote more efficient transfection procedures. In addition, the penton base has been shown to mediate the release of choline from plasma membrane vesicles with an optimum at pH 5.5–6.0, conditions which imply that the penton base could interact with the lipid bilayer of endocytic vesicles (SETH 1994).

Notwithstanding all these studies, it seems very likely that there are more complex requirements for the internalisation of the virion. Electron microscopic analysis by thin-sectioning of cells infected with Ad2*ts*1, a temperature-sensitive mutant with a defective protease (WEBER 1976), clearly shows that although the virus particle can be assembled at the restrictive temperature with unprocessed structural proteins pVI, pVII and pVIII and that they do attach in the normal way and enter into endosomic vesicles, the particles cannot be internalised (see Fig. 1). Thus at 30 min after uptake, with wild-type virus between 40% and 60% of the inoculum virus can be seen inside cells with 20%–35% in the cytoplasm, whereas with the mutant virus, although 80% of the virus inoculum has been taken up, only 0.4% can be visualised in the cytoplasm and most are sequestered in endosomic vesicles with two or more particles in each (S. PATTERSON and W.C. RUSSELL, unpublished observations). Since intact fibre and penton base are integral components of the structure of the mutant virus, this finding clearly implies that,

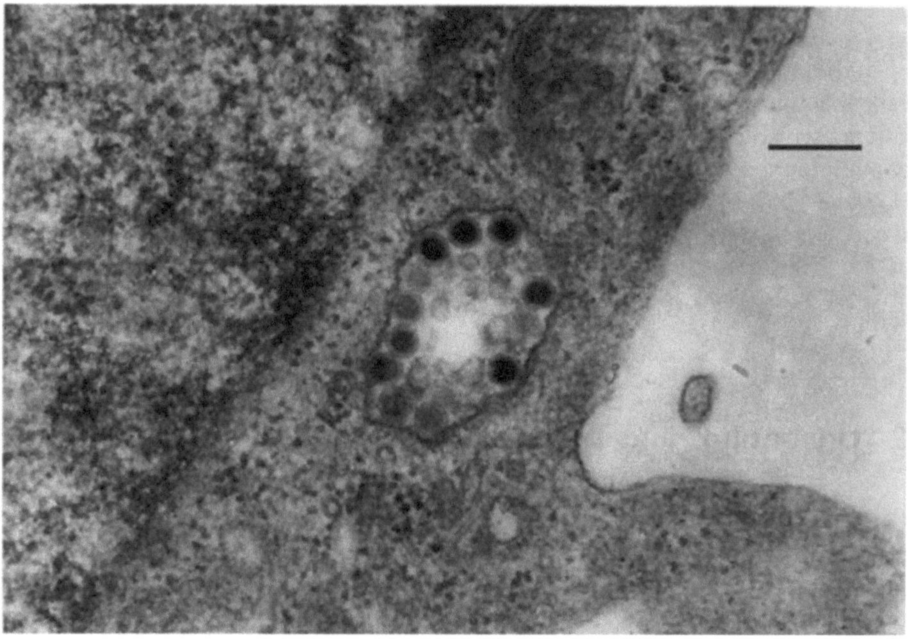

**Fig. 1.** Accumulation of adenovirus type 2 (Ad2) *ts*1 particles inside a cytoplasmic vacuole. The virus was purified from cells infected at the non-permissive temperature (39°C) using standard procedures (WINTERS and RUSSELL 1971) and then dialysed against tissue culture medium containing 5% calf serum. HeLa cell monolayers were infected at a multiplicity of approximately 1–5 x $10^3$ particles/cell and the cells kept on ice for 1 h before overlaying with warm medium and maintained at 37°C for 30 min. The cells were subsequently fixed for 1 h at 20°C in 3% glutaraldehyde in 0.1 *M* cacodylate buffer, pH 7.2, containing 5% (w/v) sucrose. After rinsing in buffer and staining for 3 h in 2% (w/v) aqueous uranyl acetate, the cells were then processed for thin-section electron microscopy, as described previously (PATTERSON and BINGHAM 1976). *Bar*, 160 nm

although the fibre can function as a first step to promote internalisation of the particle and the penton base can apparently facilitate permeabilisation of the endosomal membrane, the passage of the much larger virus particle into the cytoplasm requires the recognition of some other parameter perhaps related to the conformation of the virus particle. It is evident that the virion protease (which is defective in the mutant) plays a critical part in the maturation of the virus particle (see below) and it may be that the immature particle is not presenting either the penton base or some other structural component (e.g. VI, IX, IIIa) in the appropriate mode or the active virus protease itself may be required to promote fusion and entry into the cytoplasm. These steps are clearly operating in infection of human lymphocytes, since primary cells apparently do not elaborate virus receptors and hence are resistant to virus infection (SILVER and ANDERSON 1988). On the other hand, some human lymphoid cell lines can support adenovirus infection, albeit relatively inefficiently. In this case the cell lines appear to elaborate the appropriate cell receptors for fibre and show clustering of virus particles, but the cells only internalise the virus to any extent when they are in mitosis, presumably as a function of the different repertoire of integrins available in this phase.

From the above it will be seen that the ability of adenovirus to enter a cell and initiate infection depends on the availability of specific receptors on the cell surface which can recognise the fibre; furthermore, the availability of the appropriate integrins to react with the penton base also seems to be a prerequisite. However, other parameters not yet recognised also seem likely to be of importance, e.g. hexons have type-specific components (which can induce neutralising antibodies) situated on the variable regions at the top of the tower region (see the chapter by STEWART and BURNETT, this volume).

## 3 "Uncoating" and Transport to the Nucleus

Once taken up into the cell, the phenomenon of "uncoating" of the virus particle begins. It has been thought for some time (e.g. MORGAN et al. 1969) that the particle leaves the endosome in a more or less intact form, while more recent evidence has indicated that it may penetrate into the cytosol from the early endosome in its prelysosomal state (PASTAN et al. 1986), suggesting that lysosomal enzymes do not have a role in the uncoating process. However, events governing the passage of the virus and the eventual delivery of the virus genome into the nucleus are not at all clear, although some recent studies by GREBER et al. (1993) have made an attempt to define these steps. These investigators showed that labelled virus particles are rapidly internalised by receptor-mediated endocytosis, with 80% of the receptor-bound virus being located in early endosomes within 20 min. Their result suggested that fibre protein did not enter the cytoplasm, indicating that its removal in the endosome is one of the first events. The structure of the capsid was further dismantled in the preacidic endosome with the apparent dissociation of proteins VIII and IIIa. Although the exact location and function of VIII is unclear, it is produced by the maturation of pVIII by the adenovirus protease (ANDERSON et al. 1973) and may have a role in linking the DNA to the capsid. Both observations are consistent with a location on the inner face of the capsid, as suggested by EVERITT et al. (1975). The location of IIIa is better defined (STEWART et al. 1993). It appears to make contact with the tower regions on the outer aspect of the hexon and then span the capsid linking together hexons from group-of-nine hexons (GON) from adjacent facets. Loss of IIIa would therefore weaken the capsid structure, while the removal of VIII may be a step towards releasing the genome from the surrounding capsid.

The next stage in the endosome cycle—acidification of the organelle—appeared to provide the trigger for penetration of the virus into the cytoplasm. This event was accompanied by the loss of the majority of the pentons, protein IX and, apparently, by proteolysis of protein VI and a sub-population of the hexons (GREBER et al. 1993). Proteolysis of VI and hexon appeared to be carried out by different systems, since entry into the cytosol seemed to be a prerequisite for hexon cleavage, but not for VI cleavage. However, it is unlikely that the adenovirus-coded protease is responsible for the proteolysis of VI, as suggested by GREBER et al. (1993), since there are no further sequences on the mature polypeptide VI matching the consensus derived by WEBSTER et al. (1989) which defines the cleavage specificity of the protease.

At this stage the infecting particle resembles adenovirus cores with some of the hexons from the capsid still associated. Adenovirus cores have a relatively well defined composition consisting of the DNA with the covalently attached terminal protein (TP) and the core proteins, μ, VII, V and $IVa_2$ (REKOSH et al. 1977; RUSSELL et al. 1971; CHATTERJEE et al. 1985). The precise organisation of the cores is less clear, but it appears that VII is closely associated with the DNA, since it requires 3 *M* NaCl to dissociate the two (CUPO et al. 1987). The small μ protein also has a very basic structure and binds tightly to DNA (ANDERSON et al. 1989), while the fact that V can be more readily extracted (VAYDA et al. 1983) suggests that it is located on the outside of the core. This is consistent with the results from chemical cross-linking studies which showed that while μ, VII and V could be cross-linked to each other, the only cross-link between core and capsid proteins was between V and dimers of VI (CHATTERJEE et al. 1985).

Following the stepwise uncoating of the virus particle, the DNA genome must be delivered to the nucleus, where transcription and DNA replication takes place. While the nature of the signal directing it there is not clear, there is increasing evidence from other viral systems that protein primary structure motifs (nuclear localisation signals, NLS) play a major role. Two systems which have been studied in this respect are the influenza virus and simian virus 40 (SV40), which have similarities to adenovirus in that they involve an endosomal phase. Following release from the endosome, the matrix protein of influenza virus dissociates and the nuclear entry is then mediated by a NLS in the nucleoprotein component of the viral RNA–protein complex (DAVEY J. et al. 1985). A similar situation appears to function in SV40 with the small structural proteins Vp2/3 controlling nuclear localisation of the genome (CLEVER and KASAMATSU 1991). Proposals for human immunodeficiency virus (HIV) have also been made by BUKRINSKY et al. (1993), who suggested that following virus infection, the virus pre-integration complex of HIV-1 is actively transported to the nucleus under the influence of an NLS present in the gag matrix protein. It therefore seems probable that a similar situation exists in adenovirus, and thus one or more of the core proteins must be a prime candidate for the role of nuclear targeting. The motifs describing an NLS are somewhat diverse, although several unifying features are present (GARCIA-BUSTOS et al. 1991). They are generally short sequences (typically eight to ten amino acids in length) containing a high proportion of lysine and arginine residues and often a proline (WAGNER et al. 1990). A list of around 40 such sequences has been compiled (GARCIA-BUSTOS et al. 1991) and this contains three adenovirus proteins, the pre-terminal protein (pTP), DNA-binding protein (DBP), and an E1a protein. Since DBP and E1a proteins are not structural proteins, the sequence in pTP (between residues 362 and 373, i.e. within the TP) could play a role in directing the particle to the nucleus. This sequence is very well conserved in all of the serotypes in which the sequence of the pTP is known (Table 2).

The model for the structure of the adenovirus genome in virions proposed by WONG and HSU (1989) places the ends of the DNA with the covalently attached TP outwith the supercoiled regions. This location is consistent with the results of LEITH et al. (1989), who showed that cores can act as templates for in vitro DNA replication, suggesting that the TP must be in an accessible location. A series of experiments (SCHAAK et al. 1990) have confirmed the importance of the TP in locating the virus genome to the nuclear matrix and have also shown that it is required to maintain the integrity of early transcription events. Other studies have indicated that pTP is capable of directing other proteins to the nucleus, as at late stages of infection the pTP is known to form a complex with the adenovirus DNA polymerase and take this with it into the nucleus (ZHAO and PADMANABHAN 1988).

A more definitive family of NLS sequence has been proposed by DINGWALL and LASKEY (1991). This is bipartite in nature and can be described by a motif consisting of two basic residues separated by approximately ten amino acids from a stretch of five amino acids, three of which are basic. This motif is reasonably discriminating, being found in 50% of the nuclear proteins and only 5% of the non-nuclear proteins in the Swissprot data base. Sequences matching this motif can be found in three proteins known to be present in the adenovirus, i.e. $\mu$, protein VII and core protein V.

Protein VII and $\mu$ are known to be closely associated with the DNA, and therefore the presence of basic sequence motifs is not surprising and more likely to be related to their putative function in binding to DNA. NLS conforming to this motif are found close to the C terminus in all of the protein V sequences which have been elucidated to date (Table 3), while those found in the $\mu$ and protein VII of serotype 2 are not strictly conserved in these proteins from other serotypes.

**Table 2.** Conservation of pre-terminal protein (pTP) nuclear localisation sequence

| Serotype | Sequence |
|---|---|
| 02 | RLPVRRRRRRVP |
| 04 | RLPVRRRRRRPP |
| 05 | RLPVRRRRRRVP |
| 07 | RLPVRRRRRRVP |
| 12 | RLPVRRRRRRQP |
| 40 | RLPVRRRRRRVP |
| 41 | RLPVRRRRRRVP |

**Table 3.** Sequences from the C-terminus of core protein V which conform to the bipartite nuclear localisation signal (NLS) consensus

| Serotype | Sequence |
|---|---|
| 02 | $^{336}$RRQPVLAPISVRRVAR |
| 05 | $^{336}$RRQPVLAPISVRRVAR |
| 12 | $^{317}$RRATAPSRRRGPSRRRRR |
| 41 | $^{298}$RRTAPISVRRVTRRGR |

Thus it would appear that the major candidates for the role of directing the adenovirus DNA to the nucleus following uncoating are the pTP and protein V, with presumably the higher molar concentration and location on the surface of the cores of the latter indicating that it may have the major role. However, nuclear localisation sequences cannot be unambiguously identified from sequence analysis alone, and the precise identification of the key sequences awaits further work; indeed, the significance of the nuclear pore as a portal of entry has not yet been established (CITOVSKY and ZAMBRYSKI 1993).

## 4 Transcription and Replication

The adenovirus genome is subject to transcriptional control involving the synthesis of four major classes of early transcripts, two of these having promoters towards the 5'and 3' termini, together with a series of late transcripts, most of the latter being initiated via the major late promoter situated towards the conventional left end of the genome. The factors controlling this differential transcription are not yet evident, but it seems possible that a parameter which could play a critical role could be the conformation of the genome and the availability of the appropriate promoter sequences. From the discussion above, the evidence now indicates that the infecting virus genome is located in the nuclear matrix and very likely consists of the genome with the covalently linked TP in close association with the structural core proteins VII, V, $IV_{a2}$ and $\mu$ (RUSSELL et al. 1971; RUSSELL and PRECIOUS 1982; NERMUT 1980; NEWCOMB et al. 1984), i.e. it is unlikely that the virus DNA is completely uncoated and is present initially as naked DNA.

The topology of this virus chromatin is not well defined, but there is evidence that the adenovirus DNA is contained inside the virus capsid in a tightly supercoiled state, that the virus core proteins are located at defined positions on the adenovirus DNA and, as noted above, that the termini appear to be in a different conformational state to the rest of the genome (WONG and HSU 1989) and possibly closely associated with the nuclear matrix. Since both the early transcriptional events and the origin of viral DNA replication occur at the termini and continuing replication requires the relative movement of the nascent DNA strands on the template, it may be that the template for both transcription and replication have interacting termini either mediated by protein–protein interaction via TP (and pTP) or by transient DNA hybrid formation at the inverted terminal repeats (ITR) at the termini. Such a model would, however, involve a build-up of topological stress which would require release by topoisomerase action. Furthermore, if this model was acceptable mechanistically, it could also be envisaged that such a conformation involving a "circular" genome mediated by protein–protein interaction with a transcription and replication origin in place could exist in the virion itself. The importance of supercoiled domains in the regulation of virus infection has been strengthened by the finding that topoisomerase inhibitors have significant effects on replication, transcription and packaging of the adenovirus genome (Wong and Hsu 1990).

In view of the requirement for the tight binding to the nuclear matrix of the virus genome (via the TP) to mediate transcription (Schaak et al. 1990), a plausable scenario might involve initial transcriptional events being governed by the availability of nuclear transcriptional factors and by the template restrictions imposed by the topology of the genome. The distribution of the various core proteins may also directly influence transcriptional events, although it appears likely that transcription could occur even in the presence of these proteins in a manner analogous to that prevalent during transcription on templates containing nucleosomes (e.g. Ericsson et al. 1990; Wolffe 1994).

Following the transcription and translation of the early genes, the template must become available for virus DNA replication and the first round of replication must presumably not be constricted by the topological constraints inherent in primary transcription. This may be achieved by the interaction of some of the early virus gene products (e.g. adenovirus DBP), thereby creating progeny viral DNA which would not be subject to the same constraints as the infecting virion DNA, thereby allowing secondary transcription to take place. However, it must be pointed out that knowledge of the structure of adenovirus chromatin and the protein–DNA and protein–protein interactions occurring in the virion is rudimentary, although there seems little doubt that such interactions play a crucial part in regulating virus infections at many stages in the virus life cycle.

## 5 Virion Maturation

The steps involved in the later stages of infection in which the structural proteins are assembled into the mature virion are discussed in more detail in the chapters by D'Halluin and Schmid and Hearing (this volume), but there is now good evidence that the virus-coded protease (which is a structural component) plays a major part in the final maturation of the virion (see also chapter by Weber, this volume). The protease is synthesised late in infection and has long been known to perform maturation cleavages on six viral proteins, namely pVI, pVII, pVIII, pTP, IIIa and protein X (Anderson et al. 1973). Webster et al. (1989) used synthetic peptides to define the cleavage specificity of the protease as being (M,L,I) XGX G or (M, L, I) XGG X, and searches of these consensus sequences against the translated Ad2 genome identified plausible cleavage sites in all of the known substrates. Alignments of the sequences of all the protease substrates, which are currently available in the Swissprot and GenBank/EMBL data bases, show that the cleavage sites are well conserved (Fig. 2), any changes still falling within the consensus sequences. Houde and Weber (1990) have shown that Ad2 pVII is

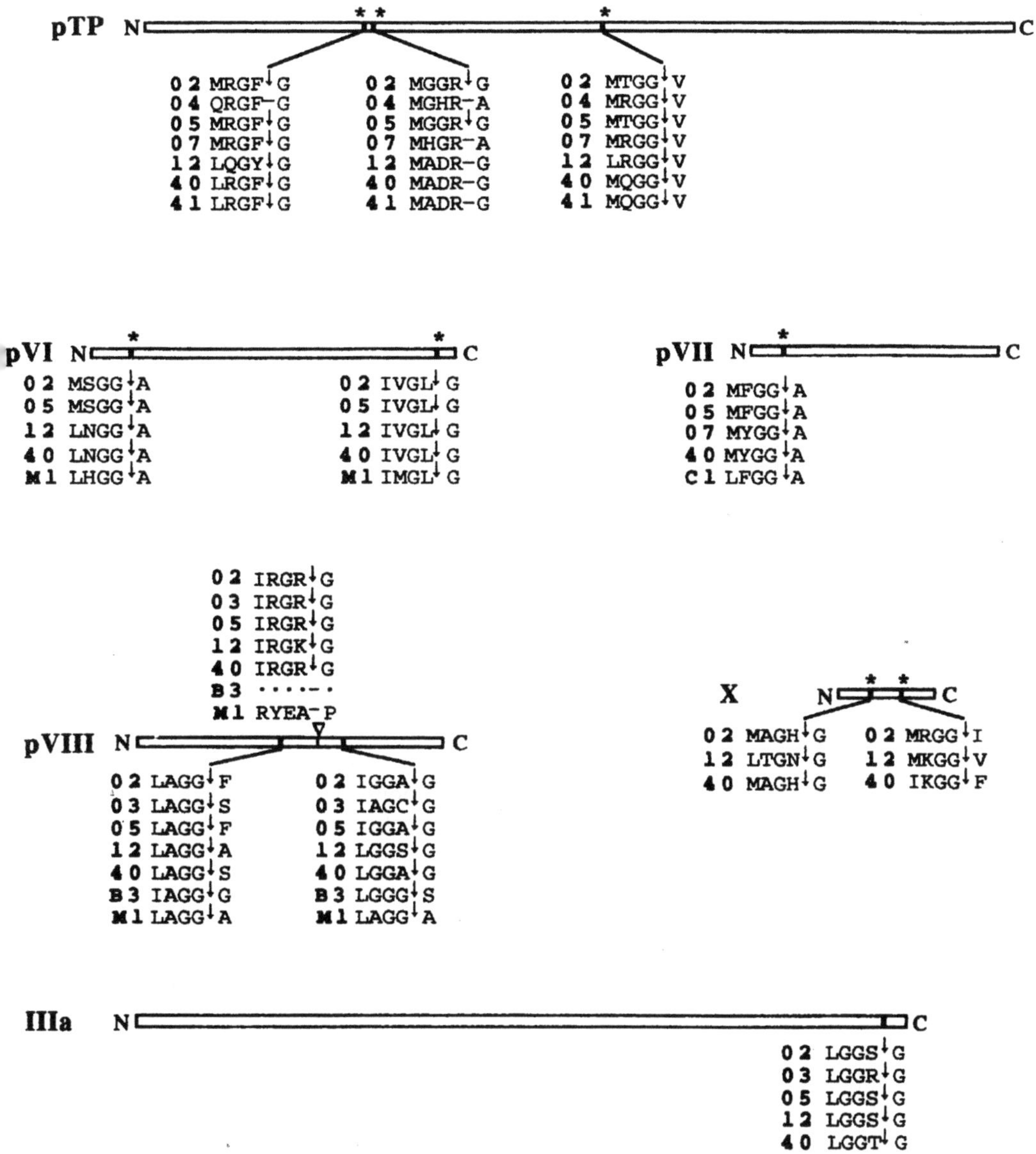


**Fig. 2.** Conservation of cleavage sites in known substrates of the adenovirus protease. Sequence alignments were prepared using the PILEUP program in the UWGCG package (DEVEREUX et al. 1984). Sequences which fit the consensus for cleavage; -, sequences which do not fit consensus for cleavage; *, sites at which cleavage is known to take place in serotype 2; ..., sequences which align with gaps. Human serotypes are denoted by *numbers*; *B3*, bovine serotype 3; *Cl*, canine serotype 1; M1, mouse serotype 1

efficiently cleaved by the protease from seven other serotypes (types 1, 3–7,9) representing four separate subgroups (B–E). These observations suggest that the roles of the protease and the precursor and mature proteins are conserved throughout the range of serotypes.

The pTP is perhaps the protease substrate which is best understood in terms of function. It acts as the protein primer for adenovirus DNA replication in a mechanism whereby the new strand is initiated from a deoxycytidine monophosphate (dCMP) linked by an ester bond to serine residue 580. The pTP forms a complex with the adenovirus DNA polymerase (pol) and this heterodimer is transported to the nucleus under the influence of the NLS in the pTP. Once in the nucleus the pTP–pol complex binds strongly to bases 9–18 of the DNA (TEMPERLEY and HAY 1992) at the origin of replication and pTP associates with the nuclear matrix (ZHAO and PADMANABHAN 1988; FREDMAN and ENGLER 1993). At a later stage of the infectious cycle the pTP is cleaved to the mature terminal protein (TP) via an intermediate TP (iTP), but the significance of these cleavages is not yet apparent. Recent work (WEBSTER et al. 1994) has established, however, the precise cleavage sites in Ad2 pTP (Fig. 2), and moreover it was shown that only pTP and iTP interact with pol and only pTP will bind to DNA, suggesting that only pTP can act as a protein primer for DNA replication. Interestingly, a seven-residue peptide (MTGGVFQ) encompassing the cleavage site which gives rise to TP was apparently not cleaved by the protease (WEBSTER et al. 1989), suggesting that the primary structure in this region of the pTP (perhaps as a result of other protein–protein interactions) may regulate the rate of TP production. Thus, since the properties of pTP and its derivatives seem to vary, these results suggest the rates of cleavage giving rise to iTP and TP may play a significant role in modulating the course of replication and thereby the infectious cycle.

Protein VI is a structural protein which, according to STEWART et al. (1983), is located on the inner face of the capsid at the apices of the icosahedral capsid. The precursor (pVI) contains two cleavage sites—one at residue 32 (types 2,5,12,41) or 33 (mouse type 1) and the other 11 residues from the C terminus. WEBSTER et al. (1993) and MANGEL et al. (1993) have shown that the 11-residue peptide derived from the C terminus is necessary for the activation of recombinant Ad2 protease. WEBSTER et al. (1993) have also demonstrated that activation requires the disulphide-bonded dimer of the peptide and provided evidence that intact pVI is also capable of activating the protease. These observations suggest that pVI has a role in the regulation of protease activity in the infected cell. As noted above, the protease-deficient mutant of Ad2 (Ad2*ts* 1) is capable of forming virus-like particles which are defective in uncoating (WEBER 1976), suggesting that capsids can be assembled from precursor proteins; indeed, it is possible that the precursor proteins are necessary for assembly. After cytoplasmic synthesis, the protease and pVI, presumably as complexes with other proteins or via scaffolding proteins (HASSON et al. 1992), are transported to the nucleus and packaged into "young" virions (see D'HALLUIN, this volume), where the concentration of pVI will be high enough to promote self-dimerisation and consequent activation of the protease. Such a regulatory mechanism should prevent premature activation of the protease and consequent precursor cleavage having a detrimental effect on capsid assembly. There is evidence for an analogous situation in HIV, where the protease itself is only active as a dimer and does not dimerise until the monomer-containing polyproteins assemble on the inner surface of the cell membrane to form the immature virion. This was elegantly demonstrated by KRAUSSLICH (1991), who engineered a second protease monomer into the polyprotein and showed that the consequent premature activation of the protease prevented formation of new virus particles as the polyproteins were self-cleaved before they could assemble at the cell membrane.

Mature VI is thought to have a role in linking core complex to the capsid (RUSSELL and PRECIOUS 1982), and recent work (MATTHEWS and RUSSELL 1994) has provided evidence that there is a strong interaction between the hexon and mature VI. This interaction appears to be about 300 times stronger than the interaction between the hexon and pVI, suggesting that proteolytic maturation alters the conformation of the VI polypeptide, facilitating tighter binding. In this regard it is noteworthy that the sequence of the peptide removed from the C terminus is very basic, with three arginine and one lysine in the 11-residue sequence of the Ad2 peptide, and it may be that proteolysis also facilitates the interaction of VI with the capsid and weakens the linkage to the core complex.

pVII and X protein are respectively precursors of protein VII (also known as the major core protein) and the μ protein which bind tightly to DNA. They are both very basic proteins and may act in a histone-like manner to neutralise the negative charge on the DNA and thereby facilitate its packaging into the virion (RUSSELL et al. 1968; ANDERSON et al. 1989).

Partial sequencing of the type 1 avian adenovirus genome (SHEPPARD and TRIST 1993) has resulted in the identification of two open reading frames which code for proteins with sequence homology to pVII and the X protein. The putative avian protein with homology to pVII is, however, much smaller, with a length of 77 residues compared to the 196 residues in Ad2 pVII. In contrast, the putative avian X homologue is 180 residues long, while Ad2 protein has only 79 residues. Both avian proteins, however, contain sites fitting the consensus for protease cleavage. Should cleavage occur at these sites, the avian VII would be very much smaller than its mammalian counterpart, while the avian μ protein would be a 15- rather than a 19-residue peptide. It is presumed that some cooperativity exists between these two proteins in facilitating charge neutralisation.

Both pVIII and IIIa are precursors for structural proteins involved in interacting with GONs and hence are critical in the formation of the icosahedral capsid. According to STEWART et al. (1991), VIII is located on the inner face of the capsid, but little is known of its function or how its properties are altered by maturation. There has been some debate about the location and role of IIIa. It was initially thought to be associated with the vertex region and in particular with the peripentonal hexons (EVERITT et al. 1973), but more recent evidence suggests that it is located at the edges of the facets and spans the capsid as noted above

(STEWART et al. 1993). Cleavage by the adenovirus protease is thought to occur close to the C terminus of the molecule (CUILLEL et al. 1990), and this is consistent with the consensus cleavage motif, but the consequences of this event are not clear.

Searches of the consensus cleavage motif for the protease against other adenovirus structural proteins reveal additional potential cleavage sites (WEBSTER et al. 1989), but to date none of these has been investigated and it may be that they are not normally available to the protease since in the mature virion they are not processed. Potential cleavage sites also exist in cellular proteins, and it has recently been demonstrated (CHEN et al. 1993) that the protease will cleave cytokeratin, promoting disruption of the cytokeratin system and thus facilitating cell lysis and the release of virus.

From the above it will be evident that the ability of adenovirus structural proteins to assemble may in many cases be tightly regulated by the virion protease; further studies of these processing pathways will undoubtedly reveal yet more subtleties in the pathway to maturation.

## 6 Epilogue

It will be seen from the review given above that there is a paucity of information concerning the role of adenovirus structural proteins in the infectious process itself. Thus one of the major structural components termed IIIa has not been studied in any detail and yet is likely to be of crucial importance in that it is clearly phosphorylated during infection and in the virus it can exist in at least six distinct phosphorylated states (M. HALL–SMITH and W.C. RUSSELL unpublished observations; see Fig. 3). The finding some years ago that the virion apparently contains a protein kinase (BLAIR and RUSSELL 1978) which phosphorylates IIIa and other structural components has not been followed up, and it seems possible that this system could play a role in packaging the virus DNA during assembly by facilitating a charge interaction between the basic proteins surrounding the virus DNA and the phosphorylated IIIa; this component is presumed to be in association with the virus capsid and also appears to have a relationship to the virus core (STEWART et al. 1993). Similarly, another major structural component, polypeptide V, has not been investigated in any depth. This core protein appears to be less tightly attached to the virus core, and it has been assumed that it may play a role analogous to H1 histones in chromatin (RUSSELL et al. 1971; MIRZA and WEBER 1982). Another adenovirus structural protein, $IVa_2$, has been little studied and it seems possible that this protein could play an important part in the encapsidation of the virus genome and perhaps in facilitating the phosphorylation events mentioned above, since it has a motif for adenosine/guanosine triphosphate (ATP/GTP) binding. There are many other interesting aspects of these structural proteins which are discussed in other chapters, and it is to be hoped that with the

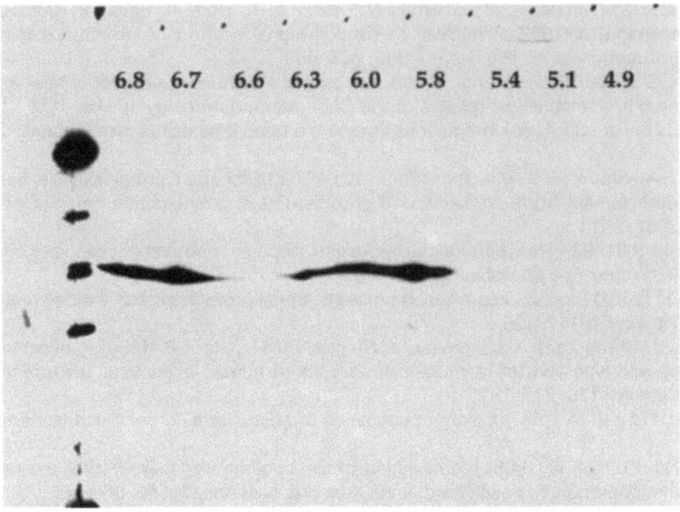


**Fig. 3.** Iso-electric focusing (IEF) using standard techniques (i.e. IEF in the first dimension and sodium deodecyl sulphate, SDS, in the second; BOULANGER et al. 1979); pattern with purified virus after in vitro phosphorylation using gamma $^{32}P$ adenosine triphosphate (ATP) with the endogenous kinase (BLAIR and RUSSELL 1978). The figures on the *horizontal axis* are the pH values monitored by direct test on these areas of the gel. An almost identical pattern was obtained using $^{32}P$-labelled virus derived from $^{32}P$ phosphate-labeled infected cells (RUSSELL and BLAIR 1977). $I^{125}$ labelled virus marker in *left-hand lane*

advent of new technologies there will be a much better understanding of all the interactions involving structural components which take place during infection.

*Acknowledgments.* We are grateful to Dr. S. Patterson, CRC, Northwick Park Hospital, Harrow for providing the electron micrographs in Fig. 1, to M. Hall-Smith for technical help in the two-dimensional gel analysis, to Margaret Wilson for typing and to Ailsa Webster and Ron Hay for discussion and permission to quote their data prior to publication. This work benefitted from the use of the SEQNET facility at Daresbury, UK.

## References

Anderson CW, Baum PR, Gesteland RF (1973) Processing of adenovirus 2 induced proteins. J Virol 12: 241–252

Anderson CW, Young ME, Flint SJ (1989) Characterisation of the adenovirus 2 virion protein, mu. Virology 172: 506–512

Bai M, Harpe B, Freimuth P (1993) Mutations that alter an Arg-Gly-Asp (RGD) sequence in the adenovirus type 2 penton base protein abolish its cell rounding activity and delay virus reproduction in flat cells. J Virol 67: 5198–5205

Belin M-T, Boulanger P (1993) Involvement of cellular adhesion sequences in the attachment of adenovirus to the Hela cell surface. J Gen Virol 74: 1485–1497

Blair GE, Russell WC (1978) Identification of a protein Kinase associated with human adenoviruses. Virology 86: 157–166

Boulanger P, Lemay P, Blair GE, Russell WC (1979) Characterisation of adenovirus protein IX. J Gen Virol 44: 783–800

Bukrinsky MI, Haggerty S, Dempsey MP, Sharova N, Adzhubel A, Spitz L, Lewis P, Goldfarb D, Emerman M, Stevenson M (1993) A nuclear localization signal within HIV-1 matrix protein that governs infection of non-dividing cells. Nature 365: 666–669
Chardonnet Y, Dales S (1970) Early events in the interaction of adenoviruses with Hela cells 1. Penetration of type 5 and intracellular release of the DNA genome. Virology 40: 462–477
Chatterjee PJ, Vayda ME, Flint SJ (1985) Interactions among the three adenovirus core proteins. J Virol 55: 379–386
Chen PH, Ornelles DA, Shenk T (1993) The adenovirus L3 23-kilodalton proteinase cleaves the amino-terminal head domain from cytokeratin 18 and disrupts the cytokeratin network of Hela cells. J Virol 67: 3507–3514
Citovsky V, Zambryski P (1993) Transport of nucleic acids through membrane channels: snaking through small holes. Annu Rev Microbiol 47: 167–197
Clever J, Kasamatsu H (1991) Simian Virus 4 VP2/3 small structural proteins harbor their own nuclear transport signal. Virology 181: 78–90
Cuillel M, Cortolezzis J, Chroboczek J, Langowski J, Rulgrok RWH, Jacrot B (1990) Purification and characterization of wild-type and ts112 mutant protein IIIa of human adenovirus 2 expressed in Escherichia coli. Virology 175: 222–231
Cupo JF, Rezanka LJ, Harpst JA (1987) Column purification of adenovirus cores. Anal Biochem 164: 267–270
Davey J, Dimmock NJ, Colman A (1985) Identification of the sequence responsible for the nuclear accumulation of the influenza virus nucleoprotein in Xenopus oocytes. Cell 40: 667–675
Defer C, Belin M-T, Caillet-Baudin M-L, Boulanger P (1990) Human adenovirus-host cell interactions: a comparative study with members of subgroups B and C. J Virol 64: 3661–3673
Devereux J, Haeberli P, Smithies D (1984) A comprehensive set of sequence analysis programs for the Vax. Nucleic Acids Res 12: 387–396
Dingwall C, Laskey RA (1991) Nuclear targeting sequences-a concensus? Trends Biochem 16: 478–481
Ericsson C, Grossbach U, Bjorkroth B, Danehalt B (1990) Presence of histone H1 on an active Balbiani Ring gene. Cell 60: 73–83
Everitt E, Sundquist B, Pettersson U, Philipson L (1973) Structural proteins of adenoviruses X. Isolation and topography of low molecular weight antigens from the virions of adenovirus type 2. Virology 62: 130–147
Everitt E, Lutter L, Philipson L (1975) Structural proteins of adenovirus XII. Location and neighbour relationships among proteins of adenovirus 2 as revealed by enzymatic iodination, immunoprecipitation and chemical cross-linking. Virology 67: 197–208
Fredman JN, Engler JA (1993) Adenovirus precursor to terminal protein interacts with the nuclear matrix in vivo and in vitro. J Virol 67: 3384–3395
Garcia-Bustos J, Heitman J, Hall MN (1991) Nuclear protein localization. Biochim Biophys Acta 1071: 83–101
Gelderblom H, Maichle-Lauppe I (1982) The fibres of fowl adenoviruses. Arch Virol 72: 289–298
Greber UF, Willetts M, Webster P, Helenius A (1993) Stepwise dismantling of adenovirus 2 during entry into cells. Cell 75: 477–486
Green NM, Wrigley NG, Russell WC, Martin SR, McLachlan AD (1983) Evidence for a repeating cross-beta sheet structure in the adenovirus fibre. EMBO J 2: 1357–1365
Hasson TB, Ornelles DA, Shenk T (1992) Adenovirus L1 52 and 55 Kilodalton proteins are present within assembling virions and colocalise with nuclear structures distinct from replication centres. J Virol 66: 6133–6142
Hennache B, Boulanger P (1977) Biochemical study of KB cell receptor for adenovirus. Biochem J 166: 237–247
Houde A, Weber JM (1990) Adenovirus type 2 precursor proteins are cleaved by proteinases of other adenoviruses. Virology 179: 485–486
Hughes RC, Mautner V (1973) Interaction of adenovirus with host-cell membranes. In: Kent P (ed) Membrane-mediated information. Medical and Technical Publishing, Lancaster, pp 104–125
Hynes RO (1992) Integrins: versatility, modulation and signalling in cell adhesion. Cell 69: 11–25
Kidd AH, Chroboczek J, Cusack S, Ruigrok RWH (1993) Adenovirus type 40 virions contain two distinct fibers. Virology 192: 73–84
Krausslich H-G (1991) Human immunodefficiency virus proteinase dimer as component of the viral polyprotein prevents particle assembly and viral infectivity. Proc Natl Acad Sci USA 88: 3213–3217
Leith IR, Hay RT, Russell WC (1989) Adenovirus subviral particles and cores can support limited DNA replication. J Gen Virol 70: 3235–3248

Mangel WF, McGrath WJ, Toledo D, Anderson CW (1993) Viral DNA and a viral peptide are cofactors of adenovirus virion proteinase activity. Nature 361: 274–275
Matthews DA, Russell WC (1994) Adenovirus proteins–protein interactions: hexon and protein VI. J Gen Virol 75: 3365–3374
Mirza MA, Weber J (1982) Structure of adenovirus chromatin. Biochim Biophys Acta 696: 76–86
Morgan C, Rosenkrantz HS, Mednis B (1969) Structure and development of viruses as observed in the electron microscope. X. Entry and uncoating of adenovirus. J Virol 4: 777–796
Nermut MV (1980) The architecture of adenoviruses: recent virus and problems. Arch Virol 64: 175–196
Newcomb WM, Boring JW, Brown JC (1984) Ion etching of human adenovirus 2: structure of the core. J Virol 51: 52–56
Pastan I, Seth P, Fitzgerald D, Willingham M (1986) Adenovirus entry into cells: some new observations on an old problem. In: Notkins AL, Oldstone MBA (eds) Concepts in viral pathogenesis, Vol 2. Springer, Berlin Heidelberg NewYork, pp 141–146
Patterson S, Bingham RW (1976) Electron microscope observations on the entry of avian infectious bronchitis virus into susceptible cells. Arch Virol 52: 191–200
Patterson S, Russell WC (1983) Ultrastructural and immunofluorescence studies of early events in adenovirus—Hela cell interactions. J Gen Virol 64: 1091–1099
Pereira HG (1958) A protein factor responsible for the early cytopathic effect of adenoviruses. Virology 6: 601–611
Philipson L, Lonberg-Holm K, Pettersson V (1968) Virus-receptor interaction in an adenovirus system. J Virol 2: 1664–1675
Rekosh DMK, Russell WC, Bellett AJD, Robinson AJ (1977) Identification of a protein linked to the ends of adenovirus DNA. Cell 11: 283–295
Ruigrok RWH, Barge A, Albiges-Rizo C, Dayan S (1990) Structure of adenovirus fibre II. Morphology of single fibres. J Mol Biol 215: 589–596
Russell WC, Blair GE (1977) Polypeptide phosphorylation in adenovirus-infected cells. J Gen Virol 34: 19–35
Russell WC, Precious BL (1982) Nucleic acid-binding properties of adenovirus structural polypeptides. J Gen Virol 63: 69–79
Russell WC, Hayashi K, Sanderson PJ, Pereira HG (1967) Adenovirus antigens—a study of their properties and sequential development in infection. J Gen Virol 1: 495–507
Russell WC, Laver WG, Sanderson PG (1968) Internal components of adenovirus. Nature 219: 1127–1130
Russell WC, McIntosh K, Skehel JJ (1971) The preparation and properties of adenovirus cores. J Gen Virol 11: 35–46
Schaak J, Ho WY-W, Freimuth P, Shenk T (1990) Adenovirus terminal protein mediates both nuclear matrix association and efficient transcription of adenovirus DNA. Genes Dev 4: 1197–1208
Seth P (1994) Adenovirus-dependent release of choline from plasma membrane vesicles at an acidic pH is mediated by the penton base protein. J Virol 68: 1204–1206
Seth P, Fitz Gerald P, Ginsberg H, Willingham M, Pastan T (1986) Pathway of adenovirus entry into cells In: Crawell RL, Lonberg-Holm K (eds) Virus attachment and entry into cells. American Society for Microbiology, Washington DC, pp 191–195
Sheppard M, Trist H (1993) The identification of genes for the major core proteins of fowl adenovirus serotype 10. Arch Virol 132: 443–449
Signas G, Akusjärvi G, Pettersson U (1985) Adenovirus 3 fibre polypeptide gene: complications for the structure of the fibre protein. J Virol 53: 672–678
Silver L, Anderson CW (1988) Interaction of human adenovirus serotype 2 with human lymphoid cells. Virology 165: 377–387
Stewart PL, Burnett RM, Cryklaff M, Fuller SD (1991) Image reconstruction reveals the complex molecular organisation of the adenovirus. Cell 67: 145–154
Stewart PL, Fuller SD, Burnett R (1993) Difference imaging of adenovirus: bridging the resolution gap between X-ray crystallography and electron microscopy. EMBO J 12: 2589–2599
Stouten PFW, Sander C, Ruigrok RWH, Cusack S (1992) New triple helical model for the shaft of the adenovirus fiber. J Mol Biol 226: 1073–1084
Svensson V, Persson R (1984) Entry of adenovirus 2 into Hela cells. J Virol 51: 687–694
Svensson V, Persson R, Everitt E (1981) Virus-receptor interaction in the adenovirus system. I. Identification of virion attachment proteins of the Hela cell plasma membrane. J Virol 38: 70–81
Temperley SM, Hay RT (1992) Recognition of adenovirus type 2 origin of DNA replication by virally encoded DNA polymerase and preterminal proteins. EMBO J 11: 761–768
Valentine RC, Pereira HG (1965) Antigens and the structure of the adenovirus. J Mol Biol 13: 13–26

Vayda ME, Rogers AE, Flint SJ (1983) The structure of nucleoprotein cores released by adenovirions. Nucleic Acids Res 11: 441–460

Wagner P, Kunz J, Koller A, Hall MN (1990) Active transport of proteins into the nucleus. FEBS Lett 275: 1–5

Watson G, Burdon MG, Russell WC (1988) An antigenic analysis of the adenovirus type 2 fibre polypeptide. J Gen Virol 69: 525–535

Weber J (1976) Genetic analysis of adenovirus type 2. III Temperature sensitivity of processing of viral proteins. J Virol 17: 462–471

Webster A, Russell S, Talbot P, Russell WC, Kemp GD (1989) Characterisation of the adenovirus proteinase: substrate specificity. J Gen Virol 70: 3224–3234

Webster A, Hay RT, Kemp G (1993) The adenovirus protease is activated by a virus-coded disulphide-linked peptide. Cell 72: 97–104

Webster A, Leith IA, Hay RT (1994) Activation of adenovirus-coded protease and processing of preterminal protein. J Virol 68: 7292–7300

Wickham TJ, Mathias P, Cheresh DA, Nemerow GR (1993) Integrins $\alpha v\beta 3$ and $\alpha v\beta 5$ promote adenovirus internalisation but not virus attachment. Cell 73: 309–319

Winters WD, Russell WC (1971) Studies on the assembly of adenoviruses in vitro. J Gen Virol 10: 181–194

Wolffe AP (1994) Transcription: in tune with the histones. Cell 77: 13–16

Wong M-L, Hsu M-T (1989) Linear adenovirus DNA is organised into supercoiled domains in viral particles. Nucleic Acids Res 17: 3535–3550

Wong M-L, Hsu M-T (1990) Involvement of topoisomerases in replication, transcription and packaging of the linear adenovirus genome. J Virol 64: 691–699

Zhao L, Padmanabhan R (1988) Nuclear transport of adenovirus DNA polymerase is facilitated by interaction with preterminal protein. Cell 55: 1005–1015

# Nuclear Organization of Replication and Gene Expression in Adenovirus-Infected Cells

E. BRIDGE and U. PETTERSSON



## 1 Introduction

The eukaryotic cell nucleus performs a series of tasks including replication and packaging of DNA and transcription, splicing, polyadenylation, and transport of RNA. Productive infection of cells by viruses is the result of complex interactions between host and viral components that effectively subvert much of the host cell machinery for replication and gene expression of the virus. In a sense, the virus is able to reprogram the host cell, resulting in dramatic changes in host gene expression and allowing for the efficient production of viral particles. Biochemical and genetic analyses of adenovirus-infected cells have provided an enormous amount of information about the interaction of host and viral factors during the infectious program (see DOERFLER 1986). Virus infection also alters structural and morphological characteristics of host nuclei (MARTINEZ-POLOMO and GRANBOULAN 1967; MARTINEZ-POLOMO et al. 1967; PHILLIPS and RASKAS 1972; VLAK et al. 1975; MOYNE et al. 1978). Adenovirus and herpesvirus replicate and accumulate DNA within distinct nuclear compartments, and host factors have also been shown to localize to these compartments (WALTON et al. 1989; WILCOCK and LANE 1991). Part of the viral reprogramming process is the formation of new virus-dependent nuclear structures and the alteration of existing nuclear structures. The role of the architecture and structural components of the nucleus in the replication and expression of both host and viral genomes is only beginning to be explored.

Department of Medical Genetics, Biomedical Center, Uppsala University, Box 589, 75 123 Uppsala, Sweden

The nucleus is organized into a series of structural domains, of which the nucleolus is the best characterized. Much of the nuclear interior is occupied by chromatin in various degrees of condensation. Interchromatin regions contain several characteristic structures, including perichromatin fibrils (PF) and granules, interchromatin granules (IG), and other nuclear bodies including coiled bodies (CB) (BRASCH and OCHS 1992). The relationship of these structures to the framework of the nucleoskeleton is not understood. Recent experiments have also demonstrated that cellular biochemical process such as replication and transcription occur in a series of focally concentrated domains. Replication domains appear as about 200 scattered foci in S-phase cells that have been labeled with a short pulse of bromodeoxyuridine (BrdU) and then examined by confocal fluorescence microscopy (NAKAMURA et al. 1986; NAKAYASU and BEREZNEY 1989; COOK 1991; M.H. FOX et al. 1991; O'KEEFE et al. 1992). The size and distribution of these foci changes in an ordered fashion as cells progress through S phase (NAKAMURA et al. 1986; NAKAYASU and BEREZNEY 1989). Similar focal sites for transcription were observed when bromouridine (BrU) was incorporated into permeabilized HeLa cells encapsulated in agarose microbeads (JACKSON et al. 1993). Most transcription foci were separate from replication foci during late S phase, but the structural relationship of transcription and replication may be more complex during the early S phase. Splicing factors are located in discrete nuclear domains, suggesting that aspects of post-transcriptional RNA processing may also take place in defined nuclear compartments (reviewed by LAMOND and CARMO-FONSECA 1993; SPECTOR 1993). Finally, confocal microscopy has been used to show that specific messages are distributed nonrandomly through the nucleus; they appear as dots or sometimes as elongated tracks (reviewed by ROSBASH and SINGER 1993; XING and LAWRENCE 1993). Taken together, these data underscore the complexity of nuclear organization. Understanding how viruses interact with and alter the nuclear architecture will be a major challenge. This review will focus on the organization of the host cell nucleus during adenovirus infection.

## 2 Nuclear Organization of Viral DNA Replication

Nuclei of infected cells progressively accumulate new induced structures as a consequence of viral replication. The earliest ultrastructural changes detected are small, irregularly shaped masses of thin fibrils (PUVION-DUTILLEUL and PUVION 1990a) that rapidly increase in size and become pleomorphic, appearing as crescents, rings and spheres. These fibrillar inclusions are labeled by antisera against the viral 72K DNA-binding protein (PUVION-DUTILLEUL et al. 1984; PUVION-DUTILLEUL and PUVION 1990b), and similar structures are observed by light microscopy when cells are fixed and labeled with anti-72K (BOSHER et al. 1992; BRIDGE et al. 1993; REICH et al. 1983; VOELKERDING and KLESSING 1986) or adenovirus DNA polymerase (SASAGURI et al. 1987). 72K is one of three viral proteins that are essential for in vitro viral DNA synthesis (CHALLBERG and KELLY 1989). The presence of the 72-kDa protein in discrete nuclear centers raises the possibility that these centers may be sites of viral DNA replication in vivo. Localization of DNA synthesis has been examined in infected cells by BrdU or $^3$H-thymidine incorporation. Several groups have reported that DNA synthesis sites colocalize with centers containing the 72-kDa protein (BOSHER et al. 1992; PUVION-DUTILLEUL and PUVION 1990a, b); however, continuous replication activity occurs in regions with little or no 72K protein (PUVION-DUTILLEUL and PUVION 1990a, b). These results can be explained by a progressive change in the functional activity of 72K centers during infection. PUVION-DUTILLEUL and PUVION (1990a, b) have shown that 72K-containing fibrillar masses represent sites of active replication in their earliest forms, but as the infection proceeds and the 72K centers become larger and pleomorphic, a 5-min pulse of $^3$H-thymidine labels only a subset of these sites. Replication is continuous in the less dense fibrillar granular network that surrounds these sites; this region is designated the peripheral replicative (PR) zone. Newly synthesized viral DNA rapidly accumulates in 72K sites following a 30-min chase (PUVION-DUTILLEUL and PUVION 1990b), suggesting that the length of the pulse may also account for different results concerning the localization of DNA synthesis; longer pulses will tend to mark accumulation as well as synthesis sites.

72K centers are sites of viral DNA accumulation (FOX et al. 1990; VOELKERDING and KLESSIG 1986; WALTON et al. 1989), but in situ hybridization studies have shown differential localization of viral single-stranded (ss) and double-stranded (ds) DNA relative to 72K (PUVION-DUTILLEUL 1993; PUVION-DUTILLEUL and PICHARD 1992; PUVION-DUTILLEUL and PUVION 1990a). Both types of DNA are found in the early 72K centers, but later in infection the larger 72K centers contain mostly ssDNA and substantially less dsDNA. PR zones contain primarily dsDNA and correspondingly less ssDNA. MURTI et al. (1990) have also described differential localization of the two other viral replication proteins, adenovirus polymerase (adPol) and terminal protein (TP), relative to 72K. This group detects regions in late phase-infected cells which label heavily with adPol and TP and have less 72K. The 72K centers contain only small amounts of adPol and TP and label very strongly with an antisera that is specific for ssDNA. These authors suggest that replication initiates in locations that contain adPol and TP, while elongation, which requires the 72K protein, takes place in 72K centers.

Taken together, the data suggest that initially (Fig.1A), 72K-containing, early replicative sites are active for DNA synthesis. These structures label with $^3$H-thymidine after a 5-min pulse and contain both ss- and dsDNA. As the infectious program continues, 72K centers enlarge, become pleomorphic, and are intermittently active for DNA synthesis; these structures contain the majority of viral ssDNA molecules (Fig.1B). Continuous replication is displaced from these sites to the surrounding PR zone (Fig.1B), which contains ds viral DNA. Still later, the 72K-containing centers become smaller and more homogeneous in shape, appearing as globular inclusions rather than crescents and rings (Fig.1C). These centers are eventually located at the outer edge of a region termed the viral genome storage site (PUVION-DUTILLEUL 1993; PUVION-DUTILLEUL and PICHARD 1992;

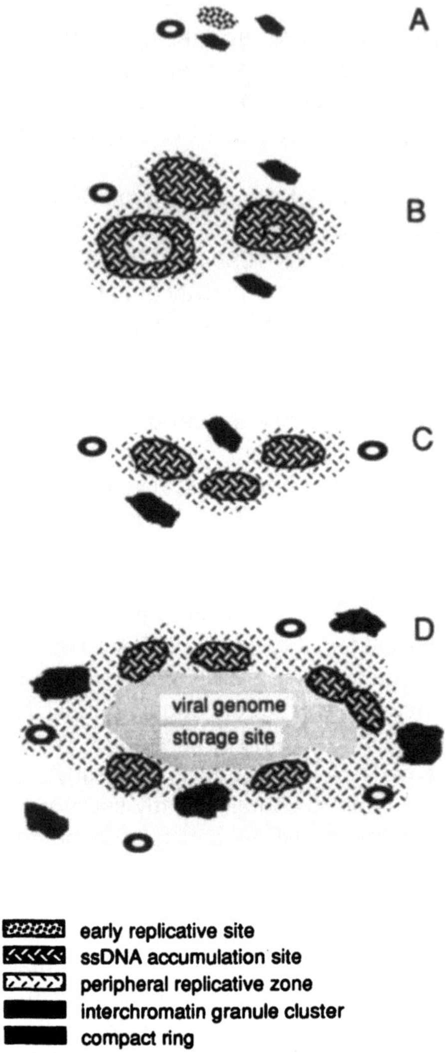


**Fig. 1 A–D.** Alterations of nuclear substructure induced by adenovirus infection. **A** Early stages of nuclear transformation by adenovirus Replicative foci consist of small areas of densely packed fibrils. **B, C** Intermediate stages of nuclear transformation. As DNA synthesis continues, replicative foci separate into two distinct regions: the single-standard (ss) DNA accumulation sites, which are strongly labeled with anti-72K, and the peripheral replicative zone, which contains double-stranded (ds) DNA and comparatively little 72K. Initially ssDNA accumulation sites are large and pleomorphic (**B**); later they become smaller and more homogeneous in shape (**C**). **D** Late stages of nuclear transformation. A large viral inclusion body consisting of inactive dsDNA appears; this is named the viral genome storage site. Interchromatin granules and replicative foci (including both ssDNA accumulation sites and the peripheral replicative zone) are located at the periphery of the viral genome storage site. During the late phase, clusters of interchromatin granules in infected cells are enlarged compared to those of uninfected cells. Virus-induced compact ring structures are also present in infected nuclei and become more numerous as the late phase continues. Adapted from PUVION-DUTILLEUL et al. (1992)

PUVION-DUTILLEUL and PUVION 1990a), which is detected in infected nuclei after 20 h post-infection (pi; Fig.1D). The genome storage site contains ds viral DNA that is inactive for both DNA replication and transcription; the 72K protein is only rarely found at this site. The genome storage site is frequently associated with crystalline arrays of virus, and it has been suggested that DNA from this compartment may be incorporated into virions (PUVION-DUTILLEUL and PICHARD 1992; PUVION-DITILLEUL and PUVION 1990a).

Several cellular factors are also important for viral replicatiopn. The transcription factors nuclear factor I (NF I) and NF III and a topoisomerase, NF II, are essential for viral replication in vitro (CHALLBERG and KELLY 1989). A recent study by BOSHER et al. (1992) showed that in adenovirus type 2 (Ad2) infections NF I colocalized with the 72K protein in immunofluorescence studies, demonstrating that both cellular and viral proteins required for replication are present in the same sites. In Ad4 infections, which do not require NFI for replication, no colocalization with 72K was observed. These results suggest that host proteins needed for viral replication are recruited from their normal locations to sites where viral replication is taking place. In herpesvirus-infected cells, several other host antigens involved in cellular DNA replication including, retinoblastoma (Rb), p53, proliferating cell nuclear antigen (PCNA), ssDNA-binding protein, DNA ligase 1, and DNA pol $\alpha$ are recompartmentalized to sites of viral DNA synthesis (WILCOCK and LANE 1991). The distribution of these and other replication antigens in adenovirus-infected cells remains to be determined. The protein composition and complexity of the 72K staining centers as they form and evolve is also not known, but should provide important clues concerning the interaction of host and viral factors during in vivo viral DNA synthesis.

## 3 Nuclear Organization of Transcription

Transcription of adenovirus genes is temporally regulated (for reviews see FLINT 1986; SHARP 1984). Prior to the onset of DNA synthesis, transcription occurs primarily from the viral early regions. After the onset of DNA replication, transcription of the late regions increases dramatically. How is transcription of the template coordinated with replication of the viral DNA? The location of transcription in adenovirus infected cells has been examined in electron microscopy studies of infected cells subjected to a 3-min pulse of $^3$H-uridine (PUVION-DUTILLEUL and PUVION 1991; PUVION-DUTILLEUL et al. 1992). At early stages, transcription was present at the early replicative sites. At intermediate times transcription occurs at PR zones, and the 72K-containing ssDNA accumulation sites are only labeled rarely. At later stages when the nucleus has acquired a large genome storage site, transcription is less intense and limited to the PR zone and the surrounding fibrillar compartment. Following a 1-h chase of infected cells pulsed for 3 min at 17 h pi (PUVION-DUTILLEUL et al.1992), label is present in PR zones, interchromatin granules (IG), virus-induced compact ring structures, and ribosome-containing areas of the cytoplasm. 72K centers are only slightly labeled after the chase. After a 2-h chase the pattern is similar, but the IG are frequently only labeled at the periphery. In situ hybridization to viral RNA with adenovirus DNA probes marked similar structures (PUVION-DUTILLEUL and PUVION 1991; PUVION-DUTILLEUL et al. 1992), indicating that viral RNA is being detected in the transcription assays. In the early stages, viral RNA was detected only at sites of transcription in the early replicative centers. At 17 h pi, RNA was present in the transcriptionally active PR zones and was also present in some, but not all, IG and in the cytoplasm. At 24 h pi, nearly all of the large clusters of IG as well as the virus-induced compact ring structures contained RNA.

These data show that transcription and DNA replication of the virus take place in overlapping compartments. This is consistent with electron microscopy studies showing replication and transcription occurring on the same viral DNA molecule (MATSUGUCHI et al. 1979; WOLGEMUTH and HSU 1981). However, DNA replication can also occur at late-stage 72K centers, whereas little or no transcription is observed in these sites. Likewise, data from POMBO et al. (1994) indicate that at intermediate times in the late stage transcriptionally active regions overlap, but are not coincident with regions undergoing replication. Thus, initially, viral replication and transcription occur together in the viral early replicative sites. After separation of the PR zone from the 72K-containing ssDNA accumulation sites, transcription occurs almost exclusively in the PR zone, where it partially overlaps with areas of ongoing DNA synthesis. Following transcription, viral RNA is detected in IG and compact rings both by in situ hybridizations and after pulse-chase experiments (PUVION-DUTILLEUL and Puvion 1991; PUVION-DUTILLEUL et al. 1992). The data suggest that these nuclear structures may be involved in post-transcriptional biogenesis or intranuclear trafficking of RNA as it moves through the nucleus and into the cytoplasmic compartment.

Adenovirus infection is known to inhibit cellular DNA synthesis (DOLPH et al. 1988; HUANG and SCHNEIDER 1991), but, in contrast, infection seems to have little effect on the transcription of cellular genes (BABICH et al. 1983; BELTZ and FLINT 1979; YODER and BERGET 1985). Thus transcriptional analysis of adenovirus-infected cells will presumably detect both viral and host RNA synthesis. The coordination of host and viral transcriptional activities in infected cells is at present not known, but transcriptional studies combined with in situ hybridizations to viral and cellular RNA may provide some clues.

## 4 Nuclear Organization of Post-transcriptional Processing

Adenovirus RNA is modified by a series of post-transcriptional processing events that are characteristic of eukaryotic mRNA biogenesis; viral transcripts are capped, polyadenylated, spliced, and subjected to internal methylation (FLINT 1986). Polyadenylation and splicing of RNA have been the subject of intense research. In vitro systems have been used to investigate the mechanisms of these reactions and to define their components (reviewed by GREEN 1991; MOORE et al. 1993; WAHLE and KELLER 1992). The temporal order of adenovirus transcripts processing has been determined by kinetic studies (NEVINS and DARNELL 1978), which showed that adenovirus transcripts are polyadenylated before they are spliced. However, more recent results from the study of other systems suggest that there are several examples of splicing of nascent transcripts (BEYER and OSHEIM 1988; LEMAIRE and THUMMEL 1990), indicating that there may be some flexibility in the order of these post-transcriptional events. Studies of regulated complex transcription units suggest that commitment to a particular spliced form of a message can influence which polyadenylation site is chosen (ADAMI and NEVINS 1988; EMESON et al. 1989; LEFF et al. 1987). Likewise, other examples exist where the selection of polyadenylation (poly A) sites directs the selection of particular splice sites (DANNER and LEDER 1985; GALLI et al. 1988). Such results suggest a degree of interaction between splicing and polyadenylating processes.

Although considerable progress has been made in biochemical analyses of post-transcriptional processing events, less is known about their nuclear organization. Most approaches to studying the nuclear organization of RNA biogenesis have focused on immunofluorescence detection of antibodies directed against the components of post-transcriptional reactions and on the use of in situ hybridization to determine the location of different RNA forms. Splicing factors have been shown to accumulate in several distinct nuclear regions including IG, PF, and CB, (CARMO-FONSECA et al. 1991a, b, 1992, 1993; FAKAN et al. 1984; NYMAN et al. 1986; SPECTOR et al. 1991; ZAMORE and GREEN 1991; ZHANG et al. 1992). This distribution pattern is complicated still further by the observation that different splicing factors occupy different subsets of nuclear compartments. Small nuclear ribonucleoprotein (snRNP) splicing factors are found throughout the nucleoplasm, but concentrate in the aforementioned IG, PF, and CB structures. In situ detection of U1 snRNP-specific proteins and RNA suggests that this particle is not concentrated as dramatically in nuclear substructures as the other snRNP particles (CARMO-FONSECA et al. 1991a, b, 1992); significant U1 snRNP is distributed throughout non-nucleolar regions of the nucleus. The splicing factor SC-35 is concentrated in PF and IG, but is not present in CB (SPECTOR et al. 1991). In contrast, the U2AF splicing factor is present in CB, as well as in the other snRNP-containing nuclear structures (ZAMORE and GREEN 1991; ZHANG et al. 1992). U2AF also shows a widespread nuclear distribution similar to that of U1. Finally, a recent electron microscopy (EM) study has examined the distribution of U1 and U2 snRNA and shown that U1 snRNA accumulates in a novel IG-associated fibrillar domain, whereas U2 snRNA does not (VISA et al. 1993a). The functional role of the nuclear compartments that contain splicing factors is the subject of current debate (LAMOND and CARMO-FONSECA 1993; SPECTOR 1993). Although the concentration of splicing components in discrete nuclear compartments raises the possibility that they are involved in splicing, it is also possible that the compartments are important for aspects of splicesome assembly before the splicing reaction occurs, for recycling of splicing factors after the splicing reaction is complete, or for storage of splicing factors.

We have examined the distribution of splicing factors in adenovirus infected cells (BRIDGE et al. 1993; see also Fig. 2). As infection proceeds, we find that the normal distribution pattern of snRNP in HeLa cell nuclei is altered. Initially snRNP is present in IG, CB, and as diffusely staining nuclear protein, as has been reported previously (panel a). snRNP is found associated with 72K centers in a subset of the cells containing early-stage replication sites (data not shown). After the onset of the viral replication, when large 72K protein-containing ssDNA accumulation sites appear in the nucleus (panel d), we find that snRNP usually surrounds the 72Kcenters but does not localize at these sites (compare

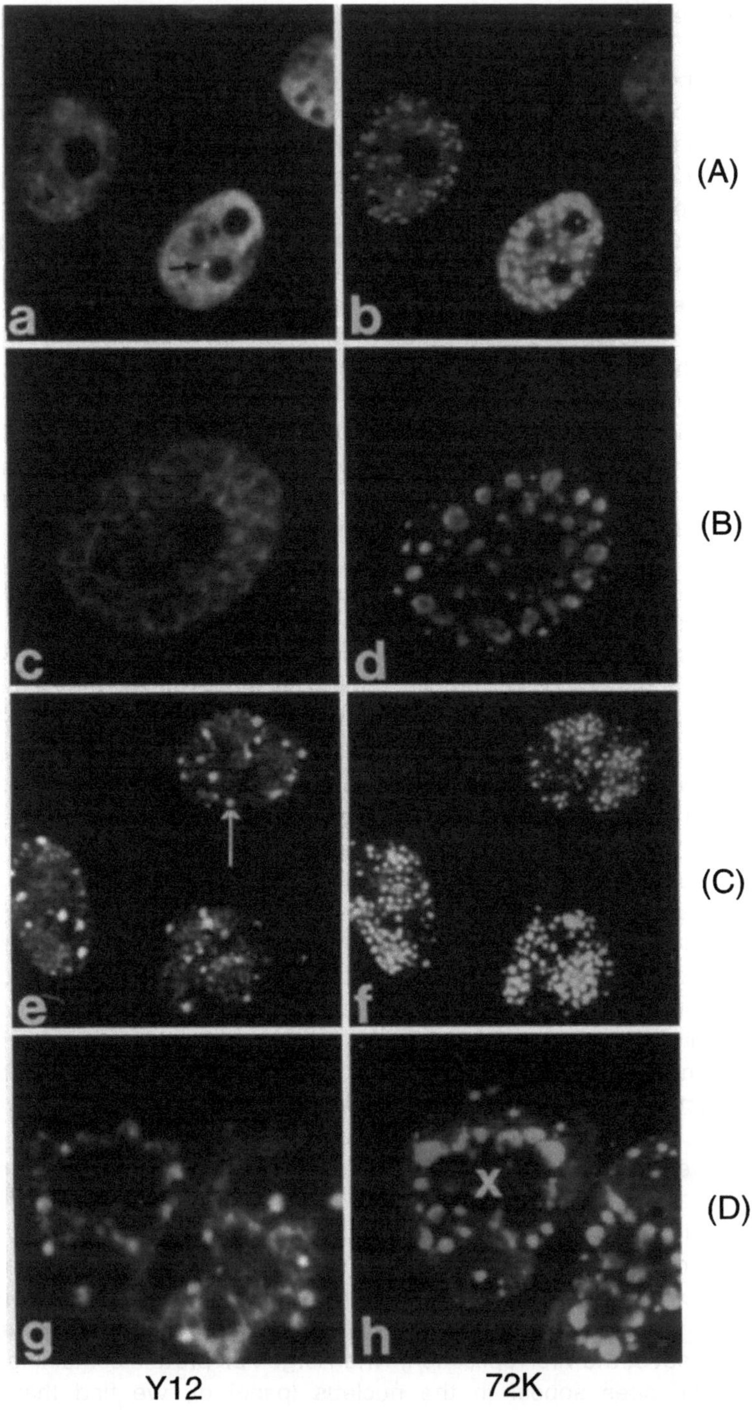
a
b
c
d
e
f
g
h
x
(A)
(B)
(C)
(D)
Y12
72K

panels c and d). As the viral late phase continues, splicing snRNP accumulates in large globular foci that are likely to correspond to IG clusters (panel e). We find that the late-phase snRNP clusters are separate from ssDNA accumulation sites identified with anti-72K (compare panels e and f). We have treated cells with hydroxyurea at various times after infection to inhibit aspects of the virus infectious program (BRIDGE et al. 1993). Preventing viral DNA synthesis limits the viral infection to the early phase. Under these conditions we find no alteration of the snRNP-staining pattern in infected cells relative to that of uninfected cells. Previous studies have shown that inhibiting viral DNA synthesis after the onset of the late phase does not prevent the continued expression of viral late genes (THOMAS and MATHEWS 1980). When hydroxyurea is added to cells at 10 h pi and cells are analyzed at various times after the addition of the drug, we find that the percentage of cells with snRNP in late-phase characteristic structures and the percentage of cells expressing the viral late protein penton continue to increase, even though the percentage of cells containing large replication centers does not change. Furthermore, there is good correlation between the appearance of late-phase snRNP clusters and the presence of the viral penton protein within individual cells. Taken together, the data suggest that the accumulation of snRNP in late-phase clusters is correlated with late gene expressioin rather than with DNA synthesis or early gene expression.

Other groups report that snRNP and additional factors involved in RNA biogenesis are localized to virus-induced inclusions containing ssDNA and RNA (JIMÉNEZ-GARCIA and SPECTOR 1993) and viral replication centers (WALTON et al. 1989). Our data (E. BRIDGE et al. 1995) offer several explanations for the apparent discrepancies between our work and that of these authors. First, we find that the monoclonal antibody (mab) used by JIMÉNEZ-GARCIA and SPECTOR (1993) to identify the non-snRNP splicing factor SC-35 in western blot analysis detects proteins from infected extracts that are not present

**Fig. 2a–h.** Distribution of small nuclear ribonucleprotein (snRNP) and 72K protein in adenovirus-infected cells. HeLa cells were infected with adenovirus type 5 (Ad5) and fixed and labeled for immunofluorescence at various times after infection as described by BRIDGE et al. (1993). SnRNP was localized with monoclonal antibody (mab) Y12, and 72K was localized with a polyclonal rabbit antisera against the viral 72K protein. *Letters in parentheses* indicate the corresponding stages from Fig.1. In stage *A* (**a, b**), snRNP localization is similar to that of uninfected cells; present in coiled bodies (*black arrow*), interchromatin granules (*black arrowhead*), and as a diffuse nuclear antigen; 72K is present in a large number of small centers which are likely to correspond to the early replicative sites identified in the electron microscopy studies of Puvion-Dutilleul and colleagues (**b**). In stage B, the 72K centers become large and pleomorphic, assuming the shape of rings, dots, and crescents (**d**). At this stage snRNP is generally excluded from the 72K staining single-stranded (ss) DNA accumulation sites; snRNP is instead concentrated in the region surrounding the 72K centers (compare **c** and **d**). At least some of snRNP labeling in cells at this stage is likely to be present in the peripheral replicative (PR) zone described in the text and Fig.1. Infected cells with 72K labeling patterns corresponding to stage *C* (f) and many cells from stage *B* show concentration of snRNP in specific sites that are likely to correspond to enlarged clusters of interchromatin granules (**e**, *white arrow*), BRIDGE et al 1993). During later stages of nuclear transformation (*D*), 72K centers surround a centrally located area that is devoid of 72K staining (**h**, *white X*). We interpret this area as the viral genome stroage site described by PUVION-DUTILLEUL and colleagues; the identity of this region is not, however, established at the level of light microscopy. SnRNP clusters are also frequently located at the edge of this site (**g**) in uninfected extracts. Furthermore, the SC-35 mab detects a protein in a western blot that was previously immunoprecipitated from infected cells by anti-72-kDa protein. These data suggest that mab SC-35 cross-reacts with the viral 72-kDa protein. This likely accounts forthe location of mab SC-35 at ssDNA accumulation sites (Jiménez-Garcia and Spector 1993) and its complete colocalization with 72K antibodies in immunofluorescence (E. Bridge et al. 1995). Thus the actual location of the SC-35 splicing factor in infected cells is not known. Second, we find that snRNP can be associated with 72K centers, but we have only observed this in a relatively small subset of the cells in the infected culture. We find that the percentage of cells showing association of snRNP with 72K centers changes as the infection progresses, reaching a peak of approximately 16–20% of the cells in the culture at 12 h pi (multiplicity of infection, MOI, 20 FFU/cell) and then declining. SnRNP containing late-phase clusters are first detected at 12–14 h pi and then accumulate throughout the late phase until by 20–24 h pi they are present in virtually 100% of the cells in the infected culture. We suggest that snRNP may transiently associate with 72K-containing ssDNA accumulation sites (Jiménez-Garcia and Spector 1993; Walton et al. 1989), prior to its incorporation into late-phase snRNP clusters.

It is necessary to consider the dynamic nature of viral replication inclusions that was discussed in previous sections to put these data in perspective. Puvion-Dutilleul and colleagues have shown that the earliest forms of viral replication inclusions contain RNA, ss- and dsDNA, and 72K. These inclusions are involved in both transcription and replication. As replication proceeds, ssDNA together with 72K are rapidly partitioned to a separate compartment that can be quite heterogeneous in size and shape, consisting of spheres rings and crescents; dsDNA is present in the surrounding PR zone. Replication occurs in both of these compartments, but significant transcription occurs only in the PR zone. Later in infection, 72-kDa protein centers become smaller and more homogeneous; still later they surround a large area of dsDNA that is inactive for both replication and transcription and has been called the genome storage site. During these stages (Fig. 1B–D), the PR zone is present at the outskirts of the 72K containing ssDNA accumulation sites. Thus, the development of the PR zone signifies a shift in transcriptional activity away from sites of ssDNA accumulation. This shift in the localization transcription might also be associated with a shift in the localization of RNA processing factors. Although the PR zone has been described at the level of EM, it is less clear what this region corresponds to at the level of light microscopy. In contrast, 72K-containing viral inclusions are easily identified by both EM and light microscopy and can be used to assess the progress of the viral infection in an individual cell.

We suggest that at the onset of DNA synthesis, snRNP can be associated with the early replicative sites (Fig. 1A; Fig. 2, panel b) that contain RNA, DNA and 72K and are sites of transcription and replication. Our unpublished data indicate that early forms of the 72K centers can sometimes be surrounded by a halo or crescent of snRNP staining; possibly this represents the early stages of PR zone formation. These structures look very similar to the earliest forms of snRNP, RNA-, and ssDNA-containing inclusions detected by JIMÉNEZ-GARCIA and SPECTOR (1993). When the 72K-containing ssDNA accumulation centers become larger and pleomorphic, we find that snRNP is usually excluded from the 72K centers. Viewed in cross-section, the snRNP staining appears as a series of interconnected rings, with the empty areas in the staining corresponding to the location of 72K (Fig. 2 compare panels c and d). The location of the PR zone in cells at this stage is not entirely clear from light micrographs, but since it is known to surround the 72K-containing ssDNA accumulation sites, at least a portion of the snRNP staining is likely to be present in the PR zones. The next stage is accumulation of snRNP in late-phase clusters; this is seen in some cells with pleomorphic 72K centers and in cells at later stages which have smaller, more homogeneous 72K centers (Fig. 2, panel f). SnRNP-containing clusters are still present in cells containing a large, centrally located space that is surrounded by 72K containing centers but is not labeled by anti-72k (Fig. 2, panels g and h). We interpret this structure as the genome storage site described by Puvion-Dutilleul and colleagues (Fig. 1D); it is rarely detected at 18 h pi and frequently detected at 24 h pi under our infection conditions.

Defining the nuclear organization of post-transcriptional processing also requires an analysis of the location of viral RNA. In situ hybridization and EM experiments described above show that at 17 h pi viral RNA is present in the transcriptionally active PR zone, viral-induced compact ring structures, and in IG clusters, as well as in ribosome-containing areas of the cytoplasm. Although JIMÉNEZ-GARCIA and SPECTOR (1993) detected RNA at ssDNA accumulation sites, Puvion-Dutilleul and colleagues reported that the presence of RNA in ssDNA accumulation centers is comparatively rare at late stages (PUVION-DUTILLEUL and PUVION 1991; PUVION-DUTILLEUL et al. 1992). Since JIMÉNEZ-GARCIA and SPECTOR (1993) also detected Pol II at ssDNA accumulation sites, it is possible that they have focused on a stage prior to the separation of the transcriptionally active PR zone from the ssDNA accumulation sites. The available data suggest that RNA is present at the site of transcription, as would be expected, and that there is also RNA present in several nuclear structures which are not associated with transcription, namely compact rings and IG (Fig. 1). The post-transriptional association of RNA with these structures makes them excellent candidates for structures involved in either processing or intranuclear trafficking of viral RNA.

No information is currently available concerning the comparative localization of intron, exon, and polyA sequences of viral RNA with compact rings and IG. In uninfected cells polyA is observed in IG when detected in in situ hybridizations using oligo(dT) as a probe (CARTER et al. 1991, 1993; Visa et al. 1993b). Unspliced RNA that is microinjected into cells is also detected in IG (WANG et al. 1991). However, others report that the concentration of splicing factors in IG is increased under conditions that inhibit transcription and presumably also splicing (CARMO-FONSECA et al. 1992). Thus the functional role of the IG structure is still not clear. The observation that viral RNA is detected in late-phase IG structures that are also likely to contain splicing factors is quite interesting (BRIDGE et al. 1993; PUVION-DUTILLEUL and PUVION 1991; PUVION-DUITILLEUL et al. 1992). It will now be critical to determine the nature of the viral RNA present in different nuclear compartments with probes for intron, exon, and polyA sequences of specific viral messages.

## 5 Interaction of Adenovirus with the Nucleolus

The nucleolus is a specialized nuclear compartment responsible for the transcription and processing of cellular ribosomal RNA (rRNA; reviewed by SCHEER et al. 1993). Nucleoli contain RNA polymerase I (pol I, the polymerase responsible for transcribing rRNA), U3 snRNP (which is essential for processing of rRNA), and fibrillarin, nucleolin, and other phosphoproteins. Adenovirus infection has been shown to interfere with the biogenesis of host rRNA (CASTIGLIA and FLINT 1983; LEDINKO 1972; RASKAS et al. 1970). This may in part be the result of the inhibition of cellular protein synthesis induced by adenovirus infection (CASTIGLIA and FLINT 1983); inhibition of cellular protein synthesis by cyclohexmide also inhibits rRNA production.

In addition to the observation that adenovirus infection affects rRNA production, several groups also report that nucleolar proteins are redistributed in adenovirus-infected cells. Bodnar and colleagues (FOX et al. 1990; WALTON et al. 1989) report that nucleolar proteins B23, NOR90 (nucleolar organizing region, 90), pol I, NuF1 (nucleolar factor 1), and antigen AF/CDC6 are present in viral 72K containing centers. Antisera against U3 snRNP labeled nucleoli in both infected and uninfected cells, while the nucleolar snRNP antigen Th was released from the nucleolus in infected cells to show a diffuse nucleoplasmic staining pattern. These data are consistent with the idea that adenovirus infection disrupts the organization of the nucleolus. The observation that several nucleolar antigens are present in viral 72K containing sites has led to the suggestion that adenovirus replication occurs in pseudonucleoli. Further experiments are needed to substantiate this hypothesis. More recently, the distribution of fibrillarin has been studied in adenovirus-infected cells by EM (PUVION-DUTILLEUL and CHRISTENSEN 1993). Fibrillarin was found in extranucleolar fibrillar spots of 0.15–0.3 $\mu$ in diameter. During early stages of nuclear transformation these spots were in close proximity to ssDNA accumulation sites; at later stages they were present in the PR zone. Residual staining of fibrillar nucleolar regions was also observed. The morphology of the nucleolus itself was found to be altered by adenovirus infection; nucleolar granules became condensed and compact, while the nucleolar fibrils migrated to the border of the nucleolar structure. The role of these alterations in nucleolar organization for the virus infectious program is not understood, but such alterations are likely to be involved in at least some aspects of altered rRNA gene expression in infected cells.

## 6 Discussion and Perspectives

The results discussed above present a preliminary picture of functional organization in adenovirus-infected nuclei, summarized in Fig.1 and Table 1. During the earliest stages of detectable nuclear alteration induced by adenovirus, replication and transcription are tightly integrated at the early replicative sites. Viral RNA and DNA are detected at these sites by in situ hybridization; the viral 72K ssDNA-binding protein is also present. At least some of these sites are associated with concentrations of cellular splicing factors (JIMÉNEZ-GARCIA and SPECTOR 1993; E. BRIDGE and U. PETERSSON unpublished data), raising the possibility that post-transcriptional processing may also occur here. This focal concentration of viral nucleic acids, replication, and gene expression activities is then rapidly compartmentalized as infection progresses. 72K, together with ss viral DNA, is partitioned to a separate compartment that is only intermittently active for DNA synthesis (see Fig. 1B), while continuous replication and transcription are shifted to the surrounding PR zone. Our results suggest that splicing factors are also likely to be compartmentalized to the PR zone during this stage (BRIDGE et al. 1993; POMBO et al. 1994), but this is controversial (JIMÉNEZ-GARCIA and SPECTOR 1993). POMBO et al. (1994) have examined replication and transcription simultaneously by confocal microscopy in cells corresponding primarily to stage B (Fig. 1); they found that although these activities are overlapping, they are not coincident. Their results support a model in which transcription occurs at the periphery of viral replication, followed by segregation of transcribing and replicating templates to separate regions. Such a model would also nicely accommodate observations from EM studies indicating that transcription and replication can occur simultaneously from a single template (MATSUGUCHI et al. 1979; WOLGEMUTH and HSU 1981) and results from biochemical investigations indicating that there

**Table 1.** Subcellular localization of viral nucleic acids, replication, and transcriptional activities and post-transcriptional processing factors in adenovirus-infected HeLa cells

| | Replication | Viral DNA | 72K | Transcription | Viral RNA | Splicing factors |
|---|---|---|---|---|---|---|
| Nucleus | | | | | | |
| Early replicative site | + | + | + | + | + | +? |
| Peripheral replicative zone | + | + | – | + | + | +? |
| ssDNA accumulatiion site | + | + | + | – | –? | –? |
| Viral genome storage site | – | + | – | – | – | ND |
| Late-stage IG cluster | – | – | – | – | + | +? |
| Compact ring | – | – | – | – | + | ND |
| Cytoplasm | | | | | | |
| Ribosome-containing areas | – | – | – | – | + | – |

ss, single-stranded; IG, interchromation granule; ND, not determined +? and –? indicate that different results have been obtained by different laboratories. In these cases, + and – indicate the current view of the authors. Adapted from Puvion-Dutilleul et al. (1992).

are separate replication and transcription complexes (Brison et al. 1977, 1979; Wilhelm et al. 1976). Furthermore, there is a good correspondence between the location of splicing factors and transcription in cells showing intense transcriptional activity; this supports the view that pre-RNA processing occurs at sites of transcription (Pombo et al. 1994).

During stages B–D (Fig.1), replication occurs at both ssDNA accumulation sites and the PR zone. Transcription occurs in the PR zone, but, interestingly, splicing factors are progressively compartmentalized into late phase-specific centers that are likely to correspond to enlarged clusters of IG (Fig. 2, panel e). Viral RNA is detected in the transcriptionally active PR zone during stages B–D (Fig.1), but at late times in infection RNA also progressively accumulates in IG structures. Compact rings (Fig.1) are another type of nuclear structure containing viral RNA. The presence of RNA in IG and compact rings, structures not associated with active transcription, raises the possibility that they are important for some aspect of post-transcriptional processing or intranuclear trafficking of RNA. Little is known about the compact ring structures other than the observation that they contain viral RNA. Jiménez-Garcia et al. (1993) have recently found that the viral Pol III transcript VA RNA I is localized in numerous small nuclear dots. It was suggested that these might correspond to the compact ring structures described by Puvion-Dutilleul et al. (1992); this has yet to be confirmed by EM studies. More investigations have been made to assess the role of IG. Several studies of uninfected cells have shown that poly A sequences are associated with IG (Carter et al. 1991,1993; Visa et al. 1993b). RNA tracks of some cellular genes have been shown to lie in close proximity to splicing factor containing IG (Carter et al. 1993; Xing et al. 1993). These data support the idea that IG might be involved in an aspect of intranuclear RNA trafficking. Adenovirus-infected cells should be an excellent model system for further testing this hypothesis. The location of splicing factors in IG also suggests a role in some aspect of post-transcriptional processing. This could be a location where introns are degraded, or splicing factors may need to cycle through this organelle as part of the normal process of spliceosome formation or turnover.

Progression of adenovirus infection interferes with cellular processes at several levels. Host DNA synthesis is dramatically inhibited (Hodge and Scharf 1969; Pina and Green 1969), and there is nearly complete inhibition of cellular protein synthesis during the viral late phase (Dolph et al. 1988; Huang and Schneider 1991). Although adenovirus infection has not been shown to inhibit cellular transcription, the export of newly synthesized cellular messages from the nucleus to the cytoplasm is dramatically reduced during the viral late phase (Beltz and Flint 1979; reviewed in Flint 1986). Cellular messages appear to be polyadenylated and spliced, but they are not transported to the cytoplasm.The location of host transcripts in adenovirus-infected cells has not been determined. Although the mechanism by which viral messages are selectively transported to the cytoplasm during the late phase is not known, it has been proposed that viral messages may be synthesized at intranuclear sites that are particularly favorable for subsequent transport of mRNA (FLINT 1986). Likewise, SANDLER and KETNER (1991) propose that nuclear events in viral late gene expression, from transcription to nuclear export, may be coordinated in physical assemblies, late gene expression machines. The association of viral RNA with IG structures in late phase-infected cells is interesting in this regard. If IG are involved in some aspect of nuclear RNA trafficking, as suggested above, then preferential association of viral RNA with these structures could interfere with their access to cellular RNA, resulting in improper nuclear trafficking. A comparison of cellular RNA localization in infected and uninfected cells could provide some clues as to the nature of the transport defect of host RNA. In situ hybridizations comparing cellular and viral messages could also be used to determine what aspects of intranuclear trafficking these messages do and do not have in common.

The controls regulating the viral early to late phase transition have long been enigmatic. Drugs or mutations that block viral DNA synthesis limit the infection to the early phase; expression of the viral late genes is not observed (reviewed in TOOZE 1982). This indicates that viral replication is required in some way for late gene expression. Superinfection experiments (THOMAS and MATHEWS 1980), in which cells are infected by one virus allowed to progress to the late phase and then superinfected with a second, genetically distinct virus, show that the presence of late-phase *trans*-acting factors is not sufficient to allow late gene expression of the superinfecting virus. Rather, the second viral template must itself be replicated before late gene expression is observed. These results suggest that DNA synthesis is required in *cis* for late gene expression; only replicated viral DNA molecules can serve as templates for late transcription. This raises the question of whether replicated viral DNA is an obligatory template for late gene expression. One attractive hypothesis to account for the *cis*-dependence of late gene expression on DNA replication is that the nuclear location or compartmentalization of replicated templates specifies the transition from early to late gene expression. The results presented above describe the intranuclear compartmentalization of viral nucleic acids within individual cells. The nature of the viral RNA produced during the stages described in Fig. 1 has not been determined. Early replicative sites are associated with both replication and transcription, but it is not known if early or late genes or both are being transcribed at this stage, since thus far only probes corresponding to the entire adenovirus genome have been used. It is possible that only viral early genes are transcribed at early replicative sites and that late gene expression is only initiated after separation of ssDNA accumulation sites from the peripheral replicative zone. This is a shift in the organization of viral templates that is observed during infection and might correspond to the late gene expression transition. The viral late phase has been defined in cultures of infected cells as the period following the onset of viral DNA replication. Is this also the criteria for defining the late phase within an individual cell? An in situ hybridization analysis of viral early and late messages in cells at different stages of the viral infectious cycle should help to address this question.

In summary, adenovirus-infected cells have been particularly useful as a model system for studying the biochemistry and the genetic elements of eukaryotic gene expression. The application of sensitive techniques for visualizing the subcellular organization of proteins and nucleic acids now raises the possibility that adenovirus-infected cells will again serve as a critical model for understanding the cell biology of eukaryotic gene expression. In addition to simply comparing the nuclear organization between uninfected and infected cells, the adenovirus system also offers an opportunity to study the cell biology of the host during normal infections, nonpermissive infections, and during infection by viral mutants. The possibility of integrating the large body of biochemical and genetic information available from these systems together with new information describing subcellular organization underscores the enormous potential of adenovirus-infected cells as a tool for studying the role of nuclear architecture in aspects of replication and gene expression.

## References

Adami G, Nevins JR (1988) Splice site selection dominates over poly(A) site choice in RNA production from complex adenovirus transcription units. EMBO J 7: 2107–2116

Babich A, Feldman LT, Nevins JR, Darnell JE Jr, Weinberger C (1983) Effect of adenovirus on metabolism of specific host mRNAs: transport control and specific translational discrimination. Mol Cell Biol 3: 1212–1221

Beltz GA, Flint SJ (1979) Inhibition of HeLa cell protein synthesis during adenovirus infection: restriction of cellular messenger RNA sequences to the nucleus. J Mol Biol 131: 353–373

Beyer AL, Osheim YN (1988) Splice site selection, rate of splicing and alternative splicing on nascent transcripts. Genes Dev 2: 754–765

Bosher J, Dawson A, Hay RT (1992) Nuclear factor I is specifically targeted to discrete subnuclear sites in adenovirus type 2-infected cells. J Virol 66: 3140–3150

Brasch K, Ochs RL (1992) Nuclear bodies (NBs): a newly "rediscovered" organelle. Exp Cell Res 202: 211–223

Bridge E, Carmo-Fonseca M, Lamond AI, Pettersson U (1993) Nuclear organization of splicing small nuclear ribonucleoproteins in adenovirus-infected cells. J Virol 67: 5792–5802

Bridge E, Xia D-X, Carmo-Fonseca M, Cardinali B, Lamond AI, Pettersson U (1995) Dynamic organization of splicing factors in adenovirus-infected cells. J Virol 69: 281–290

Brison O, Kédinger C, Wilhelm J (1977) Enzymatic properties of viral replication complexes isolated from adenovirus type 2-infected HeLa cell nuclei. J Virol 24: 423–435

Brison O, Kédinger C, Chambon P (1979) Adenovirus DNA template for late transcription is not a replicative intermediate. J Virol 32: 91–97

Carmo-Fonseca M, Tollervey D, Pepperkok R, Barabino S, Merdes A, Brunner C, Zamore PD, Green MR, Hurt E, Lamond AI (1991a) Mammalian nuclei contain foci which are highly enriched in components of the pre-mRNA splicing machinery. EMBO J 10: 195–206

Carmo-Fonseca M, Pepperkok R, Sproat BS, Ansorge W, Swanson MS, Lamond AI (1991b) In vivo detection of snRNP-organelles in the nuclei of mammalian cells. EMBO J 10:1863–1873

Carmo-Fonseca M, Pepperkok R, Carvalho MT, Lamond AI (1992) Transcription dependent localization of the U1, U2, U4/U6 and U5 snRNPs in coiled bodies. J Cell Biol 117: 1–14

Carmo-Fonseca M, Ferreira J, Lamond AI (1993) Assembly of snRNP-containing coiled bodies is regulated in interphase and mitosis—evidence that the coiled body is a kinetic nuclear structure. J Cell Boil 120: 841–852

Carter KC, Taneja KL, Lawrence JB (1991) Discrete nuclear domains of poly (A) RNA and their relationship to the functional organization of the nucleus. J Cell Biol 115: 1191–1202.
Carter KC, Bowman D, Carrington W, Fogarty K, McNeil JA, Fay FS, Lawrence JB (1993) A three-dimensional view of precursor messenger RNA metabolism within the mammalian nucleus. Science 259: 1330–1335
Castiglia CL, Flint SJ (1983) Effects of adenovirus infection on RNA synthesis and maturation in HeLa cells. Mol Cell Biol 3: 662–671
Challberg M, Kelly T (1989) Animal virus DNA replication. Annu Rev Biochem 58: 671–717
Cook PR (1991) The nucleoskeleton and the topology of replication. Cell 66: 627–635
Danner D, Leder P (1985) Role of an RNA cleavage/poly (A) addition site in the production of membrane-bound and secreted IgM mRNA. Proc Natl Acad Sci USA 82: 8658–8662
Doerfler W (ed) (1986) Adenovirus DNA: the viral genome and its expression. Nijhoff, Boston
Dolph PJ, Racaniello V, Villamarin A, Palladino F, Schneider RJ (1988) The adenovirus tripartite leader may eliminate the requirement for cap-binding protein complex during translation initiation. J Virol 62: 2059–2066
Emeson RB, Hedjran F, Yeakley JM, Guise JW, Rosenfeld MG (1989) Alternative production of calcitonin and CGRP mRNA is regulated at the calcitonin-specific splice acceptor. Nature 341: 76–80
Fakan S, Leser G, Martin TE (1984) Ultrastructural distribution of nuclear ribonucleoproteins as visualized by immunocytochemistry on thin sections. J Cell Biol 98: 358–363
Flint SJ (1986) Regulation of adenovirus mRNA formation. Adv Virus Res 31: 169–228
Fox E, Moen PT Jr, Bodnar JW (1990) Replication of minute virus of mice DNA in adenovirus-infected or adenovirus-transformed cells. Virology 176: 403–412
Fox MH, Arndt-Jovin DJ, Jovin TM, Baumann PH, Robert-Nicoud M (1991) Spatial and temporal distribution of DNA replication sites localized by immunofluorescence and confocal microscopy in mouse fibroblasts. J Cell Sci 99: 247–253
Galli G, Guise J, Tucker PW, Nevins JR (1988) Poly(A) site choice rather than splice site choice governs the regulated production of IgM heavy-chain RNAs. Proc Natl Acad Sci USA 85: 2439–2443
Green MR (1991) Biochemical mechanisms of constitutive and regulated pre-mRNA splicing. Annu Rev Cell Biol 7: 559–599
Hodge LD, Scharf MD (1969) Effect of adenovirus on host cell DNA synthesis in synchronized cells. Virology 37: 554–564
Huang J, Schneider RJ (1991) Adenovirus inhibition of cellular protein synthesis involved inactivation of cap-binding protein. Cell 65: 271–280
Jackson DA, Bassim Hassan A, Errington RJ, Cook PR (1993) Visualization of focal sites of transcription within human nuclei. EMBO J 12: 1059–1065
Jiménez-Garcia LF, Spector DL (1993) In vivo evidence that transcription of splicing are coordinated by a recruiting mechanism. Cell 73: 47–59
Jiménez-Garcia LF, Green SR, Mathews MB, Spector DL (1993) Organization of the double-stranded RNA-activated protein kinase DAI and virus-associated VA RNA I in adenovirus-2-infected HeLa cells. J Cell Sci 106: 11–22
Lamond AI, Carmo-Fonseca M (1993) The coiled body. Trends Cell Biol 3: 198–204
Ledinko N (1972) Nucleolar ribosomal precursor RNA and protein synthesis in HEK cultures infected with Ad12. Virology 49: 79–89
Leff SE, Evans RM, Rosenfeld MG (1987) Splice commitment dictates neuron-specific alternative RNA processing in calcitonin/CGRP gene expression. Cell 48: 517–524
LeMaire MF, Thummel CS (1990) Splicing precedes polyadenylation during *Drosophila* E74A transcription. Mol Cell Biol 10: 6059–6063
Martinez-Palomo A, Granboulan N (1967) Electron microscopy of adenovirus 12 replication. II. High-resolution autoradiography of infected KB cells labeled with tritiated thymidine. J Virol 1: 1010–1018
Martinez-Palomo A, Le Buis J, Bernhard W (1967) Electron microscopy of adenovirus 12 replication. I. Fine structural changes in the nucleus of KB cells. J Virol 1: 817–829
Matsuguchi M, Puvion-Dutilleul F, Moyne G (1979) Late transcription and simultaneous replication of simian adenovirus 7 DNA as revealed by spreading lytically infected cell cultures. J Gen Virol 42: 443–456
Moore MJ, Query CC, Sharp P (1993) Splicing precursors to mRNAs by the spliceosome. In: Gesteland RF, Atkins JF (eds) The RNA world. Cold Spring Harbor Laboratory, New York, pp 303–357
Moyne G, Pichard E, Bernard W (1978) Localization of simian adenovirus 7 (SA 7) transcription and replication in lytic infection. An ultracytochemical and autoradiographical study J Gen Virol 40: 77–92

Murti KG, Davis DS, Kitchingman GR (1990) Localization of adenovirus-encoded DNA replication proteins in the nucleus by immunogold electron microscopy. J Gen Virol 71: 2847–2857
Nakamura H, Morita T, Sato C (1986) Structural organization of replicon domains during DNA synthetic phase in the mammalian nucleus. Exp Cell Res 165: 291–297
Nakayasu H, Berezney R (1989) Mapping replication sites in the eukaryotic cell nucleus. J Cell Biol 108: 1–11
Nevins JR, Darnell JE Jr (1978) Steps in the processing of Ad2 mRNA: poly(A)+nuclear sequences are conserved and poly(A) addition precedes splicing. Cell 15: 1477–1493
Nyman U, Hallman H, Hadlaczky G, Pettersson I, Sharp G, Ringertz NR (1986) Intranuclear localization of snRNP antigens. J Cell Biol 102: 137–144
O'Keefe RT, Henderson SC, Spector DL (1992) Dynamic organization of DNA replication in mammalian cell nuclei: spatially and temporally defined replication of chromosome-specific a satellite DNA sequences. J Cell Biol 116: 1095–1110
Phillips DM, Raskas HJ (1972) Ultrastructural changes in KB cultures infected with adenovirus type 2. Virology 48: 156–169
Pina M, Green M (1969) Biochemical studies on adenovirus multiplication XIV. Macromolecule and enzyme synthesis in cells replicating oncogenic and non-oncogenic human adenoviruses. Virology 38: 573–586
Pombo A, Ferreira J, Bridge E, Carmo-Fonesca M (1994) Adenovirus replication and transcription are spatially separated in the nucleus of infected cells. EMBO J 13: 5075–5085
Puvion-Dutilleul F (1993) Protocol of electron microscope in situ nucleic hybridization for the exclusive detection of double-stranded DNA sequences in cells containing large amounts of homologous single-stranded DNA and RNA sequence: application to adenovirus type 5 infected HeLa cells. Micros Res Tech 25: 2–11
Puvion-Dutilleul F, Christensen ME (1993) Alterations of fibrillarin distribution and nucleolar ultrastructure induced by adenovirus infection. Eur J Cell Biol 61: 168–176
Puvion-Duitilleul F, Pichard E (1992) Segregation of viral double-stranded and single-stranded DNA molecules in nuclei of adenovirus infected cells as revealed by electron microscope in situ hybridization. Biol Cell 76: 139–150
Puvion-Dutilleul F, Puvion E (1990a) Analysis by in situ hybridization and autoradiography of sites of replication and storage of single- and double-stranded adenovirus type 5 DNA in lytically infected HeLa cells. J Struct Biol 103: 280–289
Puvion-Dutilleul F, Puvion E (1990b) Replicating single-stranded adenovirus type 5 DNA molecules accumulate within well-delimited intranuclear areas of lytically infected HeLa cells. Eur J Cell Biol 52: 379–388
Puvion-Dutilleul F, Puvion E (1991) Sites of transcription of adenovirus type 5 genomes in relation to early viral DNA replication in infected HeLa cells. A high resolution in situ hybridization and autoradiographical study. Biol Cell 71: 135–147.
Puvion-Dutilleul F, Pédron J, Cajean-Feroldi C (1984) Identification of intranuclear structures containing the 72K DNA-binding protein of human adenovirus type 5. Eur J Cell Biol 34: 313–322
Puvion-Dutilleul F, Roussev R, Puvion E (1992) Distribution of viral RNA molecules during the adenovirus type 5 infectious cycle in HeLa cells. J Struct Biol 108: 209–220
Raskas HJ, Thomas DC, Green M (1970) Biochemical studies on adenovirus multiplication. XVII. Ribosome synthesis in uninfected and infected KB cells. Virology 40: 893–902
Reich NC, Sarnow P, Duprey E, Levine AJ (1983) Monoclonal antibodies which recognize native and denatured forms of the adenovirus DNA-binding protein. Virology 128: 480–484
Rosbash M, Singer RH (1993) RNA travel: tracks from DNA to cytoplasm. Cell 75: 399–401
Sandler AB, Ketner G (1991) The metabolism of host RNAs in cells infected by an adenovirus E4 mutant. Virology 181: 319–326
Sasaguri Y, Sanford T, Aguirre P, Padmanabhan R (1987) Immunological analysis of 140-kDa adenovirus-encoded DNA polymerase in adenovirus type 2-infected HeLa cells using antibodies raised against the protein expressed in Escherichia Coli. Virology 160: 389–399
Scheer U, Thiry M, Goessens G (1993) Structure, function and assembly of the nucleous. Trends Cell Biol 3: 236–241
Sharp PA (1984) Adenovirus transcription. In: Ginsberg H (ed) The adenoviruses. Plenum, New York, pp 173–204
Spector DL (1993) Nuclear organization of pre-RNA processing. Curr Opin Cell Biol 5: 442–448
Spector DL, Fu X-D, Maniatis T (1991) Associations between distinct pre-mRNA splicing components and the cell nucleus. EMBO J 10: 3467–3481
Thomas GP, Mathews MB (1980) DNA replication and the early to late transition in adenovirus infection. Cell 22: 523–533

Tooze J (ed) (1982) The molecular biology of tumor viruses, vol 2. DNA tumor viruses, 2nd rev edn. Cold Spring Harbor Laboratory, New York

Visa N, Puvion-Dutilleul F, Bachellerie JP, Puvion E (1993a) Intranuclear distribution of U1 and U2 snRNAs visualized by high resolution in situ hybridization: revelation of a novel compartment containing U1 but not U2 snRNA in HeLa. Eur J Cell Biol 60: 308–321

Visa N, Puvion-Dutilleul F, Harper F, Bachellerie J-P, Puvion E (1993b) Intranuclear distribution of poly(A) RNA determined by electron microscope hybridization. Exp Cell Res 208: 19–34

Vlak JM, Rozij THH, Spies F (1975) Localization of adenovirus DNA replication in KB cells. Virology 65: 535–545

Voelkerding K, Klessig DF (1986) Identification of two nuclear subclasses of the adenovirus type 5-encoded DNA-binding protein. J Virol 60: 353–362

Wahle E, Keller W (1992) The biochemistry of 3′-end cleavage and polyadenylation of messenger RNA precursors. Annu Rev Biochem 61: 419–440

Walton TH, Moen PT Jr, Fox E, Bodnar JW (1989) Interactions of minute virus of mice and adenovirus with host nucleoli. J Virol 63: 3651–3660

Wang J, Cao L, Wang Y, Pederson T (1991) Localization of pre-messenger RNA in discrete nuclear sites. Proc Natl Acad Sci USA 88: 7391–7395

Wilcock D, Lane DP (1991) Localization of p53, retinoblastoma and host replication proteins at sites of viral replication in herpes-infected cells. Nature 349: 429–431

Wilhelm J, Brison O, Kédinger C, Chambon P (1976) Characterization of adenovirus type 2 transcriptional complexes isolated from infected HeLa cell nuclei. J Virol 19: 61–81

Wolgemuth DJ, Hsu M-T (1981) Visualization of nascent RNA transcripts and simultaneous transcription and replication in viral nucleoprotein complexes from adenovirus 2-infected HeLa cells. J Mol Biol 147: 247–268

Xing Y, Lawrence JB (1993) Nuclear RNA tracks: structural basis for transcription and splicing? Trends Cell Biol 3: 346–353

Xing Y, Johnson CV, Dobner PR, Lawrence JB (1993) Higher level organization of individual gene transcription and RNA splicing. Science 259: 1326–1329

Yoder SS, Berget SM (1985) Posttranscriptional control of DHFR gene expression during adenovirus 2 infection. J Virol 54: 72–77

Zamore PD, Green MR (1991) Biochemical characterization of U2 snRNP auxiliary factor: an essential pre-mRNA splicing factor with a novel intranuclear distribution. EMBO J 10: 207–214

Zhang M, Zamore PD, Carmo-Fonseca M, Lamond AI, Green MR (1992) Cloning and intracellular localization of the U2 small nuclear ribonuclear auxiliary factor small subunit. Proc Natl Acad Sci USA 89: 8769–8773

# Abortive Adenovirus Infection and Host Range Determinants

L.A. LUCHER



## 1 Introduction

The host range of a virus constitutes at one level the distribution of host organisms that it infects and at another level the tissues in that host that are targeted by the virus. Host range is determined by properties of both the virus and the potential host cell, due to the obligate parasitic nature of viruses. The expression of cell surface receptors for virus strongly correlates with susceptibility of the cell to infection, and cells that do not allow virus to bind are necessarily resistant. Thus host range is initially determined by the presence or absence of the appropriate interaction between extracellular host range determinants that are the virus attachment protein and the host receptor. Broad host range viruses generally bind to common molecules on the cell surface; for example, influenza binds to sialic acid-containing carbohydrates on cellular glycoproteins or

Department of Biological Sciences, Illinois State, University, Normal, IL 61761, USA glycolipids. Narrow host range viruses typically bind with high affinity to a specific protein, as exemplified by the binding of rhinovirus, poliovirus, and human immunodeficiency virus (HIV) to specific members of the human immunoglobulin super family. Cell surface interactions are not the only determinants of host range, however. Cells that support all steps in multiplication of a virus are termed permissive (with maximum virus yields) or semipermissive (with less than maximum virus yields). Nonpermissive cells allow virus to bind but do not allow complete virus multiplication, and the subsequent abortive infection is caused by the block of one or more multiplication steps at the intracellular level. Abortive infection can therefore be considered a result of the action of intracellular host range determinants. This review will explore some of the intracellular determinants that operate to block human adenovirus multiplication in both human and nonhuman cells.

## 2 Adenovirus Host Range

Adenoviruses are double-stranded DNA, nonenveloped viruses that are endemic in human populations around the world, with some species responsible for epidemic outbreaks of respiratory or ocular infections (reviewed in Straus 1984; White and Fenner 1986). Only about a dozen of the 47 known species of human adenovirus (Ad) are associated with specific diseases of the upper and lower respiratory tract, conjunctiva and cornea, gastrointestinal tract, and genitourinary tract. Many Ad infections are subclinical, however, which makes it difficult to determine the role of the other Ad species in human disease.

Ad infection typically exhibits a short incubation period of 5–10 days, rarely spreading beyond the draining lymph nodes in the vicinity of the site of primary infection (e.g., the epithelia of the pharynx, conjunctiva, or small intestine). On rare occasions, patients have died from acute infection of the lungs or liver or from widely disseminated infection. Disseminated infection occurs in patients that have incompetent cellular immunity, such as infants, transplant recipients, and acquired immunodeficiency syndrome (AIDS) patients. Autopsy after death due to disseminated Ad7 infection has shown secondary virus multiplication in a variety of organs including lung, brain, and kidney. The immune system is therefore effective in restricting Ad multiplication to specific tissues and plays an important role in determining the tissue tropism of Ad.

Ad-specific characteristics are probably also involved in determining tissue tropism, since certain Ad species are typically associated with specific diseases in different tissues. In general, adenoviruses (Ads) of lower species number are associated with ocular and respiratory diseases, while Ads of higher species number are associated with gastrointestinal infections. Severe lower respiratory or disseminated disease is usually due to Ad 1–5, –7, or –21. Ad3, –4, –7, –11, and –37 are commonly involved in ocular infections, with severe infections due to Ad8 and Ad19. Some Ad7 strains are neurotropic, and Ad3 and Ad7 have been isolated from brain in cases of meningitis and encephalitis. Ad21 is the only species associated with cystitis, while Ad37 has been found associated with some cases of urethritis and cervicitis.

The common endemic Ad species (Ad1, Ad2, Ad5) persist in tonsils and adenoids, with virus being shed continuously for several months after initial infection. Intermittent shedding occurs for many years after the acute infection is resolved and is more typically fecal shedding rather than respiratory shedding. This long period of shedding may be due to reactivation of latently infected cells. Reactivation of active Ad infection can occur upon infection with *Bordetella pertussis* or measles virus, as well as during immunosuppression in transplant patients or AIDS patients.

Ads have been isolated not only from humans, but also from a wide variety of animals (M. Ishibashi and Yasue 1984). As with human Ads, other animal Ads are usually restricted to multiplication in their natural host. Indeed, the presumptive evidence for determining that a new isolate is an Ad includes characteristics of culture in the laboratory. The virus must grow well in cultures of cells such as primary kidney or testicle cells originating from the presumptive natural host, but not in the same cell type from animals that are phylogenetically close to the presumptive natural host.

Productive Ad infection can be divided into events that occur prior to and those that occur after the onset of viral DNA replication. The earliest events include receptor binding and entry into the cell, uncoating of Ad DNA and its delivery to the nucleus, and expression of early Ad genes. The early gene products are regulatory or catalyze viral DNA replication. After DNA replication begins, the late genes encoding mostly virion structural proteins are expressed, and late infection is characterized by continued DNA replication, late gene expression, and virion assembly. Progeny yield can vary widely from cell type to cell type, even when all steps in viral multiplication occur. For this review, productive infection is defined as any infection that results in an increase in virus yield per cell over the inoculum multiplicity of infection. This will distinguish between semipermissive and abortive infections.

In general, cultured human cells support the complete Ad infectious cycle, whereas simian and rodent cells do not. Infection of monkey and rodent cells with human Ad has shown, however, that this restricted host range is not due to lack of receptors, but is due to intracellular blocks arising from inappropriate interactions between cellular and viral components. Abortive infection may arise from an early block (any early step including DNA replication) or a late block (any step after the onset of DNA replication).

For isolation and growth of human Ad in cell culture, primary human embryo kidney (HEK) cells are the most sensitive to Ad infections (White and Fenner 1986). Continuous malignant human epidermoid carcinoma cell lines such as HeLa (cervical), HEp-2 (laryngeal), or KB[1] (oral) are generally good hosts and

[1] HEp-2 and KB deposited with the American Type Culture Collection are most likely contaminated with HeLa

are used widely. Human diploid embryonic fibroblast cell lines from lung (e.g., WI-38) or tonsil are less sensitive and take longer to show cytopathic effect. Proliferating human skeletal myoblasts and differentiated myoblast culture are also permissive to Ad5, by demonstration of DNA replication and subsequent extraction of nuclease-resistant, encapsidated viral DNA (KOHTZ et al. 1991).

The fastidious enteric Ad40 and Ad41 have proved difficult to cultivate in the laboratory. Ad40 can be passaged in 293 cells (HEK transformed by Ad5, expressing Ad5 E1A and E1B proteins), but there has been variable success with passaging Ad41 in 293 cells. Ad41 has been serially passaged successfully in HEp-2, HeLa, and H1407 (human intestinal) cell lines and can also be grown in KB cells (PIENIAZEK et al. 1990a). Infection of permissive cells results in a very high ratio of particle to infectious units, however, which contributes to the relatively poor growth of Ad40 and Ad41 (BROWN et al. 1992).

Not all human cells are appropriate hosts for Ad infection, however. Ad12 infections of human melanoma cells Nki-4 and human cervical carcinoma cells $C_{4/1}$ exhibit a late block (SCHWARZ et al. 1982; KRUCZEK et al. 1981). Stratified cultures of human epidermal keratinocytes show an early block to multiplication of Ad2 in the basal cells, but a late block in the suprabasal differentiated cells (ANESKIEVICH and TAICHMAN 1988a). Neither Ad40 nor Ad41 can be serially passaged in primary human cells such as HEK or human diploid fibroblasts, and Ad40 barely grows in HeLa, due to an early block (DE JONG et al. 1983). Ad40 and Ad41 do not grow well in fetal intestinal organ culture (TIEMESSEN et al. 1993), even though it presumably contains the natural target cells of these viruses.

Ad infection of human lymphoid cells may present a special case in the action of host range determinants, since Ad can persist in lymphoid tissue in the body. Continuous lymphoid cell lines are usually permissive to Ad2 and Ad5 to varying degrees (LAVERY et al. 1987), but persistent infection can be established (FAUCON and DESGRANGES 1980). In contrast, freshly isolated peripheral blood lymphocytes (PBL) are abortively infected by Ad2 or Ad5 (SCHRANZ et al. 1979; HORVATH and WEBER 1988). It has been shown that 1%–2% of human PBL harbor group C Ad DNA, and Ad1, –2, –5, and –6 have been isolated from PBL following fusion to or cocultivation with permissive cells (HORVATH et al. 1986). It is not known how abortive infection of PBL in vitro may be related to persistence or latency in vivo.

Ads generally are poorly replicated in most simian cells (KLESSIG 1984), but the extent of infection depends on the Ad species and the particular cell type infected. Infection of primary African green monkey kidney (AGMK) cells by Ad results in the death of the cells without production of virus progeny (FOX and BAUM 1974). Continuous cell lines derived from AGMK (e.g., CV-1, Vero) exhibit different permissivities to Ad. CV-1 cells do not support replication of Ad3 or Ad31 DNA, while presenting a late block to Ad2, –4, –7, or –9 multiplication (FOX and BAUM 1974; DELSERT and D'HALLUIN 1984). Subclones of CV-1 cells adapted to different serum supplements are permissive to varying degrees for Ad2 multiplication

(SILVERMAN et al. 1989). Vero cells, however, are semipermissive to both Ad2 and Ad12, as infection results in virus yields that are only threefold lower and 60-fold lower, respectively, than from infected HeLa cells (EGGERDING and PIERCE 1986; SALEWSKI et al. 1989).

A range of responses to Ad infection of rodent cells is evident as well. Contact-inhibited secondary hamster embryo cultures (HamE) present a late block to Ad2 and Ad5 based on very low yields per cell (SHIMOJO and YAMASHITA 1968), although low yields may result from a subset of cell types productively infected. Infection of HamE cultures with Ad3, –7, –12, or –31 results in blocked viral DNA synthesis (SHIMOJO and YAMASHITA 1968). A continuous hamster kidney fibroblast line (BHK21) will not support replication of Ad12 or Ad3 DNA (DOERFLER and LUNDHOLM 1970; GROFF and DANIELL 1981), yet is semipermissive for Ad2 or Ad5 (ROWE and GRAHAM 1981). For Ad3, the block in BHK21 cannot be attributed to gross problems with the concentrations of early mRNA or its association with ribosomes (GROFF and DANIELL 1981). Ad2 is blocked at a late step in Chinese hamster ovary (CHO) cells and in a variety of cloned rat liver and kidney cell lines (EGGERDING and PIERCE 1986).

Ad12 infection of primary mouse kidney cells and of continuous mouse fibroblast lines, such as 3T3 (embryo) and L (connective tissue) cells, is blocked at DNA replication in spite of at least some early gene expression (STARZINSKI-POWITZ et al. 1982; BIRON and RASKA 1976). Ad2 encounters a late block during infection of primary mouse embryo fibroblasts and baby mouse kidney cells, although late proteins are detected in a significant percentage of infected cells (ZUCKER and FLINT 1985). In contrast, primary rat embryo fibroblasts are semipermissive to Ad2 (GALLIMORE 1974; ZUCKER and FLINT 1985).

An important consequence of abortive Ad infection of rodent cells is the oncogenic transformation of a minority of the infected cells. Transformants can also be selected from Ad-infected semipermissive cells. Since productive infection results in cell death, it is likely that transformation of semipermissive cells is elicited only by replication-defective variants of Ad in the inoculum. The increased transformation ability of some DNA replication-minus mutants (WILLIAMS 1986) lends support to this possibility.

## 3 Early Block Systems

Abortive Ad infections that result from an early block are typically recognized as such by exhibiting a lack of replicated viral DNA. An early block may result from a defect in any of the steps leading up to DNA replication, as well as defects in DNA synthesis. The early events in a productive Ad infection are summarized prior to a presentation of four abortive infection systems that exhibit an early block.

## 3.1 Early Steps in Productive Adenovirus Infection

Ad binds to and enters the cell via receptor-mediated endocytosis (reviewed in DIMMOCK 1982), but the identity of the receptor has not been established. The major component of the Ad2 receptor on HeLa cells is a glycoprotein of 40–42-kDa (SVENSSON et al. 1981); in addition, it appears that group B and group C Ads have different receptors on the KB or HeLa cell surface, since they do not compete with each other for binding (DEFER et al. 1990). Internalization of bound Ad virions is promoted by integrins (WICKHAM et al. 1993). The endocytotic vesicle becomes uncoated and is increasingly acidified as it moves to the cell interior (SETH et al. 1986). A partially disassembled subviral particle is released from the acidified endosome at least in part by an interaction between fiber and penton base with the endosome membrane. The particle moves to the nuclear pores, where Ad DNA, along with its associated viral core proteins, is deposited into the nucleus. Parental Ad DNA assembles into a cell-like chromatin structure in association with cell proteins, as determined by nuclease digestion of infected nuclei (TATE and PHILIPSON 1979; DANIELL et al. 1981). It is presumed that parental Ad DNA associates with cellular histones, since subunit spacing of the Ad DNA in infected nuclei is characteristic of the nucleosome spacing in the cell, which differs from cell type to cell type (DERY et al. 1985).

Parental Ad chromatin binds to the cell nuclear matrix, a reaction that is mediated by the viral terminal protein (TP) covalently attached to each 5' end of Ad DNA (SCHAACK and SHENK 1989). Binding to the nuclear matrix is necessary for proper transcription of several of the early genes. Expression of the early genes begins with the E1A and proceeds in an ordered sequence through the other early transcription units designated E1B, E2, E3, and E4 (PERSSON and PHILIPSON 1982). Virus-associated (VA) RNA transcription begins early, and some early transcriptional activity from the Ad major late promoter (MLP) is evident. Early MLP activity results in expression of L1 mRNA and synthesis of the L1 52, 55K and the i-leader proteins (SYMINGTON et al. 1986; LUCHER et al. 1986).

The E1A and E1B proteins perform regulatory functions during Ad multiplication (GRAND 1987). The E1A proteins activate transcription of other viral early genes, but can also repress transcription. The E1B proteins protect newly replicated viral DNA, shift mRNA transport to favor late viral message, and help regulate DNA replication. The E2 proteins catalyze DNA replication and are involved in the regulation of gene expression (CHALLBERG and KELLY 1989; LAZARIDIS et al. 1988). The E3 proteins are dispensable for infection of cultured cells, but play a role in protecting infected cells from being recognized and lysed by the host immune system (WOLD and GOODING 1991). The E4 proteins are involved in the regulation of DNA replication and late gene expression (BRIDGE et al. 1991, 1993).

Ad DNA is replicated by a strand-displacement mechanism, using protein priming at the terminal replication origins in the linear, double-stranded Ad DNA (CHALLBERG and KELLY 1989; STILLMAN 1989). Replication is initiated with the covalent attachment of deoxycytidine monophosphate (dCMP) to a serine residue in the primer protein, the viral preterminal protein (pTP), in a reaction requiring the viral DNA polymerase (Adpol). This initiation reaction is optimized by cellular transcription factors nuclear factor I (NFI)/CTF and NFIII/OTF-1/OCT-1, the viral DNA-binding protein (DBP), and the presence of TP at the 5' ends of template DNA. Elongation of the nascent chain from the pTP–dCMP initiation complex proceeds continuously, catalyzed by Adpol and requiring DBP and a cellular type I DNA topoisomerase (NFII). Recently, a HeLa protein designated template activating factor 1 (TAF-1) has been detected that in vitro replication studies suggest may aid in the dissociation of core/histone proteins from the viral chromatin for replication to proceed (MATSUMOTO et al. 1993).

## 3.2 Adenovirus Type 12 Infection of BHK21 Cells

Defects in binding, entry and movement of Ad12 to the nucleus do not appear to account for abortive infection in BHK21 cells. Reports of the efficiency of Ad12 binding to hamster cells vary widely, which may reflect the characteristics of different virus stocks. From 0.03% to 22.5% of prelabeled virus inoculum remained associated with BHK21 cells and Nil-2 cells (a Syrian hamster embryo line), respectively, although neither report compared this to association with human cells (DOERFLER 1970; ZUR HAUSEN and SOKOL 1969). A ratio of plaque-forming units per HEK cell to T antigen-inducing units per BHK21 cell of 15–30:1 has been reported (STROHL 1973). We have found that increasing the multiplicity of infection for BHK21 cells only up to threefold over that for KB cells will establish equivalent infections based on viral inhibition of cellular DNA synthesis (LUCHER 1990) or on viral DNA found in cells 2 h after infection (L.A. LUCHER, manuscript in preparation). At equivalent multiplicities of infection, viral DNA reaches the nuclei of KB and BHK21 cells in equivalent amounts (J. DAMON and L.A. LUCHER, unpublished work); about 43% of cell-associated Ad12 DNA can be found in the nuclear fraction of BHK21 cells 2 h after infection (DOERFLER et al. 1972).

Hamster cell DNA synthesis is stimulated by Ad12 infection of quiescent cells, but is shut down within 12 h following infection of growing cells (STROHL 1973; LEDINKO and FONG 1969). Stimulation of DNA synthesis in quiescent cells is an E1A function, whereas the inhibitory effect is independent of viral protein synthesis (STABEL et al. 1985; YAMASHITA et al. 1971). Effects of Ad12 on host DNA replication machinery are not likely to play a role in abortive infection, however, Ad12 cannot replicate its DNA after infection of quiescent BHK21 cells, and the same inhibitory effects on DNA replication are seen after Ad12 infection of permissive human cells (STROHL 1973; LEDINKO and FONG 1969).

In vitro studies suggest that parental Ad12 chromatin structure may also not be involved in the early block to infection in BHK21 cells. Using nuclear extracts, large fragments (1500–5600 bp) of Ad12 DNA from either terminus of the genome can be assembled in vitro into DNA–protein complexes with a periodicity resembling that of chromatin (ANGELETTI 1992). The spacing of the subunits correlates with cell species rather than cell permissivity. The mean difference in subunit size after micrococcal nuclease digestion of the complexes is 248 bp for

KB-derived complexes, but is 181 bp for both BHK21- and 983.2-derived complexes (983.2 is an Ad5-transformed hamster cell line that supports limited Ad12 DNA replication, see below; P.C. ANGELETTI and L.A. LUCHER, manuscript in preparation). Spacing differences seen with in vitro assembly likely reflect differences in human and hamster nucleosome spacing. The nucleosome spacing of parental Ad2 chromatin is also significantly closer in abortively infected monkey CV-1 cells than in HeLa or HEp-2 cells (DERY et al. 1985), but this difference has no effect on the early phase of infection in CV-1 cells (Sect. 4.2).

Within 24 h following infection, up to 50% of the cell-associated Ad12 DNA is integrated into the chromatin of BHK and Nil-2 cells (DOERFLER 1970; ZUR HAUSEN and SOKOL 1969), an observation that is significant to the ability of Ad12 to transform hamster cells. The significance of DNA integration for abortive infection is not known, however, but at the least this may remove some of the viral DNA as template for early transcription and for DNA replication.

Solution hybridization and in vitro translation studies have shown that early Ad12 mRNA synthesized during abortive infection comes mostly from the same early regions expressed during productive infection (RASKA and STROHL 1972; ORTIN et al. 1976; ESCHE et al. 1979). Nevertheless, defects in early gene expression are apparent during abortive infection. The concentrations of Ad12 E1A and E1B proteins in infected BHK21 cells are six- to 20-fold lower than in infected KB cells (LUCHER 1990). This is probably not due to differences in protein turnover, since E1A protein stability as determined by pulse-chase immunoprecipitation is very similar in infected KB and BHK21 (LUCHER 1990). Pulse-chase analysis of the E1B proteins in abortive infection has not been possible; although E1B protein accumulation is detectable by immunoblot methods, the rate of synthesis is too low to detect the proteins by immunoprecipitation. Reduced E1 protein synthesis appears rather to be the result of effects on E1 RNA, since slot blot northern analysis indicates that E1A and E1B total RNA concentrations are reduced in abortive infection (L.A. LUCHER, manuscript in preparation). Whether this reduction is due to decreased transcription or to increased RNA turnover in BHK21 cells is not yet known, but transcriptional defects have been noted for other Ad12 genes during abortive infection (see below).

A more profound defect is seen with DBP. The concentration of Ad12 DBP is 120-fold lower in infected BHK21 than infected KB cells when DNA replication is blocked in KB by hydroxyurea (LUCHER et al. 1992). (DBP concentration increases after the onset of DNA replication). DBP isolated from abortive infection binds to single-stranded DNA cellulose, as does DBP from productive infection, but its low concentration in BHK21 may interfere with DNA replication. In contrast to DBP, Adpol and pTP expression do not appear defective in abortive infection. The in vitro DNA replication activities of Adpol and pTP are comparable in extracts from Ad12-infected BHK21 and KB cells treated with hydroxyurea (CHOWRIRA and LUCHER 1990; LUCHER et al. 1992). Immunoblot analysis also shows similar levels of Adpol accumulated in productive and abortive infection (LUCHER et al. 1992). Transcription of the E2B region, containing the Adpol and pTP genes, is detectable in Ad12-infected BHK21 cells by nuclear run-on assays (ZOCK et al. 1993).

Other defects in gene expression are seen during abortive infection. Neither Ad12 VA RNA nor early L1 RNA accumulate during infection of BHK21 cells, as they are not detectable by RNase protection of total cytoplasmic RNA (JÜTTERMAN et al. 1989). Nuclear run-on experiments show that the lack of L1 RNA is due to the inactivity of the MLP during abortive infection (ZOCK et al. 1993). Both VA RNA transcription and early transcription from the MLP are activated by E1A (SHENK and FLINT 1991).

These results suggest that transcriptional control is involved in the mechanism of abortive infection. The reduced synthesis of E1B and DBP and the lack of MLP and VA RNA transcription in abortive infection may be a consequence of decreased E1A concentration. However, since Adpol and pTP appear unaffected in BHK21 cells, then low E1A levels may not significantly interfere with expression of the E2B genes. Although they are in the same transcription unit, the Ad12 DBP gene (in E2A) and the Adpol and pTP genes may be regulated differently with one regulatory mechanism more sensitive to human and hamster differences. The DBP gene of Ad5 is expressed from both the E2 early (E2E) and E2 late (E2L) promoters (at 75 and 72 map units, respectively), whereas the E2B genes are expressed only from the E2E promoter. If the Ad12 E2 unit is similar to Ad5, inactivity of the E2L promoter in BHK21 might reduce DBP expression without necessarily affecting E2B expression. It should be noted, however, that an analysis of the entire Ad12 DNA sequence, recently deposited in GenBank (SPRENGEL et al. 1994), reveals only one presumptive E2 promoter by sequence homology to Ad2/5.

Multiple upstream *cis*-acting elements regulate E1A transcription (HATFIELD and HEARING 1993; YAMAZAKI et al. 1992), so one or more of these elements may be involved in the reduction of E1A expression in BHK21. Cell-specific NFI characteristics may be important for E1A expression, since NFI stimulates transcription from the distal Ad12 E1A promoter (KOIKEDA et al. 1990). Gel retardation of a consensus NFI oligonucleotide shows more abundant, faster-migrating complexes formed with uninfected BHK21 nuclear extracts as compared to KB extracts (ANGELETTI 1992). Infection by either Ad12 or Ad2 does not affect the formation or characteristics of these complexes (P.C. ANGELETTI and L.A. LUCHER, manuscript in preparation). NFI proteins are found in multiple forms in human, rat, and hamster, probably due to alternative RNA splicing from a single gene (SANTORO et al. 1988; PAONESSA et al. 1988; GIL et al. 1988). The mobility differences between KB- and BHK21-derived complexes may thus result from differences in NFI molecular weight or net charge or may be due to binding of other proteins to the NFI–oligonucleotide complex. Nevertheless, footprint analysis of the first 400 bp of Ad12 DNA reveals identical protection patterns at the NFI site with both KB and BHK21 nuclear extracts (ANGELETTI 1992). How the characteristics of KB and BHK21 NFI may relate to E1A transcription during infection remains to be tested. When cloned into a chloramphenicol acetyltransferase (CAT) expression plasmid, the Ad12 E1A upstream region (the left inverted terminal repeat, ITR, through the proximal E1A promoter; Fig. 1) is transcriptionally active after transfection into both NIH3T3 cells and HeLa cells

```
     Minimum Origin                          NFI              NFIII

L    CTATATATAT AATATACCTT ATACTGGACT AGTGCCAATA TTAAAATGAA    50
             ** ********** **
R    CATCATCAAT AATATACCTT ATACTGGACT AGTGCCAATA TTAAAATGAA    50

        SP1?                                b-enhancer

L    GTGGGCGTAG TGTGTAATTT GATTGGGTGG AGGTGTGGCT TTGGCGTGCT   100

R    GTGGGCGTAG TGTGTAATTT GATTGGGTGG AGGTGTGGCT TTGGCGTGCT   100

L    TGTAAGTTTG GGCGGATGAG GAAGTGGGGC GCGGCGTGGG AGCCGGGCGC   150

R    TGTAAGTTTG GGCGGATGAG GAAGTGGGGC GCGGCGTGGG AGCCGGGCGC   150

               ATF?

L    GCCGGATGTG ACGTTTTAGA CGCCATTTTA CACGGAAATG ATGTTTTTTG   200

R    GCCGGATGTG ACGTGTCGTG AGAACACCGT ATTATGCCGA GTCATATGCG   200

                                         E2F-1?        enhancer

L    GGCGTTGTTT GTGCAAATTT TGTGTTTTAG GCGCGAAAAC TGAAATGCGG   250

                       E2F?

R    GAAGTGATTT TTTTCGCGGG CTTTTGGGGT TTTTTTTGTG TTGTGTAGGC   250


     element 1?                  distal E1A

L    AAGTGAAAAT TGATGACGGC AATTTTATTA TAGGCGCGGA ATATTTACCG   300

                                            E4

R    GAACTTTGGC CATTTACGTC ACTATTTTTC AGTATTAAAG GTGCGCACTT   300


L    AGGGCAGAGT GAACTCTGAG CCTCTACGTG TGGGTTTCGA TACGTGAGCG   350

                                                      ESF1

L    ACGGGGAAAC TCCACGTTGC GCTCAAAGGG CGCGTTTATT GTTCTGTCAG   400

                 proximal E1A

L    CTGATCGTTT GGGTATTTAA TGCCGCCGTG TTCGTCAAGA GGCCACTCTT   450
```

**Fig. 1.** The adenovirus type 12 (Ad12) left (*L*)- and right (*R*)-terminal sequences upstream from the E1A and E4 transcription units. The *lines* indicate the positions of control elements and the E1A and E4 TATA in the sequence underneath. Those elements identified with a *question mark* are putative sites based on sequence alone. The putative $E2F^{-1}$ site is inverted relative to the consensus sequence. The *stars* indicate the 14-bp sequence conserved in all Ad species; the minimum origin is the site recognized by the preterminal protein–viral DNA polymerase (pTP–Adpol) heterodimer. It should be noted that Ad12 strain Huie exhibits marked sequence heterogeneity within the first eight nucleotides. (Based on VAN ORMONDT and GALIBERT 1984; HOGENKAMP and ESCHE 1990; TEMPERLEY and HAY 1992; YAMAZAKI et al. 1992.)

(YAMAZAKI et al. 1992). The upstream region is also active in extracts from mouse Ehrlich ascites tumor cells used for in vitro transcription analysis (SHIBATA et al. 1989). It is not possible to determine from these studies, however, the relative activity of the upstream region in the different cells.

Nothing is known about the synthesis of Ad12 E3 or E4 proteins during abortive infection. Since E3 genes are not required for viral multiplication in culture, any defect in E3 expression in Ad12-infected BHK cells may be irrelevant. Because E4 proteins are involved in regulating transcription and DNA replication, the characteristics of E4 expression in BHK21 may be significant to abortive infection.

Even though DNA replication does not occur during abortive infection, some aspects of this process may not be defective. The Ad12 pTP–dCMP initiation complex can be synthesized in vitro by combining infected cytoplasmic extract as the source of viral proteins and mock-infected nuclear extract as the source of nuclear factors (CHOWRIRA and LUCHER 1990; CHOWRIRA et al. 1991). The initiation reaction is supported by infected BHK21 cytoplasmic plus mock BHK21 nuclear extracts, although the activity is lower than that obtained using infected KB cytoplasmic with mock KB nuclear extracts. The characteristics of the initiation complex and the template requirements for the reaction are the same whether KB or BHK21 extracts are used. Cross-mixing the infected cytoplasmic and mock nuclear extracts from KB and BHK21 cells demonstrates that the reaction rate is dependent on the source of the nuclear extract, but the amounts of active Adpol and pTP are the same from productive and abortive infection. There is also no evidence of an inhibitory factor that might account for the reduced activity of BHK21 nuclear protein. These data suggest that hamster NFI and/or NFIII can participate in initiation in vitro, although perhaps not as efficiently as KB NFI and NFIII. The conditions of these in vitro reactions, however, do not appear sensitive to the DBP concentration found in infected cytoplasmic extract. The reaction rate is the same using infected cytoplasmic extract from either KB or BHK21 cells as long as the nuclear extract is from KB; thus the severely reduced DBP concentration in BHK21 cytoplasmic extract does not affect the reaction.

Crude extracts will also support limited elongation from the initiation complex using both KB and BHK21 extracts (B.M. CHOWRIRA and L.A. LUCHER, unpublished work). Chain elongation under these conditions is not more than 26 nucleotides even with KB extracts, since the mobility of the reaction product is the same whether or not dideoxyguanosine triphosphate (ddGTP) is included (ddGTP blocks elongation at the 26th base). The ability of hamster NFII to participate in replication of Ad12 DNA in vitro is therefore not known.

Complementation experiments have demonstrated the importance of the E1 region in the mechanism of Ad12 abortive infection. Ad12 DNA replication is supported in BHK21 cells coinfected with Ad5 and in Ad2- or Ad5-transformed hamster cells expressing the E1 genes (KLIMKAIT and DOERFLER 1985, 1987; LUCHER et al. 1992). Although they do not catalyze the synthesis of DNA, the E1 proteins are involved in regulating DNA replication during infection. The E1A can stimulate both cellular and viral DNA synthesis in quiescent cells (STABEL et al. 1985;

SPINDLER et al. 1985). Many E1B 55K mutants show DNA replication defects of varying severity in human cells (SHIROKI et al. 1986; WILLIAMS 1986; ZHANG et al. 1992). The E1B 19K protein is required for viral DNA stability during infection (SUBRAMANIAN et al. 1984; HERRMANN and MATHEWS 1989).

Coinfection of BHK21 with Ad12 and Ad5 *dl* 312 (E1A-minus) or *dl* 313 (E1B-minus) mutants shows the relative contributions of E1A and E1B to Ad5 complementation (KLIMKAIT and DOERFLER 1987). Analysis of Ad12 DNA levels post-infection demonstrates that Ad5 E1A does not complement Ad12 DNA replication in BHK21, although it may help stabilize the parental viral DNA. However, Ad 5 E1B complements Ad12 DNA replication well. Analysis of Ad5 DNA levels reveals that Ad12 E1A complements *dl* 312 DNA replication very well, but Ad12 E1B does not affect *dl* 313 DNA replication (*dl* 313 is a leaky, DNA replication-negative mutant that exhibits modest levels of DNA replication at the multiplicity of infection used in this study). These data suggest that Ad5 E1A is not involved in the complementation phenomenon, but rather that Ad5 E1B can correct a defect in the activity of Ad12 E1B. This is consistent with an earlier observation that Ad12 can complement the replication defects in both Ad5 *hr* 3 (E1A-minus) and *hr* 6 (E1B-minus) mutants in coinfected HeLa cells, but can only complement *hr* 3 in BHK21 cells (ROWE and GRAHAM 1981). Nevertheless, hamster cell lines transformed by and expressing the Ad5 E1B region alone do not support Ad12 DNA replication (KLIMKAIT and DOERFLER 1987). The relative importance of E1B 55K and 19K to complementation is not yet known.

During complementation, the Ad5 E1 may act directly on the Ad12 genome by substituting for Ad12 E1 proteins. Ad12 recombinants in which the E1A and/or E1B are replaced with Ad5 E1A and/or E1B are viable (SAWADA et al. 1988), confirming that Ad5 E1A and E1B proteins can act directly on Ad12 DNA, at least in human cells (multiplication of these recombinants was not tested in hamster cells). Alternatively, Ad5 E1 may act indirectly by making the hamster environment suitable for the proper interaction of the Ad12 E1 proteins with the viral DNA and the required cellular components. At least one transformed cell line alleviates the block to full expression of the Ad12 E1A, E1B, and DBP genes. The cell line 983.2 contains the Ad5 XhoI-C fragment (n1-5788) and expresses Ad5 E1A and E1B proteins (ROWE et al. 1984). Synthesis of Ad12 E1A, E1B, and DBP proteins in Ad12-infected 983.2 cells is comparable to that observed in infected KB cells, even with DNA replication blocked by hydroxyurea (LUCHER et al. 1992). Total Ad12 E1A and E1B RNA levels also increase to near-permissive levels during complementation in 983.2 (L.A. LUCHER, manuscript in preparation). It is unknown whether this is due to a direct or indirect effect of the Ad5 E1 proteins. Evidence for a possible indirect effect comes from gel retardation analysis, however, Protein–NFI oligonucleotide complexes formed with 983.2 extracts are more similar to KB- rather than BHK21-derived complexes in terms of mobility and quantity (ANGELETTI 1992). It is possible that the Ad5 E1 proteins affect the characteristics of the complexes formed with 983.2 extracts.

Complementation of the Ad12 early block has revealed the existence of an additional late block to complete multiplication in hamster cells (KLIMKAIT and

DOERFLER 1985). In BHK cells coinfected with Ad12 and Ad2 or Ad5, the DNA of both viruses will replicate but only Ad2 or Ad5 virions will be produced. Ad12 infection of Ad2- or Ad5-transformed BHK results in Ad12 DNA replication, but no virions are produced. In addition, Ad12 DNA replication in 983.2 cells remains at least 200-fold lower than in infected KB cells (L.A. LUCHER, manuscript in preparation). This late block will be covered in Sect. 4.3.

## 3.3 Adenovirus Type 40 Infection of HeLa Cells

The early defect in Ad40 infection of HeLa cells involves the E1B region but not the E1A. Ad40 complements the multiplication defect of Ad2 E1A-minus mutants in HeLa, but not of E1B-minus mutants of either Ad2 or Ad12. Multiplication of Ad40 is fully rescued by heterologous Ad2 or Ad12 E1B expressed in coinfected HeLa and KB cells or in cells that express integrated E1B genes, e.g., such as 293 or E1B-transformed KB18 cells (MAUTNER et al. 1989, 1990; HASHIMOTO et al. 1991). Complementation is achieved by E1B 55K alone, but not 19K alone, and E1A is not necessary for complementation. These results demonstrate how Ad40 behaves as a DNA replication-negative E1B 55K mutant and underscore the regulatory role of the E1B 55K protein in DNA replication. Since this complementation results in complete multiplication of Ad40, the early block in this system is different from the DNA replication block of Ad12 in BHK21.

The nature of the Ad40 defect appears to involve the control of E1B gene transcription early during infection. Ad40 E1B mRNA is not detectable by northern blotting of cytoplasmic RNA isolated at early times following infection of either nonpermissive HeLa or of complementing Ad2 E1-transformed KB16 cells (MAUTNER et al. 1990). Neither a tenfold increase in multiplicity of infection nor the use of 1-β-D-arabinofuranosylcytosine to accumulate early RNA results in detectable quantities of E1B mRNA in either cell line. The defect is not likely to be in RNA transport, since E1 RNA is not found accumulated in the nucleus at early times following infection. At late times following infection of KB16 cells, the 22S mRNA encoding both the 55K and 19K proteins and the 13S mRNA encoding only the 19K protein are both easily detectable by northern blot analysis. The 19K protein, but not the 55K protein, can be immunoprecipitated from late infection extracts using antipeptide antisera. (Although the 55K protein-specific antisera demonstrated high titers against the peptides, these antisera may not have recognized the intact 55K protein). The lack of early, detectable Ad40 E1B mRNA even during complementation is highly unusual and indicates that complementation by 293, KB16, or KB18 cells is probably due to substitution of the heterologous E1B for the Ad40 E1B early during infection. It has been suggested (MAUTNER et al. 1990) that the natural target cells in the gut contain either tissue-specific factors that activate Ad40 E1B gene expression early during infection or a cellular E1B-like protein(s) that substitutes for Ad40 E1B protein(s).

## 3.4 Adenovirus Type 41 Infection of WI-38 Cells

Infection of WI-38 cells by Ad41 provides an example of conditional abortive infection. Ad41 cannot be cultivated in primary human cell cultures due to blocked DNA replication. However, Ad41 DNA replication will proceed in WI-38 cells when the serum concentration in the medium is dropped from 5%–10% down to 0.5%–1%, suggesting the presence of an inhibitory serum factor (PIENIAZEK et al. 1990b). Using low serum also alleviates the block to Ad41 multiplication in other primary cells such as HEK and Detroit 551. Upon overcoming blocked DNA replication in low serum, Ad41 will complete its multiplication cycle; growth is successful enough that Ad41 stocks can be prepared in WI-38 as long as low serum is used for infection. Ad41 DNA replication is not stimulated by low serum during infection of permissive HEp-2 or 293 cells, nor does low serum stimulate Ad5 DNA replication in WI-38. The mechanism of serum inhibition does not involve conditional defects in either E1A or E1B activity, since Ad41 can overcome the multiplication defects of coinfected Ad5 *dl* 312 and *dl* 313 in the presence of 10% serum. In addition, Ad41 DNA replication remains defective in 10% serum during coinfection with *dl* 312 or *dl* 313. The nature of the serum inhibitory factor is not known, nor is it known why it affects Ad41 multiplication in primary cells but not continuous cell lines. The presence of this inhibitor may provide an explanation for the general inability to cultivate Ad41 in primary human cells using typical culture conditions.

## 3.5 Adenovirus Type 2 Infection of Human Continuous Lymphoid Cell Lines

Ad2 or Ad5 infection of a variety of continuous human T and B lymphoid cell lines results in the productive infection of a subpopulation of cells, which generally give yields per cell on the same order as seen with HeLa (LAVERY et al. 1987; SILVER and ANDERSON 1988). Depending on the cell line, from 1% to 60% of lymphoid cells show evidence of infection by immunofluorescence detection of Ad virion protein. Persistent infection can be readily established in these cells. Further analysis of Ad2 infection of a T cell line (MOLT-3) and a B cell line (Raji) has revealed another example of conditional abortive infection (SILVER and ANDERSON 1988). Although almost all MOLT-3 and Raji cells exposed to Ad2 will bind virus, no more than 5% of the cells will show evidence of infection by DBP or hexon immunofluorescence. Most of the adsorbed Ad2 remains at the cell surface; in Raji cells, virions are most clearly concentrated in discrete areas suggesting receptor caps. In synchronized cell populations, the percentage of cells positive for DBP immunofluorescence and negative for surface virion capping correlate very well with the percentage of mitotic cells. Since receptor capping is suppressed during mitosis (BOURGUIGNON et al. 1983), only mitotic lymphoid cells appear to be susceptible to Ad2 infection. The block in resting cells is thus very early, disrupting internalization of the bound Ad2, and conditional, being relieved at mitosis. Thus conditional abortive infection may play a role in the establishment of persistent infection in lymphoid cells.

# 4 Late Block Systems

Abortive Ad infections that result in a late block are infections in which DNA replication is observed, yet no net virus yields result from the infection. Problems with late gene expression, accumulation of DNA or structural proteins, or virion assembly may be involved. Late events in productive Ad infection are summarized prior to a presentation of four abortive infection systems that exhibit a late block.

## 4.1 Late Phase of Productive Infection

Late genes are expressed after the onset of DNA replication by a shift in transcription programming that will also result in a reduction in many early messages. A change in the structure of viral DNA which occurs only when the DNA is undergoing replication is necessary for this early-to-late switch in transcription (THOMAS and MATHEWS 1980). The regulatory switch in transcription may involve the binding of sequence-specific cellular proteins, the binding activities of which are induced late in infection (JANSEN-DURR et al. 1989; MONDESERT et al. 1992). The DBP is involved in downregulating early gene expression as late gene expression increases.

All late transcripts are grouped into five families, designated L1–L5 (although L1 is also expressed early during infection), and are synthesized under the control of the MLP (PERSSON and PHILIPSON 1982). A complex pattern of splicing is utilized to generate the multiple, 3' coterminal mRNA within each family. All transcripts contain the same tripartite leader sequence, but some transcripts include additional leader segments. The E1B 55K protein is required for the efficient transport of late viral mRNA to the cytoplasm (LEPPARD and SHENK 1989). Ad transcripts are translated preferentially over host transcripts late in infection, due to the presence of the tripartite leader and to the actions of VA RNA and a late nonstructural 100K protein (MATHEWS 1990).

Structural proteins are synthesized and transported back into the nucleus, where assembly of progeny virions takes place (PHILIPSON 1983). Progeny DNA molecules assemble into nucleoprotein structures containing the terminal protein covalently bound at the 5' ends, plus the basic core proteins V, VII, and $\mu$. Assembly of the progeny virions begins with formation of empty capsids and proceeds through a series of intermediates that encapsidate viral nucleoprotein and undergo proteolytic processing to form the final mature particle. The complete virion contains the core plus seven other structural proteins.

## 4.2 Adenovirus Type 2 Infection of Monkey Cells

Ad adsorption, early gene expression, and viral DNA replication proceed as efficiently in monkey as in human cells; however, late gene expression is problematic during abortive infection (reviewed in KLESSIG 1984). The concentration of most late proteins is two- to tenfold lower in infected monkey cells, which is paralleled by a similar reduction in the concentration of late mRNA. In vitro translation, mRNA microinjection, and cell reconstruction studies show that monkey ribosomes can translate Ad2 late mRNA synthesized in infected human cells.

At least three factors combine to reduce late mRNA concentrations in infected monkey cells such as CV-1. Transcription from the Ad2 MLP is reduced two- to tenfold, but RNA turnover is not increased over that seen in productive infection (JOHNSTON et al. 1985). Premature termination of transcription at a site 7 kb downstream is increased two- to threefold (JOHNSTON et al. 1985). Finally, transcription termination at the attenuator signal, located 182–188 nucleotides downstream from the MLP transcription start site, is also increased in CV-1 cells two- to threefold over that seen in HeLa cells (SEIBERG et al. 1989b). The increase in attenuated RNA is seen both in infected cells and during in vitro transcription using whole cell extracts. CV-1 cells contain a factor which, when added to HeLa extracts, increases the concentration of attenuated RNA obtained by in vitro transcription. The attenuated RNA can be folded into the expected G, C-rich stem and loop structure, which can be bound in vitro by purified DBP (SEIBERG et al. 1989a).

The greatest single defect in late gene expression, however, is in the synthesis of fiber protein (KLESSIG 1984). The concentration of fiber protein in abortive infection is 100- to 200-fold lower than in productive infection, although the fiber RNA concentration is only five- to tenfold lower. Further reductions in fiber protein synthesis are due to translational defects. In productive infection, 10%–25% of the fiber mRNA contain one of the ancillary leaders x, y, or z, in addition to the common tripartite leader found on all late mRNA. In abortive infection little to no x or y leader is found on fiber mRNA. The absence of the y leader decreases translation initiation twofold in abortively infected cells, but the elongation rate during translation of all fiber mRNA is decreased about threefold, a reduction that is specific for fiber (SILVERMAN and KLESSIG 1989). Combined with the lower mRNA concentration, this decrease in elongation may be enough to explain the severe reduction in fiber protein synthesis in monkey cells, even though the reduction in translation cannot be fully explained by the presence or absence of the x or y leader.

The importance of the fiber mRNA leader to infection success is also illustrated by a series of CV-1 subclones which show increased permissiveness to Ad2 multiplication (SILVERMAN et al. 1989). These cell lines show a correlation between permissivity and fiber protein synthesis; the increases in fiber protein are consistently due not to increased mRNA concentration, but to an increased proportion of the fiber mRNA containing the x or y leaders. The structure of the leader is also involved in compartmentalization of the fiber mRNA, which may be important for infection. This is suggested by the observation that fiber mRNA with the z leader is compartmentalized differently during productive and abortive infection (SILVERMAN et al. 1989). Furthermore, fiber mRNA isolated from abortively infected cells can be efficiently translated in vitro by monkey or human extracts (QUINLAN and KLESSIG 1982) and also after microinjection into monkey cells (K.P. ANDERSON et al. 1985). Thus fiber mRNA translation can be increased by disrupting or circumventing the intracellular framework that would be responsible for mRNA compartmentalization.

The late block to Ad2 multiplication in monkey cells is relieved by coinfection with SV40 or by infection of SV40 large T antigen-expressing monkey cells (KLESSIG 1984). SV40 large T antigen is a multifunctional protein that is required for viral DNA replication and is involved in the negative and positive control of viral and cellular gene expression (FANNING et al. 1989). Normal levels of Ad2 late proteins are synthesized during complementation infection, due to effects on late mRNA metabolism. In the presence of large T antigen the Ad2 MLP activity is increased, the synthesis of Ad2 attenuated RNA is decreased to normal levels, and fiber translation efficiency is probably increased as well (JOHNSTON et al. 1985; SEIBERG et al. 1989b).

## 4.3 Adenovirus Type 12 Infection of Ad5-Transformed Hamster Cells

Even though Ad2 or Ad5 E1 genes will complement the Ad12 DNA replication defect in BHK21 cells, complete multiplication of Ad12 does not occur because late proteins are not synthesized in detectable amounts. Nuclear run-on and RNase protection assays show the inactivity of Ad12 MLP in Ad12-infected BHK21 cells, but transcription from the MLP can be shown in BHK21 cells coinfected with Ad12 and Ad2 and during Ad12 infection of some Ad5-transformed cell lines (ZOCK et al. 1993). Several Ad12-specific late mRNA species are made under these conditions (KLIMKAIT and DOERFLER 1985, 1987). Coinfection with *dl*312 or *dl*313 shows that Ad5 E1B, but not E1A, is responsible for the stimulation of Ad12 late gene expression, although this may be due to the ability of the E1B to stimulate Ad12 DNA replication rather than a direct effect on the MLP. Nevertheless, the ability of Ad2 to activate the Ad12 MLP in BHK21 cells has been shown by transient expression assays (WEYER and DOERFLER 1985). The Ad12 MLP cloned into a CAT expression vector shows similar basal activities in transfected HeLa and BHK21. The Ad12 MLP-CAT plasmid is activated by both Ad2 and Ad12 infection in HeLa cells, but in BHK21 cells it is activated by Ad2 infection but not by Ad12 infection. In contrast, the Ad2 MLP cloned into a CAT expression vector is activated in BHK21 and HeLa cells by both Ad2 and Ad12 infection. This indicates that Ad12 protein (probably the E1) can properly interact with BHK21 proteins and Ad2 DNA, but not with BHK21 proteins and Ad12 DNA.

The Ad12 MLP activity is inhibited by a downstream sequence that lies between +320 and +352 nucleotides. Removal of this "mitigator" sequence from the Ad12 MLP-CAT plasmid increases the basal activity of the plasmid in uninfected BHK21 and HeLa during transient expression (Zock and Doerfler 1990). More importantly, in the absence of the mitigator, the Ad12 MLP can be activated in BHK21 by Ad12 infection as well as by Ad2 infection, utilizing the authentic transcription initiation site. Thus the mitigator prevents proper interaction among Ad12 proteins, BHK21 proteins, and Ad12 DNA to support efficient transcription of the MLP. Sequences upstream of the MLP start site also influence promoter activity, and removal of bases between –180 and –41 allow a modest increase in Ad12- and Ad5-induced MLP activity in the presence of the mitigator sequence, although basal activity is not significantly affected (Zock et al. 1993).

The mechanism of action of the mitigator is not yet clear. A similar (but not identical) sequence in the Ad2 MLP is not involved in regulating Ad2 MLP activity during transient expression of CAT (Zock et al. 1993). Insertion of the Ad12 mitigator into the Ad2 MLP does not interfere with its activity in transfected BHK21 cells. Conversion by site-directed mutagenesis of the Ad12 mitigator to the corresponding Ad2 sequence does not activate the Ad12 MLP in BHK cells. The Ad12 mitigator and its Ad2 equivalent can be bound by nuclear proteins, as evidenced by gel retardation experiments (Zock et al. 1993). There is preliminary evident that the factor YY1 may be involved in the binding. There is, however, no significant difference in specific binding to the mitigator fragment when comparing uninfected HeLa with BHK21 extracts, and the mobilities of the complexes are the same with the Ad2 equivalent DNA sequence and with extracts from Ad12-infected cells.

Although the mitigator clearly affects the activity of Ad12 MLP expressed from plasmid in vivo, it does not interfere with MLP activity during in vitro transcription using nuclear extracts from either HeLa cells or BHK21 cells. This suggests the importance of intranuclear DNA structure in the suppression of MLP activity during abortive infection. It is possible that whatever change in DNA conformation is necessary for the early-to-late switch in transcription programming does not occur during complementation of the early Ad12 block, even though DNA replication does occur.

The extent of late gene transcription in Ad2- or Ad5-transformed cells depends on the integrated Ad2/5 sequences. A BHK21 cell line transformed by the Ad2 E1 region alone does not support synthesis of VA RNA or transcription from the MLP (Jütterman et al. 1989). Transformed cell lines that contain Ad5 DNA from the region between 30 and 40 map units (the L1 and part of the L2 coding regions) in addition to the E1 genes support both MLP activity and VA RNA synthesis (Klimkait and Doerfler 1985; Jütterman et al. 1989). The stimulation of late Ad12 transcription in BHK21 cells by coinfecting *dl* 312 may therefore be due to the concerted action of the E1B and other functions encoded in the Ad5 genome. Even when late transcription is supported, however, complete Ad12 multiplication cannot occur in hamster cells due to a remaining block in the synthesis of late Ad12 proteins (Klimkait and Doerfler 1985, 1987). This block may be due to the inactivity or instability of the late mRNA or to defective translation.

## 4.4 Adenovirus Type 41 Infection of 293 Cells

Different groups have reported variable success in growing Ad41 in 293 cells. In an investigation of sequential cultivation of Ad41 using 293 cells, one study has detected a loss of greater than 90% of infectivity in the first passage and a 100% loss in the second passage (PIENIAZEK et al. 1990a). High yields of incomplete particles, as detected by CsCl density, were obtained not only from 293 infection but also from infection of permissive HEp-2 cells. Antibody analysis of the "complete" particles from 293 infection showed that only traces of the core protein V could be detected in these particles, whereas complete particles from HEp-2 infection contained the expected amounts of protein V. This is not due to a synthesis defect in 293 cells, because protein V is made in both infected 293 and infected HEp-2. This abortive system thus presents an assembly defect, since protein V is found in normal assembly intermediates as well as the mature Ad virion (PHILIPSON 1983). These results were obtained with two strains of Ad41 (Tak) and four different lots of 293 cells (which all supported multiplication of wild-type Ad5, Ad5 *dl* 312, and Ad5 *dl* 313). Since some groups can grow Ad41 on 293 but others cannot, it is not yet clear how applicable these observations are in general.

## 4.5 Adenovirus Type 2 Infection of Human Peripheral Blood Lymphocytes

Because Ad can be latently carried in human PBL (HORVATH et al. 1986), abortive infection of PBL is of interest, even though little is known about the nature of the late block. Infection of human PBL by Ad2 has been compared to productive infection of HEp-2 cells (HORVATH and WEBER 1988). Ad2 can be bound by 25%–40% of collected PBL, including both T and non-T cells, whether freshly isolated, quiescent in short-term culture, or activated by phytohemagglutinin (PHA). These cells exhibit no significant difference in Ad receptor density as compared to HEp-2 cells, when differences in surface area are taken into account. The extent of Ad2 uncoating is 40-fold lower in PBL, but DNA replication does occur. PHA stimulation increases the yield of replicated DNA and allows the assembly of capsid structures detectable by gradient centrifugation of infected extracts, but net virus production is not achieved under these conditions. The late blocks in both resting and PHA-stimulated PBL are encountered at different points in the infectious cycle, the late block in resting PBL being earlier than in stimulated PBL. The number of PBL that can bind Ad2 is greater than the 1%–2% of PBL found to harbor Ad DNA (HORVATH et al. 1986), but the relationship between abortive infection and latency is not known. It should be noted that in another study, binding of Ad2 to freshly isolated PBL could not be demonstrated (SILVER and ANDERSON 1988); these authors proposed that stimulation may be easily and inadvertently induced by collection and handling protocols, allowing PBL to bind Ad2.

## 5 Genetic Analysis of Host Range Determinants

Ad mutants have proved invaluable for elucidating the requirements for productive infection. Host range mutants quite frequently have lost the ability to multiply in permissive human cells and must be propagated in cell lines containing a functional version of the mutated gene(s) or by coinfection with helper virus (WILLIAMS 1986). For example E1-negative host range mutants cannot grow in HeLa or KB, but can be grown in 293 cells. Of interest in understanding host range determinants are those mutants that have acquired an expanded host range ($hr^E$), in that they not only multiply in human cells but also in cells that restrict the growth of wild-type Ad.

The best-characterized $hr^E$ mutants are those Ad2 and Ad5 mutants that are competent to replicate in monkey cells such as CV-1. A total of nine mutants of Ad2 (*hr*400–*hr*403), Ad5 (*hr*404), and the Ad2:SV40 recombinant Ad2$^+$ND3 (*hr*600–*hr*603) all contain the identical point mutation in the DBP gene that changes the histidine at position 130 to a tyrosine (KLESSIG and GRODZICKER 1979; BROUGH et al. 1985; KRUIJER et al. 1981; C.W. ANDERSON et al. 1983). The discovery of these mutants provided the first indication that DBP functions in a capacity to control late gene expression. These mutants replicate in monkey cells with an efficiency equivalent to wild-type Ad2 in human cells and to Ad2 in monkey cells coinfected with SV40. Late proteins are synthesized at normal levels during the productive infection of CV-1 cells by these $hr^E$ Ads, at least in part because late mRNA levels are also normal (JOHNSTON et al. 1985). MLP transcriptional activity is equivalent in CV-1 cells infected by *hr*400 and coinfected by wild-type Ad2 and SV40; although MLP activity in these infections is lower than is seen in Ad2-infected HeLa cells, the activity is sufficient to help overcome abortive infection.

DBP concentration is not altered during abortive infection, so the mutation in these $hr^E$ viruses affects DBP function in a way that ultimately results in a restoration of normal late gene expression. The activities of DBP are divided between two domains that can be separated by mild chymotryptic digestion (KLEIN et al. 1979). Mutations in the C terminal 44K domain affect early gene expression, DNA replication, nucleic acid binding, and morphological transformation; the 26K N terminal domain contains the $hr^E$ mutation (BROUGH et al. 1985). The host range effect probably results from changes in the way in which DBP interacts with species-specific components of the cellular machinery, which then affect the function of the C terminal domain (KLESSIG and GRODZICKER 1979). Consistent with this model is the observation that DBP purified from *hr*400 and wild-type Ad2 bind to attenuated RNA in vitro equally well (SEIBERG et al. 1989a). The simplicity of these mutants means that adaptive growth in semi- or nonpermissive cells does not necessarily require large alterations in the virus (as compared to Ad12 $hr^E$; see below).

Recombinants between Ad2 and SV40 may acquire an $hr^E$ phenotype in monkey cells. Four nondefective recombinants designated $Ad2^+ND$ (numbered 1, 2a, 2b, and 4) contain varying amounts of the SV40 large T antigen gene between the E3 promoter and the fiber gene, replacing most or all of the Ad2 E3 region (KLESSIG 1984). These recombinants synthesize a portion of large T antigen by itself or as a fusion protein. Besides replicating in human cells, they also grow in AGMK cells and their derivatives (except $Ad2^+ND4$). The portion of the large T antigen that is required for the helper effect is found within the extreme C-terminal 35 amino acids and includes phosphoserines 676, 677, and 679 and phosphothreonine 701. This "Ad helper" domain lies outside other large T antigen domains with the following functions: DNA replication, nuclear localization, origin DNA binding, ATPase activity, and binding to the cellular proteins p53, AP2, DNApol alpha, or pRb (FANNING et al. 1989; VILLAREAL and FAN 1989). A deletion mutant mapped to the Ad helper domain cannot grow due to severely reduced late protein synthesis (KLESSIG 1984). Thus the Ad helper domain affects the metabolism of late Ad2 mRNA in monkey cells as a probable consequence of a normal role in regulating SV40 mRNA metabolism.

$Ad2^+ND4$ is unusual in that it grows in AGMK, but not in CV-1 or BSC-1 cells without a coinfecting helper Ad (KLESSIG 1984). The portion of the large T antigen gene inserted into $Ad2^+ND4$ includes splice signals from the SV40 early transcription unit, which are placed upstream of the fiber gene in the recombinant. In addition, a cryptic splice acceptor in this gene is utilized to a significant extent in the recombinant. Thus splicing regulation is altered to the extent that fiber protein synthesis is insufficient. This further emphasizes that processing of fiber preRNA is a significant contributor to the late block in monkey cells. It is unknown why this recombinant can grow in AGMK but not in two cell lines derived from AGMK, but it is probably due to differences in splicing machinery that allow AGMK to avoid improper fiber RNA splicing.

Ad12 $hr^E$ mutants have been isolated that are adapted to grow in the nonpermissive human tumor cell lines $C_{4/1}$ and Nki-4 (KRUCZEK et al. 1981; SCHWARZ et al. 1982). Each of several mutants isolated contained insertions at the right end of the genome, which arise from duplications of either left-terminal or right-terminal sequences (Fig. 2). The duplicated sequences were added to a base from nine to 14 nucleotides in from the right end, within a 14-bp sequence that is conserved in the ITR of all Ad species. In some mutants adapted to Nki-4, the right-hand sequence from nucleotides 11–352 is repeated up to five or six times. Ad12 $hr^E$ mutants have also been isolated that multiply efficiently in semi-permissive Vero cells (WERNER and ZUR HAUSEN 1978; SALEWSKI et al. 1989). These mutants are of interest because they can also overcome the abortive infection block in primary AGMK cells. These $hr^E$ mutants have a 69-bp deletion in the transforming domain of E1A, plus terminal duplications at the right end of the genome (SALEWSKI et al. 1989). In the mutant CL-1 the terminal duplication results from the addition of nucleotides 1–180 from the left terminus to the fourth

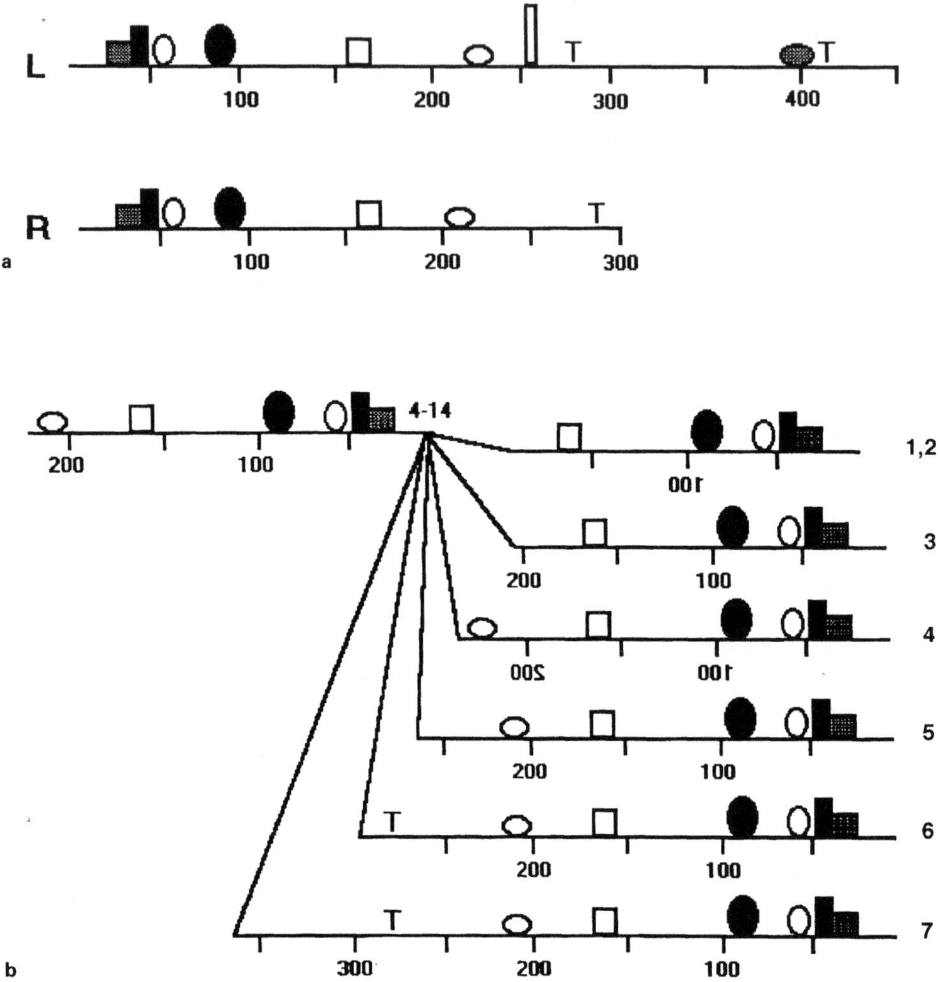


**Fig. 2.** **a** The relative positions of control elements contained in the left (*L*) and right (*R*) termini of adenovirus type 12 (Ad12). The control elements shown by sequence in Fig. 1 are represented here by *boxes* and *ellipses*. **b** Duplications of control elements at the right end of Ad12 $hr_E$ mutants selected in Vero, $C_{4/1}$, or Nki-4 cells. All duplications are due to the addition of DNA from the left terminus (reversed nucleotide numbers) or the right terminus; all additions are within four to 14 nucleotides from the right end. The $hr^E$ mutants and the nucleotides contained in the duplications are as follows: *1*, $C_{4/1}$IA, n1-192; *2*, Vero CL-1, n1-180; *3*, $C_{4/1}$II, n1-208; *4*, $C_{4/1}$IB, n1-238; *5*, $C_{4/1}$III, n1-267; *6*, Vero CS-1, n1-294; *7*, Nki-4 SI-11, n1-353. *Shaded box*, nuclear factor (NF) I; *black box*, NFIII; *black ellipse*, b-enhancer; *tall, white box*, element 1; *shaded ellipse*, ESF1; *white ellipse (vertical)* SPI; *white ellipse (horizontal)*, E2F; *white square*, activating transcription factor (ATF); T, TATA. (Based on KRUCZEK et al. 1981; SCHWARZ et al. 1982; SALEWSKI et al. 1989.)

nucleotide in from the right end. The mutant CS-1 contains a duplication of the right terminal nucleotides 1–294 attached to nucleotide 11 in from the right end (Fig. 2). Further passage in Vero cells favored more duplications in the right end. The duplications in all these $hr^E$ isolates include the NFI and NFIII sites, the b-enhancer, and presumptive SP1 and ATF sites; two isolates also contain a duplicated E4 TATA sequence.

Terminal duplications arise, and revert, spontaneously in productive infection in the absence of any recognizable host selection pressure (BRUSCA and CHINNADURAI 1983; LARSEN and TIBBETTS 1985; FEJER et al. 1992), probably as a consequence of the mode of DNA replication used by Ad (HAJ-AHMAD and GRAHAM 1986). The role of these terminal duplications in allowing Ad12 to overcome the late block in these cells is unknown. Besides their role in regulating E1A expression, the NFIII site and other regulatory sequences within the left ITR and its flanking region are important for growth by a mechanism involving DNA accumulation (HATFIELD and HEARING 1991, 1993). The consistent isolation of mutants with duplications at the right end may simply be due to selection against variants in which duplications in the left terminus have interrupted the signal for packaging DNA into capsids (TIBBETTS 1977; HAMMARSKJOLD and WINBERG 1980).

What is the importance of the E1A deletion in the Vero-derived $hr^E$? The $C_{4/1}$-derived $hr^E$ mutants exhibit reduced growth in Vero that resembles wild-type Ad12 growth in these cells, suggesting that the E1A deletion in the Vero-derived $hr^E$ mutants may be important for host range. However, at least one Vero-derived $hr^E$ (CS-1) is even more complicated, since it also contains an additional 31-bp deletion in E1B that destroys the initiator AUG for translation of the 19K gene (OPALKA et al. 1992). It is therefore not yet clear how many mutations may be required for Ad12 to multiply efficiently in Vero and to overcome the host range restriction in AGMK.

Although little is known about the early and late blocks to Ad2 multiplication in stratified epidermal keratinocyte cultures, four Ad2 $hr^E$ mutants have been isolated that overcome the late block in the suprabasal cells (ANESKIEVICH and TAICHMAN 1988b). These $hr^E$ mutants were selected in epithelial squamous cell carcinoma lines and grow to equivalent yields in suprabasal keratinocytes and in HeLa. The mutants still fail to overcome the early block in basal cells of the keratinocyte culture. The details of infection by these mutants and the Ad12 $hr^E$ mutants are not known.

It is clear that cell characteristics other than the presence or absence of a receptor are important for determining the course of Ad infection. Although some characteristics may be transitory, such as those linked to the cell cycle, many seem to be constitutively expressed. There is little information, however, concerning the genetics of host permissiveness to Ad infection. Cell fusion studies have shown that for two abortive infection systems, permissivity is dominant over nonpermissivity. Monkey–human cell hybrids infected with Ad2 at 8 days following fusion are competent to support fiber synthesis (ZORN and ANDERSON 1981). Long-term changes in the fused cells are not required for complementation, since infected monkey cells fused with human cells at 18 h after infection show an increase in fiber production within 8 h. When human and hamster cells are fused 12 h prior to Ad12 infection, the interaction of human factors with Ad12 is dominant over hamster factors, since Ad12 DNA replication occurs in both nuclei of these heterokaryons (WEBER and MAK 1972). However, the initial interaction between Ad12 and the cell determines the course of infection. Nuclei of human cells infected 2 h prior to fusion with hamster cells will support Ad12 DNA replication in the resulting heterokaryons, but nuclei of hamster cells infected 2 h prior to fusion with human cells will not (WEBER and MAK 1970). In an attempt to further characterize the dominance of human cells, a panel of mouse–human hybrids was screened for permissivity to Ad12 multiplication (BIRON and RASKA 1976). Two out of eight lines tested supported Ad12 DNA replication and late protein synthesis, but not virion production. Comparison of the human chromosomes retained in the positive and negative lines shows that no single human chromosome could be assigned responsibility for permissivity to Ad12 DNA replication. At the very least, it appears that human chromosomes A3, D14, E17, and F20 may be required.

In analyzing cell characteristics that determine Ad host range it will be important to distinguish true host characteristics from those acquired from viral genes that may be present in the cells. This is clearly illustrated by the abandonment of vaccine production in Rhesus monkey kidney cells when it was discovered they can harbor SV40 in an inapparent infection (KLESSIG 1984). Several human cervical carcinoma cell lines—HeLa, SiHa, Caski, and $C_{4/1}$—contain integrated and transcriptionally expressed copies of human papillomavirus (HPV) type 16 or HPV-18 DNA (BOSHART et al. 1984; YEE et al. 1985). The E7 gene of HPV-16 can transactivate the Ad5 E2 promoter in CV-1 cells (PHELPS et al. 1988), which suggests that HPV may be able to complement some defects of Ad multiplication in the carcinoma cell lines; this is not always the case, however, since Ad12 multiplication is restricted in $C_{4/1}$ cells. Epstein-Barr virus genes present in many lymphoid cell lines may also influence Ad multiplication (LAVERY et al. 1987). An additional consideration must be that cell culture conditions may influence permissiveness, as genetic variants of the host cell are selected (OBERLEITHNER et al. 1991; RUBIN and ELLISON 1991). A series of CV-1 subclones that show increased permissiveness to Ad2 multiplication were inadvertently selected by one laboratory due to a change in the culture medium used to maintain the cell stocks (SILVERMAN et al. 1989). This points to the possibility that the same cell line maintained in different laboratories may exhibit differential permissivity to Ad. Therefore, until more is understood about the host requirements for permissivity, caution must be used when comparing the results of abortive infection studies from different laboratories. Finally, the experience of widespread contamination of cell lines with HeLa should sound a cautionary note in deciding the permissivity of a cell line.

## 6 Applications of Abortive Infection Studies

Analysis of abortive infection systems should offer insights into the cell-specific and Ad species-specific requirements for productive Ad infection. Infection of monkey cells by Ad2 may be helpful in elucidating the mechanisms that control transcription attenuation and translation. Investigating the requirement for Ad5 E1B to support Ad40 multiplication may allow us to understand the role of this protein in regulating DNA replication. Dissection of the late block in Ad12 infection of Ad5-transformed cells may further an understanding of the early-to-late switch in transcription, as well as the roles of the E1 proteins in infection. Further analysis of the Ad12 $hr^E$ mutants may explain the involvement of control sequences within and near the ITR in efficient multiplication in different cell types. Abortive infection may also offer insights into the events leading to transformation of semipermissive and nonpermissive cells. For example, Ad12 infection of BHK21 provides an opportunity to study the differential inactivation by DNA methylation of some viral genes in Ad-transformed cells (discussed in DOERFLER 1991).

Abortive infection is also important in considering the use of recombinant Ads as vaccine agents and gene delivery vehicles. Intracellular host range determinants may be more important than cell receptors in affecting how recombinant Ads function in these two roles. Recombinant Ads are increasingly tested as means of vaccination against other viral antigens (discussed in NATUK et al. 1992). Recombinant Ad vaccines have been advocated based on the known safety record of Ad4 and Ad7 vaccines in humans. Additionally, the antigenic diversity of the large number of Ads potentially provides an extended pool of virus to use for subsequent boosters.

In spite of the advantages to using recombinant Ads as vaccines, the restricted Ad host range can become a disadvantage for testing purposes because the recombinant vaccines will not replicate efficiently in convenient animal models. A limited number of animal models have been used to test the efficacy of recombinant Ad vaccines destined for ultimate use in humans (HSU et al. 1992). Chimpanzees are susceptible to enteric infection by recombinant Ad7 and Ad4, but can only be used in limited numbers. Cotton rats and hamsters support limited multiplication of Ad5 in the lungs but are less susceptible to Ad4 or Ad7, so are not completely useful for testing recombinant Ad vaccines against agents that replicate in respiratory tissue. Beagles have served as a model for recombinant Ad vaccines against human respiratory syncytial virus (RSV) and HIV (HSU et al. 1992; NATUK et al. 1992). Even though Ad does not replicate efficiently in beagle, protective neutralizing antibodies to RSV and HIV antigens are elicited upon proper booster administration. Since these poorly replicating recombinant Ad vaccines can generate a potent humoral response in beagles, it has been postulated that efficiently replicating Ad recombinant vaccines should in addition induce potent cell-mediated immunity (NATUK et al. 1992). This suggestion emphasizes the importance of testing whether improved primary and secondary immune responses to Ad recombinant vaccines will be attainable in more permissive hosts such as chimpanzees or humans. One potential approach to this question might be to use Ad $hr^E$ mutants for recombinant vaccines that will replicate efficiently in animals such as rodents. Of course, their utility may be limited to testing in animal models, since it is possible that Ad $hr^E$ mutants may exhibit increased pathogenesis in humans.

In some situations the restricted host range of Ad is a desirable feature in vaccine development. Ad5 has been used to develop a recombinant vaccine against rabies and has been tested in skunks and foxes (CHARLTON et al. 1992). The Ad-rabies vaccine underwent little to no multiplication after oral inoculation, as shown by very limited fecal and oral shedding. Nevertheless, protective antibodies were generated within 4 weeks, without the need for boosters, with no obvious pathology in the vaccinated animals. Host restriction in this case provides a means of eliminating the potential for uncontrolled spread of the vaccine virus among animal populations in the wild.

The testing of Ads as vehicles for gene therapy has increased dramatically in the last few years (e.g. S. ISHIBASHI et al. 1993; KIRSHENBAUM et al. 1993; ZABNER et al. 1993). Understanding the permissivity or nonpermissivity of the targeted cell types may help in the design or testing of Ad as gene therapy agents. Typically, E1-deletion mutants are used for gene delivery to avoid uncontrolled multiplication of the virus vehicle and possible E1-mediated transformation of the treated cells. The potential exists, however, for cell type-specific factors to influence some of the remaining Ad genes or to substitute for the deleted E1 genes.

## 7 Intracellular Determinants: A General Problem for Host Range Analysis

It is clear that the host range of Ad can be determined at the intracellular level in many cases (Table 1). Abortive infection has shown virus defects in internalization, promoter activity, the regulation of DNA replication, the regulation of translation, and the incorporation of virion structural proteins into capsids. The presence or function of cellular determinants for infection can be influenced by the cell cycle, the state of cellular differentiation, intracellular structural features, or external signals from the culture environment. Host permissivity may generally be dominant over nonpermissivity, although some interactions between Ad and the nonpermissive cell may not be reversible by the permissive cell environment.

There are numerous examples of the influence of intracellular determinants on the host range of other viruses as well, but only a few will be mentioned here. The state of cell differentiation influences infection by hamster polyomavirus (HaPV) and HPV. HaPV was discovered originally in epitheliomas of a Syrian hamster colony (DE LA ROCHE SAINT ANDRE et al. 1989). The target cell for both virus multiplication and epithelioma production is the hair root epithelium. Virus

**Table 1.** Summary of the current understanding of the blocks in adenovirus abortive infection systems presented in the text

| Adenovirus (Ad) type and cell | Condition[a] | Nature of block |
|---|---|---|
| Ad12, BHK21 | Constitutive | Transcriptional (early genes) |
| Ad40, HeLa | Constitutive | Transcriptional (E1B only) |
| Ad41, W138 | High serum | DNA replication, not involving E1 defects |
| Ad2, lymphoid lines | Nonmitotic cells | Virion internalization |
| Ad2, CV-1 | Constitutive | Transcriptional (late genes) |
| | Constitutive | Translational (fiber protein) |
| Ad12, Ad2/5-BHK21 | Constitutive | Transcriptional (late genes) |
| | Constitutive | Translational (late proteins) |
| Ad41, 293 | Constitutive | Assembly |
| Ad2, resting PBL | Nonmitotic cells | Transcriptional? (late genes) |
| Ad2, activated PBL | Mitotic cells | Assembly? |

Ad2/5-BHK, Ad2- or Ad5-transformed hamster kidney cell lines; PBL, peripheral blood leukocytes.
[a] The conditions under which the block is evident in vivo are indicated. "Constitutive" indicates that the block is observed under all conditions tested.

particles are found in the terminally differentiated cells and in the epithelioma cells, but not in proliferating cells. Thus the host cell state of differentiation affects the ability of virus to multiply. A similar situation exists with HPV, which also multiplies only in terminally differentiated epithelial cells (Villareal and Fan 1989), and is reminiscent of Ad2 infection of epidermal keratinocyte culture (Aneskievich and Taichman 1988a). An additional feature of HPV infection is the existence of strain specificity for particular anatomical locations. In addition to characteristics of differentiated cells, the cell cycle also influences infection by porcine parvovirus (PPV). Multiplication of PPV is restricted in Madin-Darby canine kidney cells, but not swine testicle cells (Oraveerakul et al. 1992). Host cell functions required for PPV multiplication are expressed transiently in the S phase of the cell cycle and in the differentiated state; these functions are probably required for trans-activation of the structural protein gene via the PPV NS protein.

Alteration of transcriptional regulatory sequences may help determine polyomavirus host range, as appears to be the case for Ad12 hr$^E$ mutants. Field isolates of polyomavirus show well-conserved gene organization and protein coding sequences even when comparing polyomaviruses isolated from different host species (Villareal and Fan 1989). However, field isolates show a very high degree of variability in the enhancer region, even when isolates are obtained from the same host individual. Because polyomavirus eventually disseminates in the host, it has been proposed that polyomavirus utilizes continual enhancer selection to optimize multiplication in different organs. Differences in the regulatory region also appear responsible for the ability of cultivated human polyoma BK virus strains to grow well in culture, whereas BK cloned directly from patients may not (Markowitz et al. 1989). Enhancer differences between HPV strains may also be responsible for tissue tropism and the fact that multiplication only occurs in terminally differentiated epithelial cells (Villareal and Fan 1989).

Interactions between virus and host proteins or processing pathways can differ in permissive and nonpermissive cells. The host range of human rhinoviruses (HRV) is restricted to higher primates and primate cell cultures. Mouse L cells are nonpermissive to HRV multiplication due to an intracellular block at the early stage of infection (YIN and LOMAX 1983). Analysis of several $hr^E$ mutants of HRV selected to grow on L cells indicates that L cells are unable to properly cleave the wild-type HRV polyprotein, resulting in loss of nonstructural proteins and blocked RNA replication. Protein–protein interactions play a role in restricting SV40 and mouse polyomavirus multiplication to monkey and mouse cells, respectively. SV40 cannot replicate its DNA after infection of mouse cells, neither can mouse polyomavirus replicate its DNA in monkey cells. The block may result from insufficient activity of large T antigen–DNApol alpha complexes derived from binding of the respective large T antigens with DNApol from the "wrong" hosts, although other cellular factors may also be involved (BRAITHWAITE et al. 1987; GERARD et al. 1987; DORNREITER et al. 1990).

In summary, many different control mechanisms operate to restrict the multiplication of Ad and other viruses to certain host species and tissues. The control of virus infection by intracellular host range determinants is probably widespread, affecting all viruses at some point in their contacts with potential hosts. It would be of interest to know the relative contributions of cell–virus surface interactions and intracellular interactions in defining the tissue distribution of viruses within a host and the host distribution of viruses in nature.

*Acknowledgment.* Research in the author's laboratory was supported by grant #R15 CA52091-01A1 from the National Institutes of Health.

*Note Added in Proof.* (1) The processing of Ad2 late mRNA in nonpermissive monkey cells has been shown for all the late mRNA families, not just the L5 family of which fiber mRNA is a member (D. Ross and E. Ziff 1994. Defective processing of human Adenovirus 2 late transcription unit mRNAs during abortive infection in monkey cells. Virology 202: 107–115). (2) The E1 deletions and right terminal amplifications found in Ad12 $hr^E$ CS-1 have been placed separately into the wild-type Ad12 background (B. Opalka, M. Reith-Witowski, H.-C. Kirch, and H.S. Holthausen 1994. Altered host range phenotype of the transformation-defective Ad12 mutant CS-1 is due to deletions in the E1 region. Intervirology 37: 36–40). The E1 deletions alone are responsible for increased replication efficiency in semi-permissive Vero cells; the effect on replication in nonpermissive AGMK cells is not known.

## References

Anderson CW, Hardy MM, Dunn JJ, Klessig DF (1983) Independent, spontaneous mutants of adenovirus type 2-simian virus 40 hybrid Ad2⁺ND3 that grow efficiently in monkey cells possess identical mutations in the adenovirus type 2 DNA-binding protein gene. J Virol 48: 31–39

Anderson KP, Wong EA, Klessig DF (1985) Microinjection of mRNA enhances the translational efficiency of human adenovirus fiber message in monkey cells. Mol Cell Biol 5: 2870–2873

Aneskievich BA, Taichman LB (1988a) Epithelium-specific response of cultured keratinocytes to infection with adenovirus type 2. J Invest Dermatol 91: 309–314

Aneskievich BA, Taichman LB (1988b) Evidence for two points of restriction in the expression of adenovirus type 2 in cultured epidermal keratinocytes. J Virol 62: 4365–4368

Angeletti PC (1992) Adenovirus 12 abortive infection: interactions of the E1A upstream region with permissive and nonpermissive cell protein. Thesis, Illinois State University

Biron KK, Raskas K Jr (1976) Adenovirus type 12 infection of defined mouse-human hybrid cell clones. Experientia 32: 38–40
Boshart M, Gissman L, Ikenberg H, Kleinheinz A, Scheurlen W, zur Hausen H (1984) A new type of papillomavirus DNA and its presence in genital cancer biopsies and in cell lines derived from cervical cancer. EMBO J 3: 1151–1157
Bourguignon LYW, Rubin RW, Krishan A (1983) Cell cycle dependent receptor capping in mouse leukemic cells. Cell Biol Int Rep 7: 109–119
Braithwaite AW, Sturzbecher H-W, Addison C, Palmer C, Rudge K, Jenkins JR (1987) Mouse p53 inhibits SV40 origin-dependent DNA replication. Nature 329: 458–460
Bridge E, Hemstrom C, Pettersson U (1991) Differential regulation of adenovirus late transcriptional units by the products of early region. Virology 183: 260–266
Bridge E, Medghalchi S, Ubol S, Leesong M, Ketner G (1993) Adenovirus early region 4 and viral DNA synthesis. Virology 193: 794–801
Brough DE, Rice SA, Sell S, Klessig DF (1985) Restricted changes in the adenovirus DNA-binding protein that lead to extended host range or temperature-sensitive phenotypes. J Virol 55: 206–212
Brown M, Wilson-Friesen HL, Doane F (1992) A block in release of progeny virus and a high particle-to-infectious unit ratio contribute to poor growth of enteric adenovirus types 40 and 41 in cell culture. J Virol 66: 3198–3205
Brusca JS, Chinnadurai G (1983) Structure and reversion of an adenovirus type 2 mutant containing a duplication of the left end of the genome at the right end. Virology 129: 381–392
Challberg MD, Kelly TJ Jr (1989) Animal virus DNA replication. Annu Rev Biochem 58: 671–717
Charlton KM, Artois M, Prevec L, Campbell JB, Casey GA, Wandeler AI, Armstrong J (1992) Oral rabies vaccination of skunks and foxes with a recombinant human adenovirus vaccine. Arch Virol 123: 169–179
Chowrira BM, Lucher LA (1990) Extracts of hamster cells abortively infected with human adenovirus type 12 are competent to support initiation of viral DNA replication. Virology 176: 289–291
Chowrira BM, Zhao J, Lucher LA (1991) Formation in vitro of the pTP-dCMP initiation complex of human adenovirus type 12. J Gen Virol 72: 427–430
Daniell E, Groff DE, Fedor MJ (1981) Adenovirus chromatin structure at different stages of infection. Mol Cell Biol 1: 1094–1105
Defer C, Belin M-T, Caillet-Boudin M-L, Boulanger P (1990) Human adenovirus-host cell interactions: comparative study with members of subgroups B and C. J Virol 64: 3661–3673
de Jong JC, Wigand R, Kidd AH, Wadell G, Kapsenberg JG, Muzerie CJ, Wermenbol AG, Firtzlaff RG (1983) Candidate adenoviruses 40 and 41: fastidious adenoviruses from human infant stool. J Med Virol 11: 215–231
de La Roche Saint Andre C, Delmas V, Bastien C, Goutebroze L, Scherneck S, Feunteun J (1989) Molecular aspects of pathogenesis in hamster polyomavirus. In: Villareal LP (ed) Common mechanisms of transformation by small DNA tumor viruses. American Society for Microbiology, Washington DC, pp 225–238
Delsert D, D'Halluin JC (1984) Genetic expression of human adenoviruses in simian cells. Evidence for interserotypic inhibition of viral DNA synthesis. Virus Res 1: 365–380
Dery CV, Toth M, Brown M, Horvath J, Allaire S, Weber JM (1985) The structure of adenovirus chromatin in infected cells. J Gen Virol 66: 2671–2684
Dimmock NJ (1982) Initial stages in infection with animal viruses. J Gen Virol 59: 1–22
Doerfler W (1970) Integration of the deoxyribonucleic acid of adenovirus type 12 into the deoxyribonucleic acid of baby hamster kidney cells. J Virol 6: 652–666
Doerfler W (1991) Abortive infection and malignant transformation by adenoviruses: integration of viral DNA and control of viral gene expression by specific patterns of DNA methylation. Adv Virus Res 39: 89–128
Doerfler W, Lundholm U (1970) Absence of replication of the DNA of adenovirus type 12 in BHK21 cells. Virology 40: 754–757
Doerfler W, Lundholm U, Hirsch-Kauffmann M (1972) Intracellular forms of adenovirus deoxyribonucleic acid. I. Evidence for a deoxyribonucleic acid-protein complex in baby hamster kidney cells infected with adenovirus type 12. J Virol 9: 297–308
Dornreiter I, Höss A, Arthur AK, Fanning E (1990) SV40 T antigen binds directly to the large subunit of purified DNA polymerase alpha. EMBO J 9: 3329–3336
Eggerding FA, Pierce WC (1986) Molecular biology of adenovirus type 2 semipermissive infections. I. Viral growth and expression of viral replicative functions during restricted adenovirus infection. Virology 148: 97–113
Esche H, Schilling R, Doerfler W (1979) In vitro translation of adenovirus type 12-specific mRNA isolated from infected and transformed cells. J Virol 30: 21–31

Fanning E, Schneider J, Arthur A, Höss A, Moarefi I, Modrow S (1989) Structure and function of SV40 large T antigen. Curr Top Microbiol Immunol 144: 9–19
Faucon N, Desgranges C (1980) Persistence of human adenovirus 5 in human cord blood lymphoblastoid cell lines transformed by Epstein-Barr virus. Infect Immun 29: 1180–1184
Fejer G, Berencsi G, Ruzsics A, Belak S, Linne T, Nasz I (1992) Multiple enlargements in the right inverted terminal repeat of the DNA of canine adenovirus type 2. Acta Microbiol Hung 39: 159–168
Fox RI, Baum SG (1974) Posttranscriptional block to adenovirus replication in nonpermissive monkey cells. Virology 60: 45–53
Gallimore PH (1974) Interactions of adenovirus type 2 with rat embryo cells. Permissiveness, transformation and in vitro characteristics of adenovirus transformed rat embryo cells. J Gen Virol 25: 263–273
Gerard RD, Guggenheimer RA, Gluzman Y (1987) Analysis of nonpermissivity in mouse cells overexpressing simian virus 40 T antigen. J Virol 61: 851–857
Gil G, Smith JR, Goldstein JL, Slaughter CA, Orth K, Brown MS, Osborne TF (1988) Multiple genes encode nuclear factor 1-like proteins that bind to the promoter for 3-hydroxy-3-methylglutaryl-coenzyme A reductase. Proc Natl Acad Sci USA 85: 8963–8967
Grand RJA (1987) The structure and functions of the adenovirus early region 1 proteins. Biochem J 241: 25–38
Groff DE, Daniell E (1981) Early RNA of adenovirus type 3 in permissive and abortive infections. J Virol 40: 620–624
Haj-Ahmad Y, Graham FL (1986) Characterization of an adenovirus type 5 mutant carrying embedded inverted terminal repeats. Virology 153: 22–34
Hammarskjold M-L, Winberg G (1980) Encapsidation of adenovirus 16 DNA is directed by a small DNA sequence at the left end of the genome. Cell 20: 787–795
Hashimoto S, Sakakibara N, Kumai H, Nakai M, Sakuma S, Chiba S, Fujinaga K (1991) Fastidious human adenovirus type 40 can propagate efficiently and produce plaques on a human cell line, A549, derived from lung carcinoma. J Virol 65: 2429–2435
Hatfield L, Hearing P (1991) Redundant elements in the adenovirus type 5 inverted terminal repeat promote bidirectional transcription in vitro and are important for virus growth in vivo. Virology 184: 265–276
Hatfield L, Hearing P (1993) The NFIII/OCT-1 binding site stimulates adenovirus DNA replication in vivo and is functionally redundant with adjacent sequences. J Virol 67: 3931–3939
Herrmann CH, Mathews MB (1989) The adenovirus E1B 19-kilodalton protein stimulates gene expression by increasing DNA levels. Mol Cell Biol 9: 5412–5423
Hogenkamp T, Esche H (1990) Nucleotide sequence of the right 10% of adenovirus type 12 DNA encoding the entire region E4. Nucleic Acids Res 18: 3065–3066
Horvath J, Weber JM (1988) Nonpermissivity of human peripheral blood lymphocytes to adenovirus type 2 infection. J Virol 62: 341–345
Horvath J, Palkonyay L, Weber J (1986) Group C adenovirus DNA sequences in human lymphoid cells. J Virol 59: 189–192
Hsu K-HL, Lubeck MD, Davis AR, Bhat RA, Selling BH, Bhat BM, Mizutani S, Murphy BR, Collins PL, Chanock RM, Hung PP (1992) Immunogenicity of recombinant adenovirus-respiratory syncytial virus vaccines with adenovirus types 4, 5, and 7 vectors in dogs and a chimpanzee. J Infect Dis 166: 769–775
Ishibashi M, Yasue H (1984) Adenoviruses of animals. In: Ginsberg HS (ed) The adenoviruses. Plenum, New York, pp 497–562
Ishibashi S, Brown MS, Goldstein JL, Gerard RD, Hammer RE, Herz J (1993) Hypercholesterolemia in low density lipoprotein receptor knockout mice and reversal by adenovirus-mediated gene delivery. J Clin Invest 92: 883–893
Jansen-Durr P, Mondesert G, Kedinger C (1989) Replication-dependent activation of the adenovirus major late promoter is mediated by the increased binding of a transcription factor to sequences in the first intron. J Virol 63: 5124–5132
Johnston JM, Anderson KP, Klessig DF (1985) Partial block to transcription of human adenovirus type 2 late genes in abortively infected monkey cells. J Virol 56: 378–385
Jüttermann R, Weyer U, Doerfler W (1989) Defect of adenovirus type 12 replication in hamster cells: absence of transcription of viral virus-associated and L1 RNAs. J Virol 63: 3535–3540
Kirshenbaum LA, MacLellan WR, Mazur W, French BA, Schneider MD (1993) Highly efficient gene transfer into adult ventricular myocytes by recombinant adenovirus. J Clin Invest 92: 381–387
Klein H, Maltzman W, Levine AJ (1979) Structure-function relationships of the adenovirus DNA-binding domain. J Biol Chem 254: 11051–11060

Klessig DF (1984) Adenovirus–simian virus 40 interactions. In: Ginsberg HS (ed) The adenoviruses. Plenum, New York, pp 399–449
Klessig DF, Grodzicker T (1979) Mutations that allow human Ad2 and Ad5 to express late genes in monkey cells map in the viral gene encoding the 72K DNA binding protein. Cell 17: 957–966
Klimkait T, Doerfler W (1985) Adenovirus types 2 and 5 functions elicit replication and late expression of adenovirus type 12 DNA in hamster cells. J Virol 55: 466–474
Klimkait T, Doerfler W (1987) E1B functions of type C adenoviruses play a role in the complementation of blocked adenovirus type 12 DNA replication and late gene transcription in hamster cells. Virology 161: 109–120
Kohtz DS, Cole F, Wong M-L, Hsu M-T (1991) Infection and inhibition of differentiation of human fetal skeletal myoblasts by adenovirus. Virology 184: 569–579
Koikeda S, Ibuki R, Sawada Y, Nagata K, Shibata H, Masamune Y, Nakanishi Y (1990) Nuclear factor I stimulates transcription of the adenovirus 12 E1a gene in a cell-free system. Biochim Biophys Acta 1048: 85–92
Kruczek I, Schwarz E, zur Hausen H (1981) Mutants of adenovirus type 12 after adaptation to growth in tumor cell lines. II. Reproducible acquisition of additional sequences after adaptation to a human cervical cancer line. Int J Cancer 27: 139–143
Kruijer W, van Schaik FMA, Sussenbach JS (1981) Structure and organization of the gene coding for the DNA binding protein of adenovirus type 5. Nucleic Acids Res 9: 4439–4457
Larsen P, Tibbetts C (1985) Spontaneous reiterations of DNA sequences near the ends of adenovirus type 3 genomes. Virology 147: 187–200
Lavery D, Fu SM, Lufkin T, Chen-Kiang S (1987) Productive infection of cultured human lymphoid cells by adenovirus. J Virol 61: 1466–1472
Lazaridis I, Babich A, Nevins JR (1988) Role of the adenovirus 72-kDa DNA binding protein in the rapid decay of early viral mRNA. Virology 165: 438–445
Ledinko N, Fong CKY (1969) Kinetics of nucleic acid synthesis in human embryonic kidney cultures infected with adenovirus 2 or 12: inhibition of cellular deoxyribonucleic acid synthesis. J Virol 4: 123–132
Leppard KN, Shenk T (1989) The adenovirus E1B 55kd protein influences mRNA transport via an intranuclear effect on RNA metabolism. EMBO J 8: 2329–2336
Lucher LA (1990) Adenovirus type 12 tumor antigen synthesis differs during infection of permissive and non-permissive cells. J Gen Virol 71: 579–583
Lucher LA, Symington JS, Green M (1986) Biosynthesis and properties of the adenovirus 2 L1-encoded 52,000- and 55,000-$M_r$ proteins. J Virol 57: 839–847
Lucher LA, Khuntirat B, Zhao J, Angeletti PC (1992) Altered expression of adenovirus 12 DNA-binding protein but not DNA polymerase during abortive infection of hamster cells. Virology 189: 187–195
Markowitz R-B, Schneringer J, Tolbert S, Kuppers A, Dynan WS (1989) Functional organization of the BK virus transcriptional control region. In: Villareal LP (ed) Common mechanisms of transformation by small DNA tumor viruses. American Society for Microbiology, Washington DC, pp 191–205
Mathews MB (1990) Control of translation in adenovirus-infected cells. Enzyme 44: 250–264
Matsumoto K, Nagata K, Ui M, Hanaoka F (1993) Template activating factor 1, a novel host factor required to stimulate the adenovirus core DNA replication. J Biol Chem 268: 10582–10587
Mautner V, MacKay N, Steinthorsdottir V (1989) Complementation of enteric adenovirus type 40 for lytic growth in tissue culture by E1b 55K function of adenovirus types 5 and 12. Virology 171: 619–622
Mautner V, MacKay N, Morris K (1990) Enteric adenovirus type 40: expression of E1B mRNA and proteins in permissive and non-permissive cells. Virology 179: 129–138
Mondesert G, Tribouley C, Kedinger C (1992) Identification of a novel downstream binding protein implicated in late-phase-specific activation of the adenovirus major late promoter. Nucleic Acids Res 20: 3881–3889
Natuk RJ, Chanda PK, Lubeck MD, Davis AR, Wilhelm J, Hjorth R, Wade MS, Bhat BM, Mizutani S, Lee S, Eichberg J, Gallo RC, Hung PP, Robert-Guroff M (1992) Adenovirus-human immunodeficiency virus (HIV) envelope recombinant vaccines elicit high-titered HIV-neutralizing antibodies in the dog model. Proc Natl Acad Sci USA 89: 7777–7781
Oberleithner H, Westphale H-J, Gassner B (1991) Alkaline stress transforms Madin-Darby canine kidney cells. Pflugers Arch 419: 418–420
Opalka B, Reith-Witowski M, Roy A, Gnauck J, Schulte-Holthausen H (1992) The transformation-defective adenovirus 12 host range mutant CS-1 lacks the E1A-specific 9.5S mRNA and contains a second deletion in E1B. Intervirology 33: 211–216
Oraveerakul K, Choi C-S, Molitor TW (1992) Restriction of porcine parvovirus replication in non-permissive cells. J Virol 66: 715–722

Ortin J, Scheidtmann K-H, Greenberg R, Westphal M, Doerfler W (1976) Transcription of the genome of adenovirus type 12. III. Maps of stable RNA from productively infected human cells and abortively infected and transformed hamster cells. J Virol 20: 355–372
Paonessa G, Gounari F, Frank R, Cortese R (1988) Purification of a NFI-like DNA-binding protein from rat liver and cloning of the corresponding cDNA. EMBO J 7: 3115–3123
Persson H, Philipson L (1982) Regulation of adenovirus gene expression. Curr Top Microbiol Immunol 97: 157–203
Phelps WC, Yee CL, Münger K, Howley PM (1988) The human papillomavirus type 16 E7 gene encodes transactivation and transformation functions similar to those of adenovirus E1A. Cell 53: 539–547
Philipson L (1983) Structure and assembly of adenoviruses. Curr Top Microbiol Immunol 109: 1–52
Pieniazek D, Pieniazek N, Macejak D, Coward J, Rayfield M, Luftig RB (1990a) Differential growth of human enteric adenovirus 41 (TAK) in continuous cell lines. Virology 174: 239–249
Pieniazek D, Pieniazek NJ, Macejak D, Luftig R (1990b) Enteric adenovirus 41 (Tak) requires low serum for growth in human primary cells. Virology 178: 72–80
Quinlan MP, Klessig DF (1982) Normal translation of human adenovirus in cell-free lysates prepared from abortively as well as productively infected monkey cells. J Virol 44: 426–430
Raska K Jr, Strohl WA (1972) The response of BHK21 cells to infection with type 12 adenovirus. VI. Synthesis of virus-specific RNA. Virology 47: 734–742
Rowe DT, Graham FL (1981) Complementation of adenovirus type 5 host range mutants by adenovirus type 12 in coinfected HeLa and BHK21 cells. J Virol 38: 191–197
Rowe DT, Branton PE, Yee S-P, Bacchetti S, Graham FL (1984) Establishment and characterization of hamster cell lines transformed by restriction endonuclease fragments of adenovirus 5. J Virol 49: 162–170
Rubin AL, Ellison BJ (1991) Induction of transformation in NIH3T3 cells by moderate growth constraint: evidence that neoplasia is driven by adaptational change. Carcinogenesis 12: 1801–1806
Salewski E, Werner G, Opalka B, Schwarz E, Schulte-Holthausen H (1989) Genome structure of adenovirus 12 host range mutants adapted to growth in simian Vero cells. Intervirology 30: 44–51
Santoro C, Mermod N, Andrews PC, Tjian R (1988) A family of human CCAAT-box-binding proteins active in transcription and DNA replication: cloning and expression of multiple cDNAs. Nature 334: 218–224
Sawada Y, Raska K Jr, Shenk T (1988) Adenovirus type 5 and adenovirus type 12 recombinant viruses containing heterologous E1 genes are viable, transform rat cells, but are not tumorigenic in rats. Virology 166: 281–284
Schaack J, Shenk T (1989) Adenovirus terminal protein mediates efficient and timely activation of viral transcription. Curr Top Microbiol Immunol 144: 185–190
Schranz V, Kulcsar G, Dan P, Horvath J, Nasz I, Barinsky IF, Ugryumov EP (1979) Interaction of human lymphocytes and viruses in vitro. Acta Microbiol Acad Sci Hung 26: 1–9
Schwarz E, Reinke C, Yamamoto N, zur Hausen H (1982) Terminal rearrangements in the genome of adenovirus type 12 mutants adapted to growth in two human cell lines. Virology 116: 284–296
Seiberg M, Aloni Y, Levine AJ (1989a) The adenovirus type 2 DNA-binding protein interacts with the major late promoter attenuated RNA. J Virol 63: 1134–1141
Seiberg M, Aloni Y, Levine AJ (1989b) Comparison of human and monkey cells for the ability to attenuate transcripts that begin at the adenovirus major late promoter. J Virol 63: 4093–4096
Seth P, Fitzgerald D, Willingham M, Pastan I (1986) Pathway of adenovirus entry into cells. In: Crowell RL, Lonberg-Holm K (eds) Virus attachment and entry into cells. American Society for Microbiology, Washington DC, pp 191–195
Shenk T, Flint J (1991) Transcriptional and transforming activities of the adenovirus E1A proteins. Adv Cancer Res 57: 47–85
Shibata H, Zheng J-H, Koikeda S, Masamune Y, Nakanishi Y (1989) Cis- and trans-acting factors for transcription of the adenovirus 12 E1A gene. Biochim Biophys Acta 1007: 184–191
Shimojo H, Yamashita T (1968) Induction of DNA synthesis by adenoviruses in contact-inhibited hamster cells. Virology 36: 422–433
Shiroki K, Ohshima K, Fukui Y, Ariga H (1986) The adenovirus type 12 early-region 1B 58,000-Mr gene product is required for viral DNA synthesis and for initiation of cell transformation. J Virol 57: 792–801
Silver L, Anderson CW (1988) Interaction of human adenovirus serotype 2 with human lymphoid cells. Virology 165: 377–387
Silverman L, Klessig DL (1989) Characterization of the translational defect to fiber synthesis in monkey cells abortively infected with human adenovirus: role of ancillary leaders. J Virol 63: 4376–4385

Silverman L, Cleghon V, Klessig DF (1989) Increased permissivity of monkey cells to human adenovirus multiplication is affected by culturing conditions and correlates with both synthesis of virion fiber protein and altered splicing of its mRNA. Virology 173: 109–119
Spindler KR, Eng CY, Berk AJ (1985) An adenovirus early region 1A protein is required for maximal viral DNA replication in growth-arrested human cells. J Virol 53: 742–750
Sprengel J, Schmitz B, Heuss-Neitzel D, Zock C, Doerfler W (1994) The nucleotide sequence of human adenovirus type 12 DNA: a comparative functional evaluation. J Virol 68: 379–389
Stabel S, Argos P, Philipson L (1985) The release of growth arrest by microinjection of adenovirus E1A DNA. EMBO J 4: 2329–2336
Starzinski-Powitz A, Schultz M, Esche H, Mukai N, Doerfler W (1982) The adenovirus type 12-mouse cell system: permissivity and analysis of integration patterns of viral DNA in tumor cells. EMBO J 1: 493–497
Stillman B (1989) Initiation of eukaryotic DNA replication in vitro. Annu Rev Cell Biol 5: 197–245
Straus SE (1984) Adenovirus infections in humans. In: Ginsberg HS (ed) The adenoviruses. Plenum, New York, pp 451–496
Strohl WA (1973) Alterations in hamster cell regulatory mechanisms resulting from abortive infection with an oncogenic adenovirus. Prog Exp Tumor Res 18: 199–239
Subramanian T, Kuppuswamy M, Gysbers J, Mak S, Chinnadurai G (1984) 19-kDa tumor antigen coded by early region E1b of adenovirus 2 is required for efficient synthesis and for protection of viral DNA. J Biol Chem 259: 11777–11783
Svensson U, Persson R, Everitt E (1981) Virus-receptor interaction in the adenovirus system. I. Identification of virion attachment proteins of the HeLa cell plasma membrane. J Virol 38: 70–81
Symington JS, Lucher LA, Brackmann KH, Virtanen A, Pettersson U, Green M (1986) Biosynthesis of adenovirus type 2 i-leader protein. J Virol 57: 848–856
Tate VE, Philipson L (1979) Parental adenovirus DNA accumulates in nucleosome-like structures in infected cells. Nucleic Acids Res 6: 2769–2785
Temperely SM, Hay RT (1992) Recognition of the adenovirus type 2 origin of DNA replication by the virally encoded DNA polymerase and preterminal proteins. EMBO J 11: 761–768
Thomas PG, Mathews MG (1980) DNA replication and the early to late transition in adenovirus infection. Cell 22: 523–533
Tibbetts C (1977) Viral DNA sequences from incomplete particles of human adenovirus type 7. Cell 17: 243–249
Tiemessen CT, Ujfalusi M, Kidd AH (1993) Subgroup F adenovirus growth in fetal intestinal organ cultures. Arch Virol 132: 193–200
Van Ormondt H, Galibert F (1984) Nucleotide sequences of adenovirus DNAs. Curr Top Microbiol Immunol 110: 73–142
Villareal LP, Fan H (1989) History and biological strategy of polyomavirus, adenovirus, and papillomavirus. In: Villareal LP (ed) Common mechanisms of transformation by small DNA tumor viruses. American Society for Microbiology, Washington DC, pp 1–17
Weber J, Mak S (1970) Multiplication of adenovirus in heterokaryocytes produced by fusion of infected permissive and nonpermissive cells. Virology 42: 540–543
Weber J, Mak S (1972) Synthesis of viral components in hybrids of differentially permissive cells infected with adenovirus type 12. Exp Cell Res 74: 423–429
Werner G, zur Hausen H (1978) Deletions and insertions in adenovirus type 12 DNA after viral replication in Vero cells. Virology 86: 66–77
Weyer U, Doerfler W (1985) Species dependence of the major late promoter in adenovirus type 12 DNA. EMBO J 4: 3015–3019
White DO, Fenner FJ (1986) Medical virology, 3rd edn. Academic, Orlando, pp 389–400
Wickham TJ, Mathias P, Cheresh DA, Namerow GR (1993) Integrins alpha v beta 3 and alpha v beta 5 promote adenovirus internalization but not virus attachment. Cell 73: 309–319
Williams JF (1986) Adenovirus genetics. In: Doerfler W (ed) Adenovirus DNA. Martinus Nijhoff, Boston, pp 247–309
Wold WSM, Gooding LR (1991) Region E3 of adenovirus: a cassette of genes involved in host immunosurveillance and virus-cell interactions. Virology 184: 1–8
Yamashita T, Moritsugu Y, Shimojo H (1971) Suppression of cellular DNA synthesis in growing cells by ultraviolet-irradiated adenoviruses. Virology 45: 687–696
Yamazaki Y, Shimada Y, Shibata-Sakurai H, Masamune Y, Nakanishi Y (1992) Multiple cis-acting elements that regulate transcription of the adenovirus 12 E1A gene. Virus Genes 6: 261–271
Yee CL, Krishnan-Hewlett I, Baker CC, Schlegel R, Howley PM (1985) Presence and expression of human papillomavirus sequences in human cervical cell lines. Am J Pathol 119: 3261–3266

Yin FH, Lomax NB (1983) Host range mutants of human rhinovirus in which nonstructural proteins are altered. J Virol 48: 410–418

Zabner J, Couture LA, Gregory RJ, Graham SM, Smith AE, Welsh MJ (1993) Adenovirus-mediated gene transfer transiently corrects the chloride transport defect in nasal epithelia of patients with cystic fibrosis. Cell 75: 207–216

Zhang S, Mak S, Branton PE (1992) Adenovirus type 12 early region 1B proteins and metabolism of early viral mRNAs. Virology 191: 793–802

Zock C, Doerfler W (1990) A mitigator sequence in the downstream region of the major late promoter of adenovirus type 12 DNA. EMBO J 9: 1615–1623

Zock C, Iselt A, Doerfler W (1993) A unique mitigator sequence determines the species specificity of the major late promoter in adenovirus type 12 DNA. J Virol 67: 682–693

Zorn GA, Anderson CW (1981) Adenovirus type 2 expresses fiber in monkey-human hybrids and reconstructed cells. J Virol 37: 759–769

Zucker ML, Flint SJ (1985) Infection and transformation of mouse cells by human adenovirus type 2. Virology 147: 126–141

zur Hausen H, Sokol F (1969) Fate of adenovirus type 12 genomes in nonpermissive cells. J Virol 4: 256–263

# Functional Similarity Between Adenovirus E1B 19-kDa Protein and Proteins Encoded by *Bcl-2* Proto-Oncogene and Epstein-Barr Virus BHRF1 Gene

T. SUBRAMANIAN, B. TARODI, and G. CHINNADURAI



## 1 Introduction

The isolation and physical mapping of a class of mutants designated large plaque (lp) mutants of adenovirus (Ad) type 2 (group C) within the E1B 19 kDa-coding region provided initial clues about the functions of E1B 19-kDa protein (CHINNADURAI et al. 1979; CHINNADURAI 1983). A class of mutants designated the cytocidal (cyt) mutants of Ad12 (group A) isolated by Takemori and colleagues (TAKEMORI et al. 1968) which also produce large plaques on infected cell monolayers was subsequently mapped to the E1B 19 kDa-coding region by intertypic mutant complementation (SUBRAMANIAN et al. 1984b). The cyt mutants of Ad12 induce extensive cytopathic effect on infected cells (TAKEMORI et al. 1968). As expected, subsequent studies revealed that several Ad2 and Ad5 mutants defective in the 19-kDa protein (19K mutants) also induced severe cytopathic effects (SUBRAMANIAN et al. 1984b; TAKEMORI et al. 1984). In accordance with an observation by Mak and colleagues (EZOE et al. 1981) that viral DNA is extensively degraded in cells infected with Ad12 cyt mutants, several Ad2/Ad5 19K mutants were also found to induce fragmentation of cellular and viral DNA (deg) in infected cells (SUBRAMANIAN et al. 1984a; PILDER et al. 1984; WHITE et al. 1984a). Genetic studies have indicated that the cyt and deg phenotypes are linked, thereby indicating that the fragmentation of cellular and viral DNA is a consequence of the

Institute for Molecular Virology, St. Louis University Medical Center, 3681 Park Avenue, St. Louis, MO 63110, USA

cytopathic effect induced by the 19K mutants (SUBRAMANIAN and CHINNADURAI 1986). The DNA fragmentation observed in cells infected with E1B 19K mutants is reminiscent of that observed during cell death induced by several cytotoxic stimuli, including that occurring during apoptosis (WYLLIE 1980). Although it has not yet been determined whether DNA fragmentation induced by 19K mutants of adenovirus occurs by an apoptotic mechanism, it is clear that the 19-kDa protein protects against a cell death program induced by viral infection, facilitating efficient virus replication by preventing premature cell death. The E1A proteins and specifically the conserved region (CR) 1 and CR2, which interact with cellular proteins p300 and pRb and induce cellular proliferation, have been implicated in the onset of the virus-induced death program (WHITE et al. 1991; MYMRYK et al. 1994). In addition, the 19-kDa protein has been shown to suppress the cytotoxic effects of certain external stimuli such as tumor necrosis factor (TNF)-$\alpha$ (GOODING et al. 1991; WHITE et al. 1992) and anti-Fas antibody (HASHIMOTO et al. 1991). Both these agents have been shown to cause cell death through apoptosis (ITOH et al. 1991; WATENABE-FUKUNAGA et al. 1992). Similarly, the 19-kDa protein also protects cells against the effects of DNA damaging agents such as the anticancer drug cisplatin (SUBRAMANIAN et al. 1993) and ultraviolet (UV) (TARODI et al. 1993). Both cisplatin (SORENSON and EASTMAN 1988) and UV (reviewed by WILLIAMS 1991; LANE 1993) induce cell death through the apoptotic pathway, possibly through activation of the p53 tumor suppressor gene (reviewed by LANE 1993). Recently, by direct gene transfer, DEBBAS and WHITE (1993) have shown that the 19-kDa protein can zovercome the growth control effects of the p53 tumor suppressor gene.

Primary baby rat kidney (BRK) cells undergo rapid cell death in culture by an apoptosis-like process (QUINLAN 1993). Transfection of BRK cells with plasmids expressing the E1A proteins induces proliferation and transient rescue of cell death. A significant fraction of these cells appear to undergo terminal growth arrest and apoptosis (RAO et al. 1992). Coexpression of E1A proteins and the 19-kDa protein rescues these cells from terminal growth arrest and apoptosis (RAO et al. 1992). Thus, the 19-kDa protein provides a survival function in virus-infected cells and also protects cells against several external cell death-inducing stimuli.

These activities of the 19K gene appear to be strikingly similar to the activities of the cellular proto-oncogene *Bcl-2*. The *Bcl-2* oncogene was isolated from a follicular lymphoma (BAKHSHI et al. 1985; TSUJIMOTO et al. 1985; CLEARY and SKLAR 1985) and has been shown to suppress cell death induced by a number of stimuli (reviewed by VAUX 1993). Recent amino acid sequence comparisons have led to the recognition of several Bcl-2-related proteins (reviewed by WILLIAMS and SMITH 1993). Among the various Bcl-2-related proteins, the BHRF1 protein (BAER et al. 1984; PEARSON et al. 1987) of the Epstein-Barr virus (EBV) has recently been shown to promote survival of B cells after withdrawal of growth factors (HENDERSON et al. 1993) and also of cells treated with DNA damaging agents (TARODI et al. 1994). We have carried out mutant complementation studies which suggest that the 19-kDa, Bcl-2, and BHRF1 proteins may be functional homologues with regards to promoting cell survival.

## 2 Complementation of 19K Mutation by *Bcl-2*

The functional similarities between the Bcl-2 and 19-kDa proteins were determined by two different approaches. First, Chinese hamster ovary (CHO) cells ectopically expressing the human Bcl-2 protein were established and the ability of these cells to suppress the DNA fragmentation induced by Ad2 19K deletion mutant dl250 was determined (TARODI et al. 1993). Infection of parental CHO cells with dl250 induced characteristic internucleosomal DNA fragmentation. The DNA fragmentation induced by dl250 could be readily suppressed by $Zn^{2+}$ ions in a manner similar to $Zn^{2+}$-mediated suppression of DNA fragmentation during apoptosis (WYLLIE 1980). In contrast, infection of CHO Bcl-2 cells with dl250 in the absence or presence of $Zn^{2+}$ ions did not induce significant DNA fragmentation compared to infection with Ad2 wild type (wt), indicating that the Bcl-2 protein expressed in these cells efficiently suppresses the effect of the 19K mutation. The functional similarity between the 19-kDa and *Bcl*-2 proteins was further substantiated by construction of an Ad2 recombinant virus expressing the human Bcl-2 protein (Ad-Bcl2). In Ad-Bcl2 recombinant virus, a cassette expressing the *Bcl-2* gene under the transcriptional control of cytomegalovirus (CMV) immediate early promoter has been substituted for the E1B region (SUBRAMANIAN et al. 1995). The Ad-Bcl2 recombinant virus, which expresses copious amounts of Bcl-2 protein, produces small plaques on human A549 cells compared to the large-plague morphology produced by dl250 or an Ad5 E1B deletion mutant dl118, defective in both the 19-kDa and 55-kDa E1B proteins (BABISS et al. 1984). The effect of Bcl-2 expression on DNA fragmentation induced by adenovirus lacking the E1B region was determined by infection of a rat kidney (NRK) cell line. As expected, mutant dl118 induced extensive fragmentation of cellular DNA, whereas cells infected with Ad2 wt or Ad-Bcl2 did not induce significant DNA fragmentation. These studies clearly show that the Bcl-2 protein expressed from the viral chromosome can efficiently substitute for the 19-kDa protein during adenovirus infection. RAO et al. (1992) have reported that the Bcl-2 protein can substitute for the 19-kDa protein, albeit at reduced levels, in transformation of primary rat kidney cells in cooperation with E1A.

## 3 Complementation of 19K Mutation by BHRF1

The interest in the protein coded by the BHRF1 open reading frame (BAER et al. 1984) of EBV arises from the suggestion that this viral protein may be a functional homologue of the Bcl-2 protein (WILLIAMS 1991). This suggestion was based on a distant linear amino acid homology between these two proteins first reported by CLEARY et al. (1986). Recent sequence alignments have revealed significant homologies between different regions of the BHRF1 and Bcl-2 proteins (WILLIAMS and SMITH 1993). We have examined whether the BHRF1 protein can also functionally substitute for the 19-kDa protein during viral infection. CHO cells expressing the 17-kDa BHRF1 protein (PEARSON et al. 1987) were isolated by transfection of a BHRF1 expression plasmid (pRcCMV-BHRF1) and tested for their ability to suppress DNA fragmentation induced by the E1B 19K deletion mutant dl 250. As expected, the BHRF1 protein was found to efficiently suppress internucleosomal DNA fragmentation induced by dl 250 (TARODI et al. 1994). Thus, the two viral proteins, BHRF1 and the 19-kDa protein, appear to be functionally similar with regard to suppression of DNA fragmentation induced by an adenovirus E1B 19K mutant.

It should be noted that in spite of the functional similarity between the 19-kDa and Bcl-2 proteins we have observed during adenovirus infection, these two proteins may not be similar with regard to conferring survival function against certain other agents. For example, the 19-kDa protein protects human cells against the cytotoxic effects of TNF-$\alpha$ (GOODING et al. 1991) while Bcl-2 does not (VANHAESEBROECK et al. 1993). The survival promoting activity of Bcl-2 has been examined against the effects of a multitude of stimuli, while the effects of the 19-kDa protein have not been examined extensively. Future studies may well reveal additional functional selectivity in these two proteins.

## 4 Sequence Homology

The functional similarities observed between the three proteins are remarkable, since the E1B 19-kDa protein does not show significant primary sequence homology with the Bcl-2 and BHRF1 proteins. However, we have observed three short regions of homology between these three proteins which may be important for their common function (Fig. 1). Two of these sequence motifs were discovered by searching for Bcl-2 and BHRF1 sequences homologous to sequences of the Ad2 19-kDa protein required for suppression of cell death induced by the DNA-damaging agent cisplatin. The sequences of the 19-kDa protein required for this activity have been localized by our mutational analysis of the 19K gene (SUBRAMANIAN et al. 1993). This search has identified a region of the Bcl-2 protein located between residues 108 and 114 which shows significant homology to the 19-kDa sequences located between residues 47 and 53 and between 90 and 99. Similarly, two regions of the Bcl-2 protein located between residues 42 and 50 and between 77 and 87 appear to be homologous to a hydrophobic region of the 19-kDa protein located between residues 90 and 99. The corresponding BHRF1 sequences are indicated in Fig. 1. We also searched for sequences within the Ad2 19-kDa protein homologous to sequences conserved among the various Bcl-2-related proteins (WILLIAMS and SMITH 1993). This comparison has revealed that a ten-amino acid region located between residues 60 and 69 has significant homology to a domain designated conserved domain I

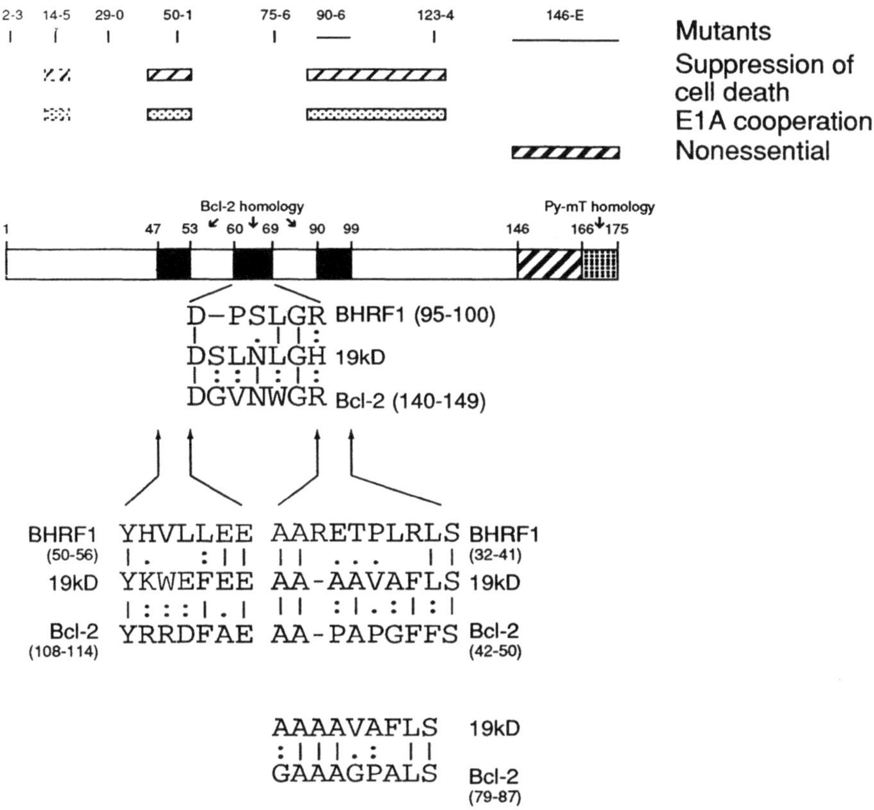


**Fig. 1.** Functional organization of the 19-kDa protein and homology to Bcl-2 and BHRFI proteins. The 19-kDa sequences involved in suppression of cisplatin-induced apoptosis and E1A cooperative transformation are based on SUBRAMANIAN et al. (1993). *Solid bars* indicate pronounced effects of various mutants on 19-kDa protein functions. The sequence similarities between the 19-kDa, Bcl-2, and BHRF1 proteins are shown at the *bottom* of the figure. Identical amino acids are indicated by *solid lines*, similar amino acids by *colons*, and distantly related amino acids by *single dots*

of the Bcl-2-related proteins (OLTVAI et al. 1993; WILLIAMS and SMITH 1993). However, we do not know whether this region of the 19-kDa protein (i.e., residues 60–69) is essential for the cell death-suppressing activity, since no 19K mutations within this domain have thus far been examined for such activity. As this motif is not conserved in the 19-kDa protein coded by Ad12, its role in the function of the 19-kDa protein is uncertain. Although both Bcl-2 and BHRF1 proteins appear to suppress the effect of Ad2 19K mutation well, the homology between different regions of Bcl-2 and the 19-kDa protein is more pronounced than between BHRF1 and the 19-kDa protein. Additional mutagenesis and domain substitution studies will be required to establish the functional importance of the sequence homologies that we have deduced.

## 5 Perspectives

The mechanism by which the 19-kDa and Bcl-2-related proteins protect against cell death is not known. Genetic studies with the nematode *Caenorhabditis elegans* indicate that a survival promoting gene, ced-9, can inhibit programmed cell death which resembles apoptosis (YUAN and HORWITZ 1990; HENGARTNER et al. 1992). The Ced-9 protein appears to be structurally related to the Bcl-2 protein. Recent gene transfer experiments have indicated that the *Bcl-2* gene can functionally substitute for the *ced-9* gene in C. *elegans* (VAUX et al. 1991). The *ced-9* gene suppresses cell death by antagonizing the activity of two other genes, *ced-3* and *ced-4*, which are activated during cell death (YUAN and HORWITZ 1990). Since *Bcl-2* can functionally substitute for *ced-9* in C. *elegans*, it is possible that *Bcl-2* may also antagonize the activities of *ced-3* and *ced-4*. Recently, it has been observed that the mammalian interleukin (IL)-1 β-converting enzyme (ICE), a homologue of Ced-3 (YUAN et al. 1993; MIURA et al. 1993), induces cell death in mammalian cells. The activity of ICE can be suppressed by Bcl-2 (MIURA et al. 1993). Similarly, Bcl-2 can also suppress cell death induced by Bax, a homologue of Bcl-2, possibly by forming Bax:Bcl-2 heterodimers (OLTVAI et al. 1993). Recently, we have identified three different cellular proteins designated Nip (19-kDa-interacting protein)1,Nip2, and Nip3 which interact with the 19-kDa protein. These proteins also interact with Bcl-2 and BHRF1 proteins, suggesting that all three proteins may facilitate cell survival possibly through common mechanisms. The Nip proteins may function as inducers of cell death in a manner similar to Ced-3, Ced-4, and Bax and the 19-kDa, Bcl-2, and BHRF1 proteins may antagonize their activities by physical association. Alternatively, the Nip proteins may function as facilitators of cell survival and the 19-kDa, Bcl-2, and BHRF1 proteins may recruit them to the subcellular compartments such as the nuclear envelope region to enhance cell survival. Interestingly, these three survival-promoting proteins are predominantly localized on the nuclear envelope region (WHITE et al. 1984b; ALNEMRI et al. 1992; MARCHINI et al. 1991; TARODI et al. 1993). Characterization of the Nip proteins will facilitate understanding of the biochemical pathways of suppression of cell death by viral and cellular proteins that promote cell survival.

Thus far, two biochemical activities of the 19-kDa protein have been reported. It has been implicated in mobilization of $Ca^{2+}$ ions, since the 19-kDa protein is required for efficient cellular proliferation of adenovirus-transformed cells in $Ca^{2+}$-deficient media (SUBRAMANIAN et al 1985). Since intracellular $Ca^{2+}$ ions appear to be important mediators of cell death (reviewed by YUAN and HORWITZ 1992), the role of the 19-kDa protein in $Ca^{2+}$ homeostasis may be important for its cell death-modulating activity. The 19-kDa proteins coded by Ad2 and Ad12 have also been reported to modulate transcriptional activities (YOSHIDA et al. 1987; SHIROKI et al. 1990). These activities of the 19-kDa protein may contribute to the cellular survival function conferred by this viral protein in conjunction with various cellular proteins.

*Acknowledgments.* This study was supported by research grants CA-33616 and CA-31719 from the National Cancer Institute.

## References

Alnemri ES, Robertson NM, Fernandes TF, Croce CM, Litwack G (1992) Overexpressed full-length human Bcl-2 extends the survival of baculovirus infected Sf9 insect cells. Proc Natl Acad Sci USA 89: 7295–7299

Babiss LE, Fisher PB, Ginsberg HS (1984) Effect on transformation of mutations in the early region 1b-encoded 21 and 55-kilodalton proteins of adenovirus. J Virol 52: 389–395

Baer R, Bankier A, Biggin M, Deininger P, Farrel P, Gibson T, Hatfull G, Hudson G, Satchwell S, Sequin C, Tuggnell P, Barrell B (1984) DNA sequence and expression of the B95-8 Epstein-Barr genome. Nature 310: 207–211

Bakhshi A, Jensen JP, Goldman P, Wright JJ (1985) Cloning the chromosomal breakpoint of t (14, 18) human lymphomas: clustering around JH on chromosome 14 and near a transcriptional unit on 18. Cell 41: 889–906

Chinnadurai G (1983) Adenovirus 1p+ locus codes for a 19kD tumor antigen that plays an essential role in cell transformation. Cell 33: 759–766

Chinnadurai G, Chinnadurai S, Brusca J (1979) Physical mapping of a large plaque mutation of adenovirus type 2. J Virol 32: 623–628

Cleary ML, Sklar J (1985) Nucleotide sequence of at(14,18) chromosomal breakpoint in follicular lymphoma and demonstration of a breakpoint cluster region near a transcriptionally active locus on chromosome 18. Proc Natl Acad Sci USA 82: 7439–7443

Clearly ML, Smith SD, Sklar J (1986) Cloning and structural analysis of cDNAs for bcl-2 and a hybrid bcl-2/immunoglobulin transcript resulting from the t (14;18) translocation. Cell 47: 19–28

Ezoe H, Lai, Fatt RB, Mak S (1981) Degradation of intracellular DNA in KB cells infected with cyt mutants of human adenovirus type 12. J Virol 40: 20–27

Debbas M, White E (1993) Wild-type p53 mediates apoptosis by E1A, which is inhibited by E1B. Genes Dev 7: 546–554

Gooding LR, Aquino L, Duerksen-Hughes PJ, Day D, Horton TM, Yei S, Wold WSM (1991) The E1B 19,000-molecular-weight protein of group C adenoviruses prevents tumor necrosis factor cytolysis of human cells but not of mouse cells. J Virol 65: 3083–3094

Hashimoto S, Ishii A, Yonehara S (1991) The E1B oncogene of adenovirus confers cellular resistance to cytotoxicity of tumor necrosis factor and monoclonal anti-Fas antibody. Int J Immunol 3: 343–351

Henderson S, Huen D, Rowe M, Dawson C, Johnson G, Rickinson A (1993) Epstein-Barr virus BHRF1 protein, a viral homologue of Bcl-2, protects human B cells from programmed cell death. Proc Nat Acad Sci USA 90: 8479–8488

Hengartner MO, Ellis RE, Horovitz HR (1992) Caenorhabditis elegans gene ced-9 protects cells from programmed cell death. Nature 356: 495–499

Itoh N, Yonehara S, Ishi A, Yonehara M, Mizushima S-I, Sameshima M, Hase A, Seto Y, Nagata S (1991) The polypeptide encoded by the cDNA for human cell surface antigen Fas can mediate apoptosis. Cell 66: 233–243

Lane DP (1993) A death in the life of p53. Nature 362: 786–787

Marchini A, Tomkinson B, Cohen JI, Kief E (1991) BHRF1, the Epstein-Barr virus gene with homology to Bcl2, is dispensable for B-lymphocyte transformation and virus replication. J Virol 65: 5991–6000

Miura M, Zhu H, Rotello R, Hartwieg EA, Yuan J (1993) Induction of apoptosis in fibroblasts by IL-1β-converting enzyme, a mammalian homolog of the C. elegans cell death gene ced-3. Cell 75: 653–660

Mymryk JS, Shire K, Bayley ST (1994) Induction of apoptosis by adenovirus type 5 E1A in rat cells requires a proliferation block. Oncogene 9: 1187–1193

Oltvai AN, Milliman CL, Korsmeyer SJ (1993) Bcl-2 heterodimerizes in vivo with a conserved homolog, bax, that accelerates programmed cell death. Cell 74: 609–619

Pearson GR, Luka J, Petti L, Sample J, Birkenbach M, Braun D, Kief E (1987) Identification of an Epstein-Barr virus early gene encoding a second component of the restricted early antigen complex. Virology 160: 151–161

Pilder S, Logan J, Shenk T (1984) Deletion of the gene encoding the adenovirus 5 early region 1B 21,000-molecular-weight polypeptide leads to degradation of viral and cellular DNA. J Virol 52: 664–671
Quinlan M (1993) E1A 12S in the absence of E1B or other cooperating oncogenes enables cells to overcome apoptosis. Oncogene 8: 3289–3296
Rao L, Debbas M, Sabbatini P, Hockenbery D, Korsmeyer S, White E (1992) The adenovirus E1A proteins induce apoptosis, which is inhibited by the E1B 19-kDa and Bcl-2 proteins. Proc Natl Acad Sci USA 89: 7742–7746
Shiroki K, Kato H, Kawai S (1990) Tandemly repeated hexamer sequences within the beta interferon promoter can function as an inducible regulatory element in activation by the adenovirus E1b 19-kilodalton protein. J Virol 64: 3063–3068
Sorenson CM, Eastman A (1988) A mechanism of cis-diamminedichloroplatinum (II) induced cytotoxicity: role of G2 arrest and DNA double-strand breaks. Cancer Res 48: 4484–4488
Subramanian T, Chinnadurai G (1986) Separation of the functions controlled by adenovirus 2 $1p^+$ locus. Virology 150: 381–389
Subramanian T. Kuppuswamy M, Gysbers J, Mak S, Chinnadurai G (1984a) 19 kDa tumor antigen coded by early region E1b of adenovirus 2 is required for efficient synthesis and for protection of viral DNA. J Biol Chem 259: 11777–11783
Subramanian T, Kuppuswamy M, Mak S, Chinnadurai G (1984b) Adenovirus cyt+ locus, which controls cell transformation and tumorigenicity, is an allele of the 1p+ locus, which codes for a 19-kilodalton tumor antigen. J Virol 52: 336–343.
Subramanian T, Kuppuswamy M, Chinnadurai G (1985) An adenovirus 2-coded tumor antigen located on the endoplasmic reticulum and nuclear envelope is required for growth of transformed cells in $Ca^{2+}$-deficient media. Mol Cell Biol 5: 3297–3300
Subramanian T, Tarodi B, Chinnadurai G (1995) p53-independent apoptotic and necrotic cell deaths induced by adenovirus infection: suppression by E1B 19 K and Bcl-2 proteins. Cell Growth Diff 6: 131–137
Subramanian T, Tarodi B, Govindarajan R, Boyd JM, Yoshida K, Chinnadurai G (1993) Mutational analysis of the transforming and apoptosis suppression activities of adenovirus E1B 175R protein. Gene 124: 173–181
Takemori N, Riggs JL, Aldrich CD (1968) Genetic studies with tumorigenic adenovirus. I. Isolation of cytocidal (cyt) mutants of adenovirus type 12. Virology 36: 575–586
Takemori N, Cladaras C, Bhat B, Conley AJ, Wold WSM (1984) Cyt gene of adenovirus type 2 and 5 is an oncogene for transforming function in early region E1b and encodes the E1B 19,000-molecular-weight polypeptide. J Virol 52: 793–805
Tarodi B, Subramanian T, Chinnadurai G (1993) Functional similarity between adenovirus E1B 19K gene and Bcl-2 oncogene: mutant complementation and suppression of cell death induced by DNA damaging agents. Int. J Oncol 3: 467–472
Tarodi B, Subramanian T, Chinnadurai G (1994) Epstein-Barr virus BHRF1 protein protects against cell death induced by DNA damaging and heterologous viral infection. Virology 201: 404–407
Tsujimoto Y, Cossman J, Jaffe E, Croce C (1985) Involvement of the Bcl-2 gene in human follicular lymphoma. Science 228: 1440–1443
Vanhaesebroeck B, Reed JC, Valck D, Grooten J, Miyashita T, Tanaka S, Beyaert R, Van Roy F, Fiers W (1993) Effect of Bcl-2 proto-oncogene expression on cellular sensitivity to tumor necrosis factor-induced cytotoxicity. Oncogene 8: 1075–1081
Vaux D (1993) Towards an understanding of the molecular mechanisms of physiological cell death. Proc Natl Acad Sci USA 90: 786–789
Vaux DL, Weissman IL, Kim SK (1991). Prevention of programmed cell death in Caenorhabditis elegans by human Bcl-2. Science 258: 1955–1957
Watanabe-Fukunaga R, Brannan CI, Copeland NG, Jenkins NA, Nagata S (1992) Lymphoproliferation disorder in mice explained by defects in Fas antigen that mediates apoptosis. Nature 356: 314–317
White E, Grodzicker T, Stillman BW (1984a) Mutations in the gene encoding the adenovirus E1B 19K tumor antigen cause degradation of chromosomal DNA. J Virol 52: 410–419
White E, Blose SH, Stillman BW (1984b) Nuclear envelope localization of an adenovirus tumor antigen maintains the integrity of cellular DNA. Mol Cell Biol 4: 2865–2875
White E, Cipriani R, Sabbatini P, Denton A (1991) Adenovirus E1b 19-kilodalton protein overcomes the cytotoxicity of Ela protein. J Virol 65: 2968–2978.
White E, Sabbatini P, Debbas M, Wold WSM, Kusher DI, Gooding L (1992) The 19-kilodalton adenovirus E1B transforming protein inhibits programmed cell death and prevents cytolysis by tumor necrosis factor α. Mol Cell Biol 12: 2570–2580
Williams GT (1991) Programmed cell death: apoptosis and oncogenesis. Cell 65: 1097–1098

Williams GT, Smith CA (1993) Molecular regulation of apoptosis: genetic control on cell death. Cell 74: 777–779

Wyllie AH (1980) Glucocorticoid-induced thymocyte apoptosis is associated with endogeneous endonuclease activation. Nature 284: 555–556

Yoshida K, Venkatesh L, Kuppuswamy M, Chinnadurai G (1987) Adenovirus transforming 19kD T antigen has an enhancer-dependent trans-activation function and relieves enhancer repression mediated by viral and cellular genes. Genes Dev 1: 645–658

Yuan J, Horvitz HR (1990) The Caenorhabditis elegans genes ced-3 and ced-4 act cell autonomously to cause programmed cell death. Dev Biol 138: 33–41

Yuan J, Horvitz HR (1992) The Caenorhabditis elegans cell death gene ced-4 encodes a novel protein and is expressed during the period of extensive programmed cell death. Development 116: 309–320

Yuan J, Shaham S, Ledoux S, Ellis HM, Horvitz HR (1993) The C. elegans cell death gene ced-3 encodes a protein similar to mammalian interleukin-1β-converting enzyme. Cell 75: 641–652

# Adenovirus Fiber

J. CHROBOCZEK[1], R.W.H. RUIGROK[2], and S. CUSACK[2]



## 1 Introduction

The adenovirus fiber protein was first described as the "cell detaching factor" by ROWE et al. (1958), as the "early cytopathic factor" by PEREIRA (1958), and as the

[1] Institut de Biologie Structurale, 41, Avenue des Martyrs, 38027 Grenoble Cedex 1, France
[2] EMBL Grenoble Outstation, c./o ILL, BP 156, 38042 Grenoble Cedex 9, France

"toxin-like factor" by Everett and Ginsberg (1958). In 1965 Valentine and Pereira observed antenna-like projections extending from the vertices of HAd2 virions, and the next year Norrby (1966) observed similar, but shorter projections on HAd3 virions. These projections were called fibers by Ginsberg and coworkers (1966), and it was shown by several laboratories that they can be purified by ion exchange chromatography and visualized on denaturing polyacrylamide gels as a distinct polypeptide, characteristic of a given adenovirus serotype.

Thirty years after its discovery, it is now known that the fiber is the structural protein of adenovirus with the crucial function of recognizing the primary cellular receptors in the initial stage of infection. Morphologically it is a trimeric, elongated protein that can be divided into a tail, a slender shaft of variable length, and a globular head (knob), this organization being the same for all serotypes so far examined. All known mammalian adenoviruses have 12 fibers, one at each vertex of the icosahedral virion, noncovalently attached to another capsid protein, the penton base. The complex consisting of penton base and fiber is called penton. Viruses belonging to subgroup F (e.g., HAd40 and HAd41) have two fibers of different length, but only one per vertex, whereas avian adenoviruses have two fibers (sometimes of different length) per vertex.

# 2 Function

There are two well-understood functions which can be directly attributed to the fiber protein. The first is the structural role in the viral capsid, and the second the interaction with cellular adenovirus receptors, the first step in viral infection. In this section we will also discuss the antigenicity of the fiber.

## 2.1 Interaction with the Penton Base

The polarity of the polypeptide chain in the fiber was unambigously determined from experiments using partial proteolysis, chemical cleavage, and anti-peptide sera. The N terminus is located in the tail region and the C terminus is in the knob (Devaux et al. 1987). Since fiber devoid of the N terminal 17 amino acids is unable to bind to the penton base, it is clear that at least this region is required for penton base binding. Furthermore, when an antibody recognizing a peptide located 30 residues from the N terminus was used, it was found that the epitope was masked in the complete virion (Weber et al. 1989). An interpretation of this is that the first 30 N-terminal fiber residues lie within the anchoring domain of the fiber in the penton base. An interesting observation is that chimeric pentons can be obtained in vitro by incubating fiber and penton base of different serotypes (Boudin and Boulanger 1982; Praszkier and Ginsberg 1987). This strongly suggests that fiber–penton base recognition is mediated by conserved structural features of both the fiber and penton base proteins. One aspect of the fiber–penton base interaction that remains to be elucidated is how the trimeric fiber is accommodated by the pentameric penton base (see Sect. 3.4.4).

## 2.2 Interaction with the Cell Receptor

Interaction of adenovirus with cell receptors is mediated by the fiber, as demonstrated by blocking of virus attachment by soluble fiber (PHILIPSON et al. 1968) and by the ability of soluble fiber (DEFER et al. 1990) as well as anti-fiber antibody (PHILIPSON et al. 1968) to inhibit infection. These studies on the role of the fiber in cell attachment were carried out predominantly on HAd2.

When the problem of fiber polarity was resolved, it was supposed that the part of fiber which interacts with the cell receptor is the terminal head. However, no experimental data existed supporting this hypothesis. Recently, it has been shown that the soluble C-terminal head fragment produced in the baculovirus expression system was able to block virus infection, confirming the hypothesis that the head domain interacts with viral receptors (LOUIS et al. 1994; Henry et al. 1994).

Adsorption studies performed in different laboratories indicate that there are between $10^3$ and $10^4$ receptors for HAd2 virions and about 100 times more for the fiber protein (PHILIPSON et al. 1968; PERSSON et al. 1983). Binding of virions is a saturable phenomenon and occurs with increasing cooperativity at temperatures between 5° and 20°C, the plateau of which is maintained up to 37°C (PERSSON et al. 1983). The positive cooperativity for HAd2 was explained as being a consequence of both the ability of the receptor sites to move laterally in the plasma membrane and the presence of 12 fibers per virions, which permits the virions to bind at low multiplicity of infections (MOI) to several receptors simultaneously. Studies with purified fiber revealed poor or no positive cooperativity, irrespective of temperature (PERSSON et al. 1983).

The molecular identity of the primary adenovirus receptors is not known. They are, at least in part, proteins, since they can be inactivated when cells are treated with proteases such as subtilisin and trypsin (PHILIPSON et al. 1968; YEH and LUFTIG 1991). Regeneration of HAd2 receptors on cell membranes in cell culture occurs in about 6 h and can be blocked by inhibitors of protein synthesis. HAd2 fiber inhibits attachment of both HAd2 and HAd5 virions, suggesting common receptors for these serotypes (PHILIPSON et al. 1968). In contrast, HAd3 attachment cannot be inhibited in the presence of HAd2, which suggests different receptors for subgroup C and subgroup B adenoviruses (A.M. DI GUILMI and J. CHROBOCZEK, unpublished results; DEFER et al. 1990). Interestingly, attachment of the unrelated coxsackievirus serotype B3 can be inhibited by HAd2 as well as by HAd2 fiber, which suggests common receptors for these two viruses (LONBERG-HOLM et al. 1976). Various and often numerous plasma membrane proteins or glycoproteins were observed to interact with ligands such as virions or fibers

(Meager et al. 1976; Hennaché and Boulanger 1977; Svensson et al. 1981; Defer et al. 1990), but these proteins were not identified or sequenced and no appropriate transfection experiments were performed. In order to really identify the adenovirus cellular receptor, eukaryotic expression libraries should be used. It might enable us to find the clones which, upon transfection, will permit the attachment of virus to nonpermissive cells

Following attachment of fiber protein to its receptor, HAd2 internalization occurs by receptor-mediated endocytosis (Chardonnet and Dales 1970a, b; Fitzgerald et al. 1983); this corresponds to the infectious pathway of entry (Varga et al. 1991). It was observed for HAd2 that the virus binds, but does not enter some cells (Silver and Anderson 1988), which could suggest that a host-derived factor in addition to the primary attachment receptor is necessary for virus entry. Penton base of HAd2 contains the sequence Arg-Gly-Asp (RGD), an integrin-binding sequence motif (Hynes 1992), suggesting a role for integrins in adenovirus infection. It was shown that HAd2 penton base protein is indeed probably a ligand for the integrins $\alpha_v\beta_3$ and/or $\alpha_v\beta_5$ (Wickham et al. 1993). Interaction of soluble HAd2 penton base with cells was blocked by an RGD-containing (but not RGE-containing) peptide and by monoclonal antibodies against integrins $\alpha_v\beta_3$ and $\alpha_v\beta_5$. Soluble penton base, the RGD-containing peptide, and the anti-integrin antobodies all inhibited HAd2 infection. Internalization studies showed that whereas fiber internalization was very slow and inefficient, penton complex and penton base were both internalized at a similar, much faster rate. These results, together with the fact that the free fiber inhibited almost totally HAd2 attachment (Wickham et al. 1993), show that the binding properties of the HAd2 penton complex are determined by the fiber, whereas the internalization properties are determined by the penton base. It thus seems that, in addition to a primary interaction between the fiber and its receptor, a second interaction between penton base and integrins must occur for the successful HAd2 infection.

Is this a general mechanism of adenovirus entry? Out of five penton base polypeptide sequences available in the data bank, three (serotypes 2, 5, 12) contain the motif RGD; so does the penton base of HAd3 (Cuzange et al. 1994). In HAd40, RGD is replaced by RGA (Davison et al. 1993). This motif is in a part of the sequence which seems to be highly hydrophilic (Hopp and Woods 1981) and as such likely to be exposed. However, the penton base sequences of HAd40 and FAV (fowl adenovirus) 10 do not contain this motif, which suggests that the above-described mechanism implicating the RGD motif in the penton base cannot be true for all adenoviruses. HAd40 has two distinct fiber genes, each penton containing one or other of the fibers (Kidd et al. 1993). FAV10, like the majority of avian adenoviruses, has two fibers in each penton base (Gelderblom and Maichle-Lauppe 1982). It could be that for these adenoviruses the second kind of fiber interacts with integrins during internalization. However, there is no RGD motif in either HAd40 fiber sequence, and the fiber sequences for avian adenoviruses are not known yet. It is possible that these viruses interact with a different second receptor.

## 2.3 Antigenicity

The fiber is one of the three capsid proteins which carry antigenic determinants that are important in the serological classification of adenoviruses (the other two are hexon and penton base). The existence of three antigenic sites on the HAd2 fiber was described. The γ determinant present on the head is a type-specific antigen (NORRBY 1969) which is probably identical to the type-specific hemagglutinin (GINSBERG 1979). This determinant seems to be the only one present on the shortest fibers of adenoviruses subgroup B (PETTERSSON and WADELL 1985). Type-specific-antisera against the fiber have been shown both to contain and to lack virus neutralizing activity, depending on the neutralization assay employed (WATSON et al. 1988; WOHLFART et al. 1985). A subgroup determinant δ (longer fibers of subgroups A, C, D) is located in the shaft, close to the junction between the fiber and the penton base, and is shared between the members of the same subgroups (WADELL and NORRBY 1969). A determinant common for subgroups C and D has been reported (PETTERSSON et al. 1968; NORRBY 1969).

The use of monoclonal antibodies makes it possible to study the antigenic determinants in more detail. This approach may reveal a larger number of epitopes and also of overlapping neutralizing antigenic epitopes (STRASSHEIM et al. 1994). The antigenic analysis of sodium dodecyl sulfate (SDS)-denatured HAd2 fiber (WATSON et al. 1988) revealed the existence of six antisera groups, some of them neutralizing and some not. Using the C-terminal part of the fiber expressed in *Escherichia coli* as an antigen and a competition assay for monoclonal antibodies that bind to this part of the fiber, the authors showed that five epitopes exist on the head domain and at least three on the rest of the fiber polypeptide. These antibodies were produced against denatured fiber, however, and additional studies are necessary on three-dimensional epitopes of native fiber.

# 3 Structure of the Fiber

## 3.1 Primary Structure

The first amino acid sequence of a fiber polypeptide was that deduced from the corresponding gene sequence of HAd2 by HÉRISSÉ et al. (1981); it contains 582 amino acid residues. This sequence was analysed in detail in the paper by GREEN et al. (1983), who suggested for the first time the organization of the fiber polypeptide in three well-defined regions. These are the N-terminal "tail" of 46 residues (GREEN et al. recognized 43 tail residues), a central shaft of 354 residues consisting of 22 pseudo-repeats, usually of 15 residues, and a C-terminal domain of 182 residues. It has subsequently been confirmed that it is indeed the tail which interacts with the penton base and that the C-terminal domain forms the distal head (knob) which has receptor-binding activity (see Sects. 2.1 and 2.2). All known human adenovirus fiber sequences, of which there are now 13 in the data base (representing subgroups A–F), conform to this same basic organization. There are also three known sequences of fiber from adenoviruses in the animal hosts (canine, murine, and bovine); these show some interesting differences from the human virus fibers which will be pointed out below. Figure 1 (see also KIDD et al. 1993) shows an alignment of all these sequences for the tail, shaft, and the head region. In subsequent paragraphs we use these alignments to discuss particular features of the three distinct regions of the fiber (sequence numbers will refer to the HAd2 fiber sequence unless otherwise stated).

```
AD11p   1  MT-KRVRLS---------DSFNPVYPYEDESTSQ--HPFINPGF
AD11a   1  MT-KRVRLS---------DSFNPVYPYEDESTSQ--HPFINPGF
AD7     1  MT-KRVRLS---------DSFNPVYPYEDESTSQ--HPFINPGF
AD21    1  MT-KRVRLS---------DSFNPVYPYEDESTSQ--HPFINPGF
AD3     1  MA-KRARLS---------TSFNPVYPYEDESSSQ--HPFINPGF
AD16    1  MA-KRARLS---------SSFNPVYPYEDESSSQ--HPFINPGF
AD2     1  M--KRARPS--------EDTFNPVYPYDTETGPPT-VPFLTPPF
AD5     1  M--KRARPS--------EDTFNPVYPYDTETGPPT-VPFLTPPF
AD4     1  MSKKRARV---------DDGFDPVYPYDADNA-PT-VPFINPPF
AD12    1  M--KRSRTQYAEETEENDDFNPVYPFDPFDTSD--VPFVTPPF
AD8     1  MT-KRLRA---------EDDFNPVYPYGYARNQN--IPFLTPPF
AD9     1  MS-KRLRV---------EDDFNPVYPYGYARNQN--IPFLTPPF
AD15    1  MS-KRLRV---------EDDFNPVYPYGYARNQN--IPFLTPPF
AD41-1  1  M--KRARL---------EDDFNPVYPYEHYN-PLD-IPFITPPF
AD40-1  1  M--KRARF---------EDDFNPVYPYEHYN-PLD-IPFITPPF
AD40-2  1  M--KRTRI---------EDDFNPVYPYDTSS-TPS-IPYVAPPF
AD41-2  1  M--KRTRI---------EDDFNPVYPYDTFS-TPS-IPYVAPPF
CANINE  1  M--KRTRSALP-------ANFDPVYPYDAPK------PSTQPPFF
BOVINE  1  M--KRS-V-P--------QDFNLVYPYKAKR------PNIMPPF
MURINE  1  M------V--EA------EALNAVYPYDLALLPEDYEKTTAPDAV
MURINE 38             QAANAAPFLNPVYPYQQPVAGDFGFPIVMPPFF

AD11p  33  ISP-NGFTQSPNGV
AD11a  33  ISP-NGFTQSPDGV
AD7    33  ISP-NGFTQSPDGV
AD21   33  ISP-NGFTQSPDGV
AD3    33  ISP-DGFTQSPNGV     END OF TAIL
AD16   33  ISS-NGFTQSPDGV
AD2    34  VSP-NGFQESPPGV
AD5    34  VSP-NGFQESPPGV
AD4    37  VFFKNGFQEKPPGV     Subgroup  A  12
AD12   40  TSS-NGLQEKPPGV        "      B  3, 7, 11, 16, 21
AD8    33  VSS-NGFQNFPPGV        "      C  2, 5
AD9    33  VSS-DGFQNFPPGV        "      D  8, 9, 15
AD15   33  VSS-DGFQNFPPGV        "      E  4
AD41-1 32  ASS-NGLQEKPPGV        "      F  40, 41
AD40-1 32  ASS-NGLQEKPPGV
AD40-2 32  VSS-DGLQENPPGV
AD41-2 32  VSS-DGLQEKPPGV
CANINE 31  NDR-KGLTESSPGT
BOVINE 27  FDR-NGFVENQEAT
MURINE 63  NSY-D-FTSIHGNT
```

**Fig. 1.** Manual alignment of all known adenovirus fiber primary sequences (except for some minor variants which have been excluded). The sequence is arranged into tail, shaft, and head and the shaft region is divided into motifs. The motifs are numbered from 1 to 22 (the last being incomplete) according to the number of motifs found in the HAd2 and HAd5 sequences; the extra motif in HAd12 is denoted *motif 19**. The nonhuman virus fiber shaft sequences between motifs 4 and 20 have not been included in the table as no meaningful alignment can be made (but see Fig. 4 for the canine and murine fiber shafts). *Bold type* indicates highly conserved residues. Sources for the sequences are given in Table 1

```
                  MOTIF 1              MOTIF 2
AD11p   46   LTLKCLTPLTTTGGS- | LQLKVGGGLTVDDTNGF
AD11a   46   LTLKCLTPLTTTGGS- | LQLKVGGGLTVDDTDGT
AD21    46   LTLNCLTPLTTTGGP- | LQLKVGGGLIVDDTDGT
AD7     46   LTLKCLTPLTTTGGS- | LQLKVGGGLTIDDTDGF
AD3     46   LSLKSVNPLTTASGS- | LQLKVGSGLTVDTTDGS
AD16    46   LTLKCVNPLTTASGP- | LQLKVGSSLTVDTIDGS
AD2     47   LSLRVSEPLDTSHGM- | LALKMGSGLTLDKAGN-
AD5     47   LSLRLSEPLVTSNGM- | LALKMGNGLSLDEAGN-
AD4     48   LSLRLADPVTTKNGE- | ITLKLGEGVDLDDSGK-
AD12    53   LALNYKDPIVTENGT- | LTLKLGDGIKLNAQGQ-
AD8     46   LSLKLADPITINNQN- | VSLKVGGGLTLGEETGK
AD9     46   LSLKLADPIAIVNGN- | VSLKVGGGLTLQDGTGK
AD15    46   LSLKLADPIAIANGN- | VSLKMGGGLTLQEGTGN
AD41-1  45   LSLKYTDPLTTKNGA- | LTLKLGTGLNIDENGD-
AD40-1  45   LSLKYTDPLTTKNGA- | LTLKLGTGLNIDKNGD-
AD40-2  45   LALKYTDPITTNAKHE | LTLKLGSNITLQ-NGL-
AD41-2  45   LALKYTDPITTNAKHE | LTLKLGSNITLE-NGL-
CANINE  44   LAVNISPPLTFSNLGA | IKLSTGAGLILKE-GK-
BOVINE  40   LAMLVEKPLTFDKEGA | LTLGVGRGIRINPAGL-
MURINE  75   LSLRLNKPLKRTAKG- | LQLLLGSGLSVNADGQ-

                  MOTIF 3
AD11p   78   LK-ENIS--ATTPLVKTGHS-
AD11a   78   LQ-ENIG--TTTPLVKTGHS-
AD7     78   LK-ENIS--AATPLVKTGHS-
AD21    78   LQ-ENIR--ATAPITKNNHS-
AD3     78   LE-ENIK--VNTPLTKSNHS-
AD16    78   LK-ENIT--AAAPLTKTNHS-
AD2     78   LTSQNVT-TVTQPLKKTKSN-
AD5     78   LTSQNVT-TVSPPLKKTKSN-
AD4     79   LIANTVN-KAIAPLSFSNNT-
AD12    84   LTASNNI-NVLEPLTNTSQG-
AD8     78   LTV-----NTEPPLHLTNNK-
AD9     78   LTV-----NADPPLQLTNNK-
AD15    78   LTV-----NTEPPLQLTNNR-
AD41-1  76   LSSDASV-EVSAPITKTNKI-
AD40-1  76   LSSDASV-EVSAPITKTNKI-
AD40-2  76   LSA-TVP-TVSPPLTNSNNS-
AD41-2  76   LSA-TVP-TVSPPLTNSNNS-
CANINE  75   LEA-----NIGPGLTTNQEGQ
BOVINE  77   LETNDLASAVFPPLASDEAGN
MURINE  107  LESSEGISEADAPLQINDGV-

                  MOTIF 4              MOTIF 5
AD2     97   ISLDTSAPLTITSGA-- | LTVATTAPLIVTSGA
AD5     97   INLEISAPLTVTSEA-- | LTVAAAAPLMVAGNT
AD4     98   ISLNMDTPLYTKDGK-- | LSLQVSPPLSILKSTILN
AD12    103  LKLSWSAPLAVKASA-- | LTLNTRAPLTTTDES
AD8     93   LGIALDAPFDVIDNK-- | ---------------
AD9     93   LGIALDAPFDVIDNK-- | ---------------
AD15    93   IGIALDAPFDVIGGK-- | ---------------
AD41-1  95   VGLNYTKPLALRSNA-- | LTLSYNAPLNVVNNN
AD40-1  95   VGLNYTKPLALQNNA-- | LTLSYNAPFNVVNNN
AD41-2  94   LGLATSAPIAVSANS-- | LTLATAAPLTVSNNQ
AD40-2  94   LGLATSAPIAVSANS-- | LTLATAAPLTVSNNQ

                  MOTIF 6              MOTIF 7
AD2     127  LSVQSQAPLTVQDSK-- | LSIATKGPITVSDGK
AD5     127  LTMQSQAPLTVHDSK-- | LSIATQGPLTVSEGK
AD12    133  LALITAPPITVESSR-- | LGLATIAPLSLDGGGN
AD41-1  125  LALNISQPVTVNANNE- | LSLLIDAPLNADTGT
AD40-1  125  LALNMSQPVTINANNE- | LSLLIDAPLNADTGT
```

| | | MOTIF 8 | MOTIF 9 |
|---|---|---|---|
| AD2 | 157 | LALQTSAPLSGSDSDT- | LTVTASPPLTTATGS |
| AD5 | 157 | LALQTSGPLTTTDSST- | LTITASPPLTTATGS |
| AD12 | 164 | LGLNLSAPLDVSNNN-- | LHLTTETPLVVNSSGA |
| AD41-1 | 156 | LRLQSAAPLGLVDKT-- | LKVLFSSPLYLDNNF |
| AD40-1 | 156 | LRLRSDAPLGLVDKT-- | LKVLFSSPLYLDNNF |

| | | MOTIF 10 | MOTIF 11 |
|---|---|---|---|
| AD2 | 188 | LGINMEDPIYVNNGK-- | IGIKISGPLQVAQNSDT |
| AD5 | 188 | LGIDLKEPIYTQNGK-- | LGLKYGAPLHVTDDLNT |
| AD12 | 195 | LSVATADPISVRNNA-- | LTLPTADPLMVSSDG-- |
| AD41-1 | 186 | LTLAIERPLALSSSRA- | VTLKYSPPLKIENEN-- |
| AD40-1 | 186 | LTLAIERPLALSSNRA- | VALKYSPPLKIENEN-- |

| | | MOTIF 12 | MOTIF 13 |
|---|---|---|---|
| AD2 | 220 | LTVVTGPGVTVEQNS-- | LRTKVAGAIGYDSSNN |
| AD5 | 220 | LTVATGPGVTINNTS-- | LQTKVTGALGFDSQGN |
| AD12 | 225 | LGISVTSPITVINGS-- | LALSTTAPLNSTGST- |
| AD41-1 | 217 | LTLSTGGPFTVSGGN-- | LNLTTSAPLSVQNNS- |
| AD40-1 | 217 | LTLSTGGPFTVSGGN-- | LNLATSAPLSVQNNS- |
| AD40-2 | 124 | LSINTGRGLVITNNA-- | VAVNPTGALGFNNTGA |
| AD41-2 | 124 | LSINAGRGLVITNNA-- | LTVNPTGALGFNNTGA |

| | | MOTIF 14 | MOTIF 15 |
|---|---|---|---|
| AD2 | 251 | MEIKTGGGMRINNNL-- | LILDVDYPFDAQTKLR |
| AD5 | 251 | MQLNVAGGLRIDSQNRR | LILDVSYPFDAQNQLN |
| AD4 | 132 | ----------------- | LALAFGSGLGLSGSA- |
| AD12 | 255 | LSLSVANPLTISQDT-- | LTVSTGNGLQVSGSQ- |
| AD41-1 | 247 | LSLVITSPLKVINSM-- | LAVGVNPPFTITDSG- |
| AD40-1 | 247 | LSL-------------- | ---GVNPPFLITDSG- |
| AD40-2 | 155 | LQLNAAGGMRVDGAN-- | LILHVAYPFEAINQ-- |
| AD41-2 | 155 | LQLNAAGGMRVDGAN-- | LILHVAYPFEAINQ-- |

| | | MOTIF 16 | MOTIF 17 |
|---|---|---|---|
| AD2 | 282 | LKL-GQQGPLYINASHN- | LDINYNRGLYLFNASNNTKKLE |
| AD5 | 284 | LRL-GQQGPLFINSAHN- | LDINYNKGLYLFTASNNSKKLE |
| AD4 | 147 | LAVQLASPLTFDDKGN- | IKITLNRGLHVTTGDAIE |
| AD12 | 285 | LVTRIGDGLTFDNGV-- | MKVNVAGGMRTSGGR------ |
| AD8 | 108 | ----------------- | LTLLAGHGLSIITKETSTLPGLVNT |
| AD9 | 108 | ----------------- | LTLLAGHGLSIITKETSTLPGLRNT |
| AD15 | 108 | ----------------- | LTLLAGHGLSIITEETSPLPGLVNT |
| AD41-1 | 277 | LAMDLGDGLALGGSK-- | LIINLGPGLQMSNGAIT----- |
| AD40-1 | 262 | LAMDLGDGLALGGSK-- | LIINLGPGLQMSNGAIT----- |

| | | MOTIF 18 | MOTIF 19 |
|---|---|---|---|
| AD2 | 319 | VSIKKSSGLNFDNTA-- | IAINAGKGLEFDTNTSESPDINP |
| AD5 | 321 | VNLSTAKGLMFDATA-- | IAINAGDGLEFGSPNA--PNTNP |
| AD4 | 181 | SNISWAKGIKFEDGA-- | IATNIGKGLEFGTSNTETGVNNAYP |
| AD12 | 315 | IILDVNYPFDASNN--- | LSLRRGLGLIYNQSTNWNLT--- |
| AD8 | 133 | ----------------- | LVVLTGKGIGTDLSNNGGNIC |
| AD9 | 133 | ----------------- | LVVLTGKGIGTESTDNGGTVC |
| AD15 | 133 | ----------------- | LVVLTGKGLGTDTTDNGGSIR-- |
| AD41-1 | 309 | LALDAALPLQYRDNQLQ | LRIGSTSGLIMSGVTQTLN---- |
| AD40-1 | 294 | LALDAALPLQYKNNQLQ | LRIGSASALIMSGVTQTLN---- |

```
                 MOTIF 19*
AD12     349  TDISTEKGLMSFGQN--

                 MOTIF 20             MOTIF 21
AD11p    95   IGLPLGAGLGTNENK--  |LCIKLGQGLTFNSNN-
AD11a    95   IGLSLGAGLGTDENK--  |LCTKLGKGLTFNSNN-
AD7      95   IGLSLGPGLGTNENK--  |LCAKLGEGLTFNSNN-
AD21     95   VELSIGNGLETQNNK--  |LCAKLGNGLKFNNGD-
AD3      95   INLPIGNGLQIEQNK--  |LCSKLGNGLTFDSSNS
AD16     95   IGLLIGSGLQTKDDK--  |
AD16     110  LCLSLGDGLVTKNDK--  |
AD16     125  LCLSLGDGLITKNDV--  |LCAKLGEGLTFNSNN-
AD2      357  IKTKIGSGIDYNENGA-  |MITKLGAGLSFDNSGA
AD5      357  LKTKIGHGLEFDSNKA-  |MVPKLGTGLSFDSTGA
AD4      221  -----------------  |IQVKLGSGLSFDSTGA
AD12     364  IALNAGQGLTFNNGQ--  |LRVKLGAGLIFDSNNN
AD8      154  -----------------  |VRVGEGGGLSFNDNGD
AD9      154  -----------------  |VRVGEGGGLSFNNDGD
AD15     154  -----------------  |VRVGEGGGLSFNEAGD
AD41-1   345  VNANTGKGLAVENNS--  |LVVKLGNGLRFDSWGS
AD40-1   330  VNANTSKGLAIENNS--  |LVVKLGNGLRFDSWGS
AD40-2   184  LTLRLENGLEVTNGGK-  |LNVKLGSGLQFDNNGR
AD41-2   184  LTLRLENGLEVTSGGK-  |LNVKLGSGLQFDSNGR
CANINE   322  IAAVIGAGLKFENNA--  |ILAKLGNGLTIRD-GA
BOVINE   759  LEVPLGQGLEIADESQ-  |VRVKLGDGLQFDSQAR
MURINE   354  ----PLNPIEVNQHGQ-  |LGIALGYGFHAHRGY-

                 MOTIF 22?        START OF HEAD

AD11p    125  ICIDDNIN--------TLWTGVNPTTANCQIMNSSE
AD11A    125  ICIDDNIN--------TLWTGINPTTANCQMMDSSE
AD7      125  ICIDDNIN--------TLWTGVNPTTANCQIMASSE
AD21     125  ICIKDSIN--------TLWTGIKP-PPNCQIVENTD
AD3      126  IALKNN----------TLWTGPKP-EANCIIEYGK-
AD16     155  ITIENN----------TLWTGAKP-SANCVIKEGE-
AD2      389  ITIGNKN-----DDKLTLWTTPDP-SPNCRIHS---
AD5      389  ITVGNKN-----NDKLTLWTTPAP-SPNCRLNA---
AD4      237  IMAGNKD-----YDKLTLWTTPDP-SPNCQILA---
AD12     395  IALGSSSNTP--YDPLTLWTTPDP-PPNCSLIQ---
AD8      170  LVAFNKK-----EDKRTLWTTPDT-SPNCRIDQ---
AD9      170  LVAFNKK-----EDKRTLWTTPDT-SPNCKIDQ---
AD15     170  LVAFNKK-----EDMRTLWTTPDP-SPNCKIIE---
AD41-1   376  ITVSP-----TTTTPTTLWTTADP-SPNATFYE---
AD40-1   361  IAVSP-----TTTTPTTLWTTADP-SPNATFYE---
AD40-2   216  ITISNRIQTRSVTSLTTIWSI-SP-TPNCSIYE---
AD41-2   216  IAISNSNRTRSVPSLTTIWSI-SP-TPNCSIYE---
CANINE   352  IEAVAPQP--SFT-PVTLWTGPDP-NVNASINGT--
BOVINE   792  ITT-APNM---VT--ETLWTG-TGSNANVTWRGY--
MURINE   381  LEL---------T-PQTLWTG-LPIGNNGTFHT---
```

```
AD11p   153 SN-DCKLILTLVKTGAL-VTAFVYVIGVSNNFNMLTTH---------
AD11a   153 SN-DCKLILTLVKTGAL-VTAFVYVIGVSNNFNMLTTY---------
AD7     153 SN-DCKLILTLVKTGGL-VTAFVYVIGVSNDFNMLTTH---------
AD21    152 TN-DGLLTLVLVKNGGL-VNGYVSLVGVSDTVNQMF-----------
AD3     150 QNPDSKLTLILVKNGGI-VNGYVTLMGASDYVNTLF-----------
AD16    179 DSPDCKLTLVLVKNGGL-INGYITLMGASEYTNTLF-----------
AD2     416 DN-DCKFTLVLTKCGSQ-VLATVAALAVS-GDLSS-------------
AD5     416 EK-DAKLTLVLTKCGSQ-ILATVSVLAVK-GSLAP-------------
AD4     264 EN-DAKLTLCLTKCDSQ-ILATVSVLVVRSGNLNP------------
AD12    425 EL-DAKLTLCLTKNGSI-VNGIVSLVGVK-GNLLN-------------
AD8     197 DK-DSKLSLVLTKCGSQ-ILANVSLIVVAGRYKIINNNTN-------
AD9     197 DK-DSKLTLVLTKCGSQ-ILANVSLIVVDGKYKIINNNTQ-------
AD15    197 DK-DSKLTLILTKCGSQ-ILGSVSLLVVKGKFSNINNTTN-------
AD41-1  403 SL-DAKVWLVLVKCNGM-VNGTISIKAQKGILLRPTASFISFVMYFY
AD40-1  388 SL-DAKVWLVLVKCNGM-VNGTISIKAQKGTLLKPTASFI-------
AD40-2  247 TQ-DANLFLCLTKNGAH-VLGTITIKGLKGALRE-------------
AD41-2  247 TQ-DANLFLCLTKNGAH-VLGTITIKGLKGALRE-------------
CANINE  382 --PVIRSFISLTRDSNL-VTVNASFTGEGSYQS--------------
BOVINE  819 TAPGSKLFLSLTRFSTGLVLGNMTIDSNASFGQY-------------
MURINE  403 KQ-DCKIFLSLTRLGPM-VHGTFMLQAPQYELTT-------------

AD11p   189 -RNINFTA---ELFFDSTGNLLTRLSSLK-TPL--
AD11a   189 -RNINFTA---ELFFDSAGNLLTSLSSLK-TPL--
AD7     189 -KNINFTA---ELFFDSTGNLLTSLSSLK-TPL--
AD21    186 -TQKSATIQL-RLYFDSSGNLLTDESNLK-IPL--
AD3     185 -KNKNVSINV-ELYFDATGHILPDSSSLK-TDL--
AD16    215 -KNNQVTIDV-NLAFDNTGQIITYLSSLK-SNL--
AD2     448 -MTGTVASVSIFLRFDQNGVLMENS-SLK-KH-YW
AD5     448 -ISGTVQSAHLIIRFDENGVLLNNS-FLD-PE-YW
AD4     297 -ITGTVSSAQVLFRFDANGVLLTEHSTLK-K--YW
AD12    457 -IQSTTTTVGVHLVFDEQGRLITSPTTALVPQASW
AD8     235 -TNPALKGFTIKLLFDKNGVLMESS-NLG-KS-YW
AD9     235 -TQPALKGFTIKLLFDENGVLMESS-NLG-KS-YW
AD15    235 -PNEADKQITVKLLFDANGVLKQGS-TMD-SS-YW
AD41-1  448 YSDGTWRKNY--PVFDNEGILANSA--------TW
AD40-1  433 YSDGTWRKNY--PVFDNEGILANSA--------TW
AD40-2  279 -MNDNA--LSVKLPFDNQGNLLNCA--LE-SS-TW
AD41-2  279 -MHDNA--LSLKLPFDNQGNLLNCA--LE-SS-TW
CANINE  412 -VSPTQSQFSLILEFNQFGQLMSTG-NLN-STTTW
BOVINE  853 -INAGHEQIECFILLDNQGNLKEGS-NLQ-G--TW
MURINE  435 ----NGMREITFS-FNSTGGLEQPA-----PVTYW
MURINE  460 GALDPPPTAKAAEIENQKRVKKRAAPDPPVEPPPK
MURINE  495 RRGDLAVLFAKVAEQAMELAKEQAVQA

AD11p   217 NHKSGQNMATGAITNAKGF--MPSTTAYPFNDNSR
AD11a   217 NHKSGQNMATGAITNAKSF--MPSTTAYPFNNNSR
AD7     217 NHKSGQNMATGALTNAKGF--MPSTTAYPFNVNSR
AD21    216 KNKSS-TATSEVLQPAEAF--MPSTTAYPFNTTTR
AD3     217 DLKYKQ--TAD-FS-ARGF--MPSTTAYPFVLPNAGT
AD16    245 NLKDNQNMATGTITSAKGF--MPSTTAYPFITYATEI
AD2     479 NFRNGNSTNANPYTNAVGF--MPNLLAYPKTQSQT
AD5     479 NFRNGDLTEGTAYTNAVGF--MPNLSAYPKSHGKT
AD4     328 GYKQGDSIDGTPYTNAVGF--MPNSTAYPKTQSST
AD12    491 GYRQGQSVSTNTVTNGLGF--MPNVSAYPRPNASE
AD8     266 NFRNQNSIMSTAYEKAIGF--MPNLVAYPKPTTGSKKY
AD9     266 NFRNENSIMSTAYEKAIGF--MPNLVAYPKPTAGSKKY
AD15    266 NYRSDNSNLSQPYKKAVGF--MPSKTAYPKQTKPTNKEISQ
AD41-1  472 GYRQGQSANTN-VSNAVEF--MPSSKRYPNQKGSE
AD40-1  457 GYRQGQSANTN-VSNAVEF--MPSSKRYPNEKGSE
AD40-2  307 --RYQE-TNAV-ASNALTF--MPNSTVYPRNKTAD
AD41-2  307 --RYQE-TNAV-ASNALTF--MPNSTVYPRNKTAH
CANINE  394 GEKPWGNNTVQVQPSHTWKLCMPNREVYSTPAATL
BOVINE  883 EVKN-----NPSASKA-AF--LPSTALYPILNESR
MURINE  522 QPPEHVNTDWADHMNLLRF--MPNTLVYPTAATIA
```

```
AD11p  250 EKENYIYGTCYYTAS-DRTAFPIDISVMLNRRAINDE---
AD11a  250 EKENYIYGTCHYTAS-DHTAFPIDISVMLNQRAIRAD---
AD7    250 EKENYIYGTCYYTAS-DHTAFPIDISVMLNQRALNNE---
AD21   248 DSENYIHGICYYMTSYDRSLVPLNISIMLNSRTISSN---
AD3    246 HNENYIFGQCYYKAS-DGALFPLEVTVMLNKRLPDS----
AD16   280 LNEDYIYGEVYYKST-NGTLFPLKVTVTLNRRMLAS----
AD2    512 AKNN-IVSQVYLHGD-KTK--PMILTITLNGTSESTETSE
AD5    512 AKSN-IVSQVYLNGD-KTK--PVTLTITLNGTQETGDTT-
AD4    361 TKNN-IVGQVYMNGD-VSK--PMLLTITLNGTDDTT----
AD12   524 AKSQ-MVSSLTYQGD-TSK--PITMKVAFNGITSLN----
AD8    305 ARD-IVYGNIYLGG--KPHQ-PVTIKTTFNQET-------
AD9    305 ARD-IVYGNIYLGG--KPDQ-PVTIKTTFNQET-------
AD15   305 AK-NKIVSNVYLGG--KIDQ-PCVIIISFNEEA-------
AD41-1 504 VQNM-ALTYTFLQGD--PHMA-ISFQSIYNHAL-------
AD40-1 489 VQNM-ALTYTFLQGD--PNMA-ISFQSIYNHAI-------
AD40-2 336 PGNM-LIQIS-------PN---ITFSVVYNEIN-------
AD41-2 336 PGNM-LIQIS-------PN---ITFSVVYNEIN-------
CANINE 479 TSCGLNSIAHDGA----PNR-SIDCMLIINKLRGAA----
BOVINE 910 GSLPGKNLVGMQAIL-GGGG-TCTVIATLNGRRSNNYP--
MURINE 555 ANLQFHDTRLSLRR---------ATLKIRLNGSPD------

AD11p  286 -TSYCIRITWSWNTGDAPEVQTSATTLVTSPFTFYYIREDD  325
AD11a  286 -TSYCIRITWSWNTGDAPEGQTSATTLVTSPFTFYYIREDD  325
AD7    286 -TSYCIRVTWSWNTGVAPEVQTSATTLVTSPFTFYYIREDD  325
AD21   285 -VAYAIQFEWNLNAKESPESNIA--TLTTSPFFFSYIREDDN 323
AD3    281 RTSYVMTFLWSLNAGLAPETTQA--TLITSPFTFSYIREDD  319
AD16   314 GMAYAMNFSWSLNAEEAPETTEV--TLITSPFFFSYIREDD  352
AD2    548 VSTYSMSFTWSWESGKYTTETFAT-----NSYTFSYIAQE-  582
AD5    547 PSAYSMSFSWDWSGHNYINEIFAT-----SSYTFSYIAQE-  581
AD4    393 -SAYSMSFSYTWTNGSYIGATFGA-----NSYTFSYIAQQ-  426
AD12   556 --GYSLTFMWSGL-SNYINQPFST-----PSCSFSYITQE-  587
AD8    334 QCEYSITFDFSWAKTYV-NVEFET-----TSFTFSYIAQE-  362
AD9    334 QCEYSITFDFSWAKTYV-NVEFET-----TSFTFSYIAQE-  362
AD15   334 DSDYSIVFYFKWYKTYE-NVQFDS-----SSFNFSYIAQE-  367
AD41-1 533 -EGYSLKFTWRVRN----NERFDI-----PCCSFSYVTEQ-  562
AD40-1 518 -EGYSLKFTWRVRN----NERFDI-----PCCSFSYVTEQ-  547
AD40-2 358 -SGYAFTFKWSAEP----GKPFHP-----PTAVFCYITEQ-  387
AD41-2 358 -SGYAFTFKWSAEP----GKPFHP-----PTAVFCYITEQ-  387
CANINE 510 --TYTLTFRFLNFNKLSSSTVFKTD-----VLTFTYVGENQ  543
BOVINE 946 -AGQSIIFVWQEFNTIARQPLNHS------TLTFSYWT---  976
MURINE 581 -SAYQLGFMLELVGTQS--------ASIVTDTISFWYYAEDY  613
```

### 3.1.1 N-Terminal Tail

The sequence of the N-terminal tail is strongly conserved and in particular contains a number of constant prolines and glycines (Fig. 1). Of the short, highly conserved peptide motifs, two are worthy of comment. Firstly, all human fiber sequences and the canine fiber contain the sequence 2-**KRλR** (bold type indicates absolutely conserved, λ indicates a small residue, G,A,T,S,V, exceptionally an L in HAd8,–9, and –15). Evidence from the work of HONG and ENGLER (1991) strongly suggests that this is a nuclear localization signal (see Sect. 4.5). The second conserved motif is 11-**FNPVYPYD/E**; in a protein data base search this peptide is only found in adenovirus fibers. CAILLET-BOUDIN (1989) speculated that this hydrophobic motif, which is usually flanked by charged residues, might be complementary to a peptide in the penton base and thus be implicated in the specific recognition between the fiber and penton base. Specifically, it was proposed that it might interact with residues 251–263 in the HAd2 penton base sequence through complementary ionic and hydrophobic interactions (CAILLET-BOUDIN 1989). This peptide of the penton base is in a region which is extremely well conserved even between such distant serotypes as human and avian adenoviruses (NEUMANN et al. 1988; SHEPPARD and TRIST 1992). It would be surprising, however, if fiber–penton base recognition depends only upon a quasi-linear relationship between two peptides, as proposed in this work.

Some exceptional features exist in the N-terminal tail sequence of the murine adenovirus fiber, which is 75 residues long, compared to the normal size of about 46 residues. This is due to there being two versions of the **FNPVYPYD/E** motif separated by an insert of 26 residues (Fig. 1). Furthermore, there is no putative nuclear localization signal in the N-terminal tail, although the sequence 477-KRVKKR, reminiscent of the normal signal, is found in a large insertion in the head domain.

### 3.1.2 Central Shaft

The major insight of the paper by GREEN et al. (1993) was the observation that the central region of the HAd2 fiber sequence contains a pseudorepeat of 15 residues (Fig. 2a). These repeats are characterised by conserved positions of hydrophobic residues and, in particular, a conserved proline or glycine (exceptionally an alanine, e.g., in HAd2 motif 13) at the eighth position of the motif. Consensus sequences for the proline and glycine repeats are shown in Fig. 3.

The positions marked $\phi$ in Fig. 3 are invariably hydrophobic (i.e., L, I, V, M, F, occasionally A, T, Y). In particular, position 1 is almost exclusively L,I,V, or M with a very strong preference of L. The same is true for position 9, although an F sometimes follows a P in position 8. Position 11 is the most variable hydrophobic position. GREEN et al. (1983) further proposed that the secondary structure corresponding to these repeats would comprise two short β-strands (Sa:1–3 and Sb:9–11, see Fig. 2b) separated by a five-residue turn containing the conserved proline or glycine (Ta:4–8) and a four-residue turn (Tb:12–15, Fig. 2b). The putative turns have many more hydrophilic residues. How this may be arranged in a three-dimensional structure to form the fiber shaft will be discussed below (Sect. 3.6). Whereas the five-residue loop Ta is almost invariable in length, the four-residue loop is often extended, particularly at the distal end of the shaft, e.g., by up to seven residues in motif 17 of HAd5 (Fig. 2a). A perhaps more significant deviation from the consensus pattern is observed for the third repeat from the tail (Fig. 1). All known sequences (except those of subgroups D viruses HAd8, HAd9, and HAd15, which have a regular third repeat) lack the normal hydrophobic residue at position 3, have extra residues before the conserved proline, and often have a lysine in the normal hydrophobic position 11. It seems unlikely that this segment can have the same structure as the normal repeats, and indeed the position of repeat 3 appears to correlate with that of a frequently occurring bend in several of the fibers observed by electron microscopy (see Sect. 3.4). The sequence of the last complete repeat before the head (motif 21 in HAd2) is noteworthy for

| | Sa | Ta | Sb | Tb |
|---|---|---|---|---|
| 47 | L S L | R L S E P | L V T | S N G M |
| 62 | L A L | K M G N G | L S L | D E A G N |
| 78 | L T S | Q N V | | |
| | T | T V S P P | L K K | T K S N |
| 97 | I N L | E I S A P | L T V | T S E A |
| 112 | L T V | A A A A P | L M V | A G N T |
| 127 | L T M | Q S Q A P | L T V | H D S K |
| 142 | L S I | A T Q G P | L T V | S E G K |
| 157 | L A L | Q T S G P | L T T | T D S S T |
| 173 | L T I | T A S P P | L T T | A T G S |
| 188 | L G I | D L K E P | I Y T | Q N G K |
| 203 | L G L | K Y G A P | L H V | T D D L N T |
| 220 | L T V | A T G P G | V T I | N N T S |
| 235 | L Q T | K V T G A | L G F | D S Q G N |
| 251 | M Q L | N V A G G | L R I | D S Q N R R |
| 268 | L I L | D V S Y P | F D A | Q N Q L N |
| 284 | L R L | G Q G P | L F I | N S A H N |
| 299 | L D I | N Y N K G | L Y L | F T A S N N S K K L E |
| 321 | V N L | S T A K G | L M F | D A T A |
| 336 | I A I | N A G D G | L E F | G S P N A P N T N P |
| 357 | L K T | K I G H G | L E F | D S N K A |
| 373 | M V P | K L G T G | L S F | D S T G A |
| a 389 | I T V | G N | | |

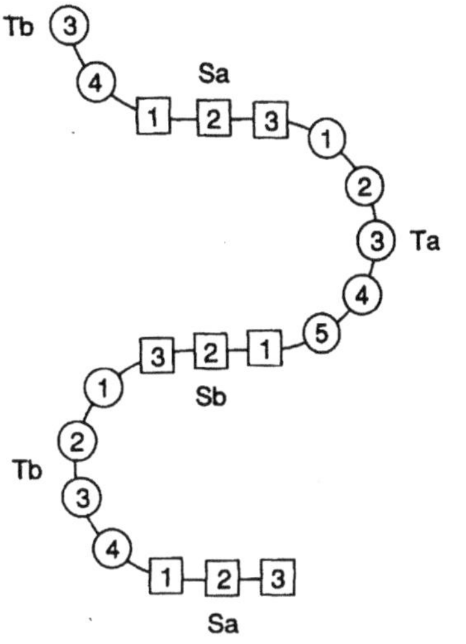


b

**Fig. 2. a** Sequence of the shaft of HAd5 fiber arranged to show the 21.5 approximately 15-residue repeats. The conserved hydrophobic residues in positions 1, 3, 9, and 11 are in *bold type*. Note also the conserved proline or glycine (exceptionally an alanine) in position 8. The *boxed* region was that used in modeling the fiber shaft by STOUTEN et al. (1992; see Sect. 3.6). **b** Predicted secondary structure of the 15-residue repeat (GREEN et al. 1983), *Sa* and *Sb* are two short β-strands, *Ta* is a five-residue turn, and *Tb* is a four-residue turn. In the cross-β model of the shaft, there is intrachain hydrogen bonding between consecutive β-strand segments (GREEN et al. 1983), but this does not give rise to a long enough shaft. In a more recent model (STOUTEN et al. 1992), a triple-helical structure is formed by interstrand hydrogen bonding, as shown in Fig. 7

ϕ-X-ϕ-X-X-X-X-P-ϕ-X-ϕ-X-X-X-X

1 2 3 4 5 6 7 8 9101112131415

Sa Ta Sb Tb

ϕ-X-ϕ-X-X-G-X-G-ϕ-X-ϕ-X-X-X-X

1 2 3 4 5 6 7 8 9101112131415

**Fig. 3.** Consensus sequences for the proline and glycine repeats in the central shaft

its strong conservation (except again for subgroup D fibers). It contains the consensus pattern **KLGXGLXFD/N** (where X is any residue), which if used in a protein data base search is only found in adenovirus fibers.

The exact number of repeats in the shaft depends somewhat on the definition of where the head starts. What is designated motif 22? in Fig. 1 appears to be an imcomplete motif, although the first putative β-strand (ϕ-X-ϕ) is conserved. Thus for example for HAd2, there are 43 putative β-strands corresponding

**Table 1.** Characteristics of adenovirus fibres of known primary sequence

| Serotype | Subgroup (human) | Repeats in shaft (*n*) | Amino acids in fiber (*n*) | Data base accession no. | Reference |
|---|---|---|---|---|---|
| HAd12 | A | 22.5 | 587 | X73487 (E) | Sprengel et al. 1993 |
| HAd3 | B | 5.5 | 319 | P04501 (S) | Signäs et. al. 1985 |
| HAd16 | B | 7.5 | 352 | – | Shieh and Tibbetts 1993<br>Shieh 1992 |
| HAd 7 | B | 5.5 | 325 | P15141 (S) | Hong et al. 1988 |
| HAd21 | B | 5.5 | 323 | – | Shieh and Tibbetts 1993<br>Shieh 1992 |
| HAd11a | B | 5.5 | 325 | LO8232 (G) | Mei et al. 1993 |
| HAd11p | B | 5.5 | 325 | – | Mei et al. 1993 |
| HAd2 | C | 21.5 | 582 | P03275 (S) | Herissé et al. 1981 |
| HAd5 | C | 21.5 | 581 | P11818 (S) | Chroboczek and Jacrot 1987 |
| HAd8 | D | 7.5 | 362 | X74660 (E) | Pring-Akerblom 1993 |
| HAd9 | D | 7.5 | 362 | X74659 (E) | Pring-Akerblom 1993 |
| HAd15H9 | D | 7.5 | 362 | X74658 (E) | Pring-Akerblom 1993 |
| HAd15HX | D | 7.5 | 367 | X74669 (E) | Eiz et al. 1993 |
| HAd4 | E | 11.5 | 426 | L19194 (G) | Gruber et al. 1993 |
| HAd40 short | F | 11.5 | 387 | – | Kidd et al. 1993 |
| HAd41 short | F | 11.5 | 387 | – | Pieniazek et al. 1990 |
| HAd40 long | F | 20.5 | 547 | P18047 (S) | Kidd and Erasmus 1989 |
| HAd41 long | F | 21.5 | 562 | P14267 (S) | Pieniazek et al. 1989 |
| Canine CAd1 | – | 18.5? | 543 | P22230 (S) | Dragulev et al. 1991 |
| Bovine BAd3 | – | 46.5 | 976 | D16839 (D) | Mittal et al. 1992, 1993 |
| Murine MAd1 | – | 17.5? | 613 | P19721 (S) | Raviprakash et al. 1989 |

S, Swissport; E, EMBL; G, Genbank; D, DDBJ.

to 21.5 repeats (including the anamolous motif 3). Using this method of accounting, Table 1 gives the number of motifs found in the known fiber sequences: it ranges from 5.5 in subgroup B and D fibers to an astonishing 46.5 in the bovine adenovirus fiber. As shown below, the number of motifs corresponds very well with the length of the fiber shaft as observed by electron microscopy. The length increment per motif was first determined to be 13.2 Å by comparing the lengths of HAd2 and HAd3 fibers (Ruigrok et al. 1990), and similar values have been found with other fibers (Table 2, see below). The bovine fiber shaft is exceptionally long with 46 complete repeats, of which 80% are glycine repeats (compared to 40% in HAd2).

The canine and murine adenovirus fiber shafts, on the other hand, contain a slightly modified organization. Large sections of these fiber shafts appear to be made up of alternating 20-residue proline segments and 16-residue glycine segments (Fig. 4). The canine sequence, for example, contains six of these 36-residue long "double" repeats. The most significant aspect of these double repeats is the occurrence in the canine sequence of a long charged residue (boxed in Fig. 4d) three residues after the conserved proline, normally a hydrophobic position. It could be that the preceeding triplet $\phi$ TY/F constitutes the

```
MURINE  75  L S L - - - - - R L N K P L K R T A K G - - L Q L L L G S G L S V N A D G Q -
MAd1    107 L E S S E G I S E A D A P L Q I N D G V - - L Q L S F G E G L S V N D H G E
        143 L E S K G K V E A V T L P L A L Q D H V - - M S L S F G Q G L Q V N D Q G Q
        179 L E A L A M V H S T S A P L K V T N N N - - L E L A L G R G L I V D D Q G Q
        215 L R L A P N L L W P E S P L A I E Q G T N H L I L F Y N Q S L D V E D G K
        252 L T L - - - - P E P F D P L T L D G G R - -
        268 L R M - - - Q L A P N S G L A V T E K G S - L G I N W G E G I Q V K E Q K
        301 I T L - - - K V T P A N G L A V T E Q G G - L N I N W G N G I K V D E Q K V T
        336 L K T S N E F A L T E N G L Y L T S P L N P
        358 I E V - - - N Q H G Q L G I A L G Y G F

CANINE  44  L A V - - - - - - - - - N I S P L T F S N L G A I K L S T G A G L I L K E G K
CAd1    74                                              L E A N I G P G L T T N Q E G Q
        90  I T V E K D S D G L T F T S P L H K I E N T - V S L S I G E G L E D E S G T
        127 L K V - N P P P P L L F S P P L A E A G G T - V S L P L Q E S M Q V T E G K
        163 L G V - - - - P - T T Y S P P L Q K T D Q Q - V S L R V G P G L T V L N G Q
        195 L Q A - V Q P P A T T Y K E P L L E T E N S - V S L K V G A G L A V Q D G A
        231 L V A - T - P P N V T F S A P L E K N G N A - V S V R V G A G L S I Q G N A
        266 L V A - T T S P T L T F A Y P L I K N N N H - I T L S A G S G L R V S G G S
        302                                             L T V A T G P G L S H I N G T
        317                                             I A A V I G A G L K F E N N A
        332                                             I L A K L G N G L T I R D G A
```

**Fig. 4.** Arrangement of the sequences of the murine (Raviprakash et al. 1989) and canine (Dragulev et al. 1991) fibers in the shaft region showing regions of proline–glycine double repeats. *Boxed* residues indicate significant deviations of amino acid type from those found in the standard shaft motif (see Sect. 3.1.2)

β-strand region. Interestingly, the canine shaft sequence ends with four consecutive "normal" glycine repeats, whereas the murine sequence has a pair of double glycine repeats.

### 3.1.3 C-Terminal Head

The head is a C-terminal domain of about 180–225 residues which we will assume (arbitarily) begins at the very well conserved 400-TLWT motif. There are a relatively small number of segments containing highly conserved residues which permit all the different head sequences to be aligned (Fig. 1). However, between these segments there is considerable variability, although members of the same subgroup show more similarity. Variability in the head is expected due to antigenic variation between serotypes and due to the fact that receptors (and hence the receptor-binding site on the head) may vary. A three-dimensional atomic structure would be a great step forward in the mapping of the functional regions of the head.

## 3.2 Evolution, Gene Duplication

The evolutionary relationship between the different human fiber proteins has not been examined systemically, although interesting observations have been made concerning the relationship between HAd2 and HAd5 (Chroboczek and Jacrot 1987) and also on the origin of HAd4 (Gruber et al. 1993). Two points will be commented upon here: first, the occurrence of gene duplication in the evolution of the shaft, and second, the relatedness of the fiber sequences as revealed by multiple sequence alignments.

Green et al. (1983), in their original analysis of the HAd2 fiber sequence, proposed that gene duplication was the most likely mechanism for the generation of the shaft repeats, although they remarked that the underlying periodicities are much weaker in the DNA sequence than in the amino acid sequence. They proposed that there were two ancestral segments with either a proline or a glycine in the conserved position 8 and that these multiplied and evolved to form the other repeats. Now that many other fiber sequences are available (Fig. 1), we can see that the first two motifs of all the fiber shafts are a proline repeat and a glycine repeat, respectively (both rather well conserved), and the third repeat is generally an anomalous proline repeat. Furthermore, the last two complete repeats (20 and 21 in Fig. 1) of all known fibers (except the murine) are both glycine segments, motif 21 being particularly well conserved. These five repeats would thus appear to constitute the minimal fiber shaft as indeed found in the short fibers of subgroups B and E. The longer fibers could have evolved by insertion after motif 3 of a variable number of other repeats either by multiplication of proline segments, glycine segments, or, especially in the case of the murine and canine fibers, double proline–glycine segments. An interesting point is that the long HAd40 fiber shaft differs from that of HAd41 by the absence of one 15-residue motif which can be unambiguously identified as motif 14 (Fig. 1), since these two fibers are otherwise about 95% identical.

The multiple alignment given in Fig. 1 can be presented as a rootless tree diagram showing the relatedness of the different fibers based on amino acid sequence identity (Higgins et al. 1992). Figure 5a shows such a diagram for the last two shaft motifs and head, while the diagram in Fig. 5b represents the tail and first two motifs. These diagrams show that the subgroup classification of the human adenoviruses based on antigenic cross-reactivity correlates well with the relatedness of fiber sequences. Members of each subgroup are clustered together (60%–100% identical), whereas members of distinct subgroups are generally only about 25%–40% identical. An interesting exception is HAd4, the head of which is 67% identical to that of HAd5, these fibers belonging to subgroups E and C, respectively. This similarity has already been pointed out (Gruber et al. 1993) and correlates with the fact that HAd4 fiber shows immunological cross-reactivity with fibers of subgroups C and D (Wadell and Norrby 1969). Interestingly, the tail of HAd4 is distinct from other subgroups (compare Fig. 5a and b), suggesting two different origins for the two ends of the fiber. One remarkable observation is that the short and long fibers of HAd40 and HAd41 are almost as distinct as if they were from different subgroups (42% identity). This strongly suggests that one or the other of the two fibers has been acquired from an external genome rather than evolved internally by gene duplication (Kidd et al. 1993). The three nonhuman viruses each form a distinct group.

## 3.3 Adenovirus Fiber Is a Trimer

The adenovirus fiber is a trimer, as shown by sedimentation analysis, stoichiometric analysis of radioactively labeled virus, electron microscopy image reconstruction, and cross-linking experiments on human serotypes 2, 3, and 5 (Sundquist et al. 1973; van Oostrum and Burnet 1985; van Oostrum et al. 1987; Chatellard and Chroboczek 1989; Ruigrok et al. 1990; Albiges-Rizo et al. 1991). Analysis of HAd2 and HAd5 fiber crystals also suggested a trimer (Devaux et al. 1990), and trimeric views of the fiber head can sometimes be observed in electron micrographs of negatively stained fibers (see Fig. 6, HAd2 and HAd3). The primary sequence similarity between all known fibers permits us to assume that all fibers are trimeric.

## 3.4 Morphology of Fiber and Penton

### 3.4.1 Human Adenoviruses

The organization of the fiber sequences in tail, shaft, and head can also be observed in electron microscopy images of negatively stained fiber: a box-like head of about 50 x 50 Å on top of a thin long shaft 28 Å wide and sometimes clearly of a beaded character (Fig. 6, HAd2, top; Ruigrok et al. 1990). Beads were also found to extend from the penton base in the image reconstruction of adenovirus by Stewart et al. (1991). In electron microscopy, the conformation of the N-terminal tail is variable. When fully stretched out, the HAd2 fiber is 373 Å long (see Table 2), but often the tail is folded back along the shaft or bent at a 90° angle perpendicular to the carbon support film, so that it does not contribute to the length but produces a stain-excluding (white) spot. These short fibers were found to be 307 Å long for HAd2. In some cases the stretched-out long fibers showed a kink at 60 Å from their N-terminal end (Ruigrok et al. 1990). This kink can also be observed in intact pentons (Fig. 6, HAd2, bottom) where the fiber comes out of the penton base straight but is bent just above it, which is probably the reason why only the first two beads of the fiber were observed in the image reconstruction of virus from cryoelectron microscopy (Stewart et al. 1991).

Electron microscopy of negatively stained intact penton shows two morphologies. The penton can lie with the base on its side (base looking like a bucket), which we call the *side view*, or with the base standing upward (base appearing round or pentagonal), which we call the *end-on-view* (see Fig. 6). In this end-on

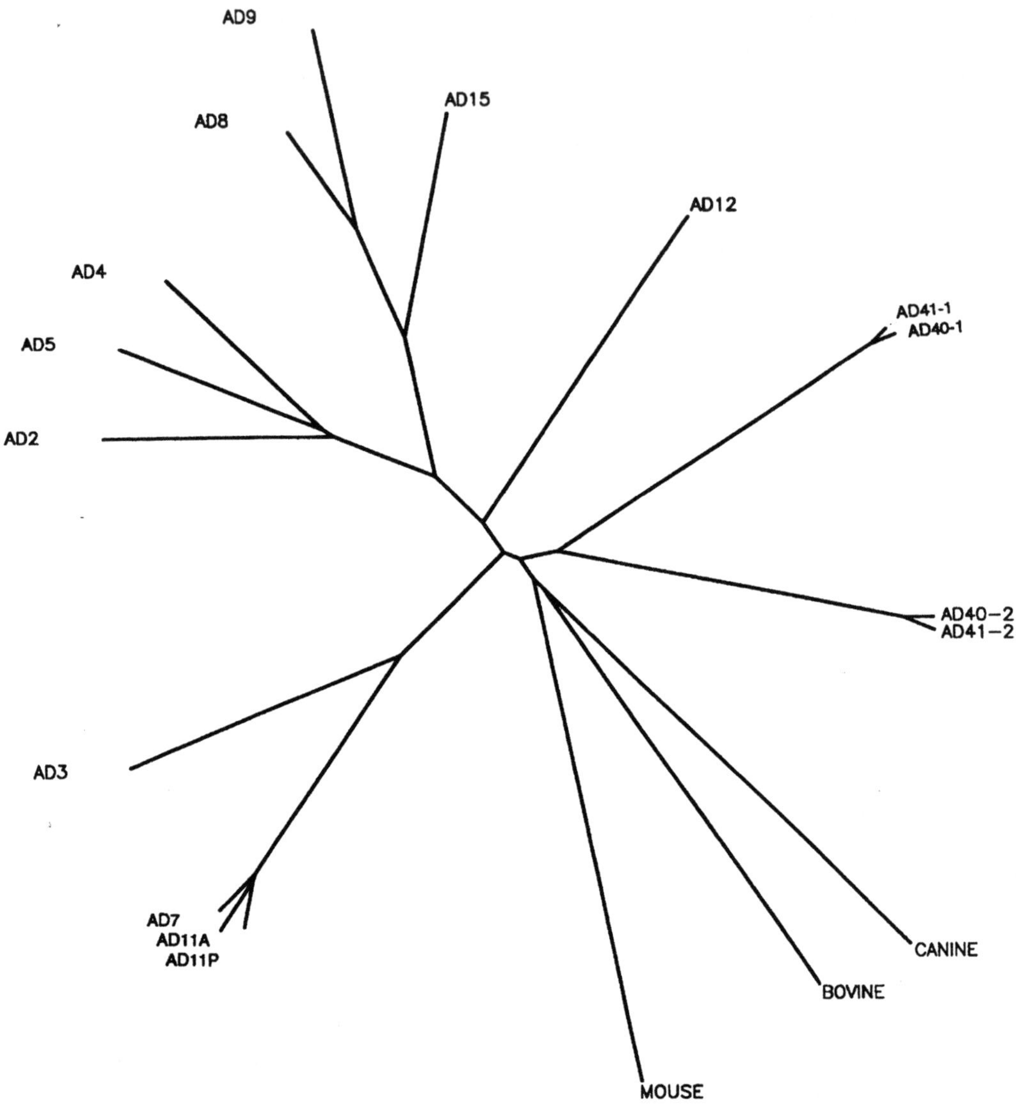


**Fig. 5a, b.** Rootless tree diagrams showing the relatedness of the known adenovirus fiber sequences derived from the alignment given in Fig. 1 using the program CLUSTALV (Higgins et al. 1992). **a** Head domain only (including the last three shaft motifs). **b** Tail domain only (including the first three shaft motifs). See Sect. 3.2

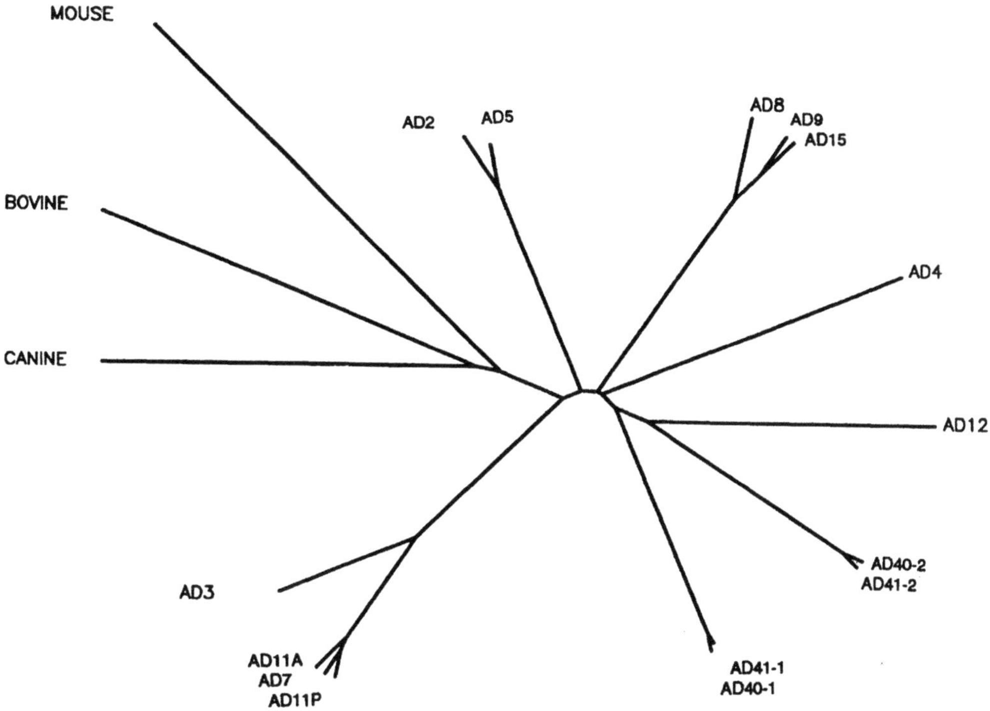


**Fig. 5b**

orientation of the penton, the fiber needs to bend in order to be able to lie flat on the carbon film. The total length of the penton in the side view is longer than that of the penton in the end-on view, the average difference being about 40 Å (Table 2).

The fiber of HAd3 is much shorter than that of types 2 and 5, only 160 Å long, correlating with the presence of only five and a half 15-residue repeats in the shaft (SIGNÄS et al. 1985). These fibers also show a bend at the same position as the HAd2 and HAd5 fibers, and two populations were observed with a length difference of 60 Å (ALBIGES-RIZO et al. 1991). Apart from the shaft length, the morphology of fiber and penton and the two views of the penton are similar to those of the subgroups C viruses. Because of its small length, the free HAd3 fiber often associates with the head onto the carbon film, the shaft standing away from the support film, showing clear trimeric views of the head (Fig. 6).

The viruses of subgroup F (HAd40 and -41) are so far unique among the human viruses in that they contain two fiber genes (PIENIAZEK et al. 1989; KIDD et al. 1993) which are both expressed and both fiber proteins are incorporated into the virion (KIDD et al. 1993, see Fig. 6). The lengths of the HAd40 fibers (derived from the lengths of the pentons in the side view orientation) are 372 and 259 Å, again related to the number of repeats in the shaft; 20.5 and 11.5 repeats, respectively. As far as

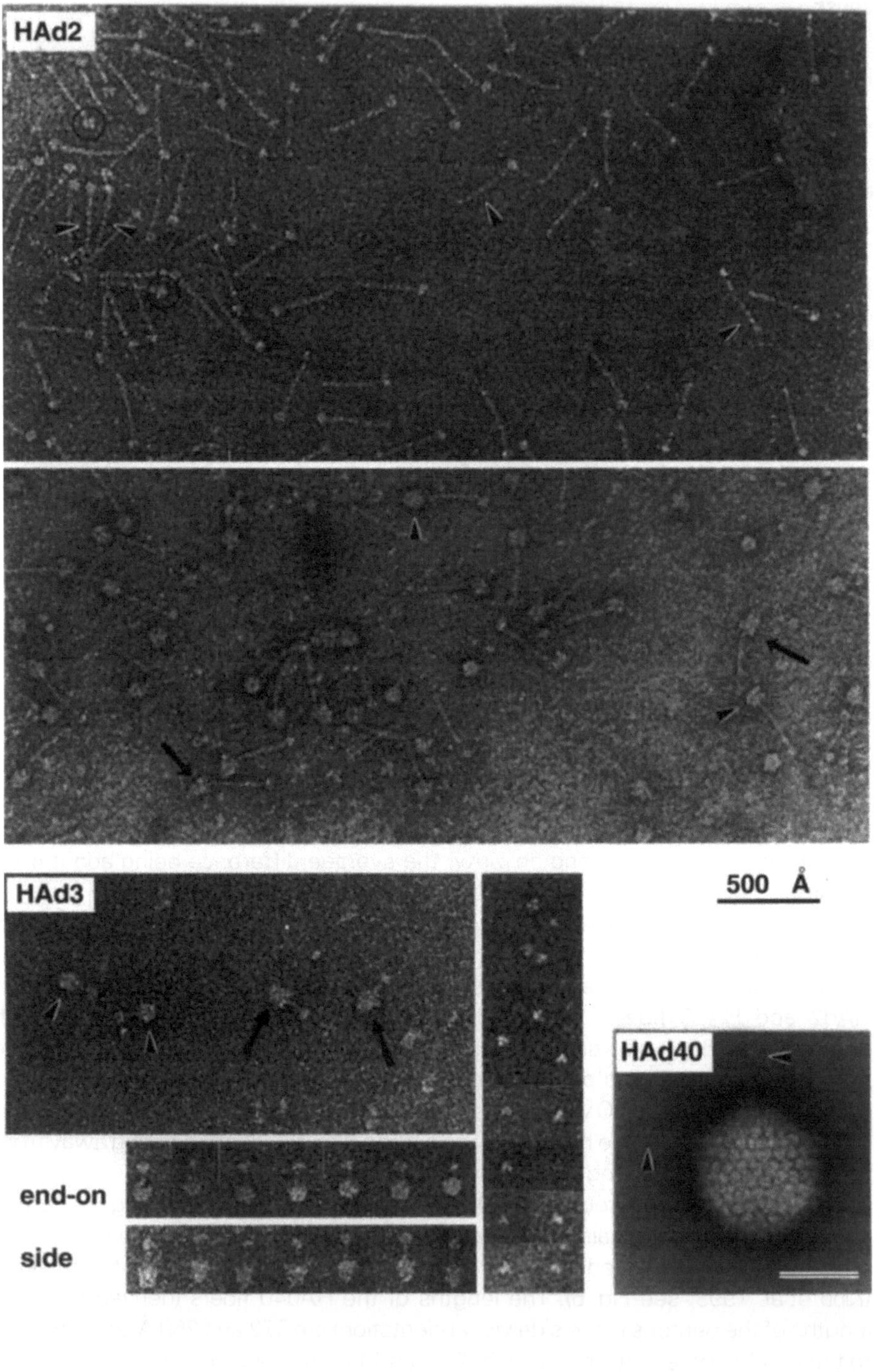
HAd2
HAd3
500 Å
HAd40
end-on
side

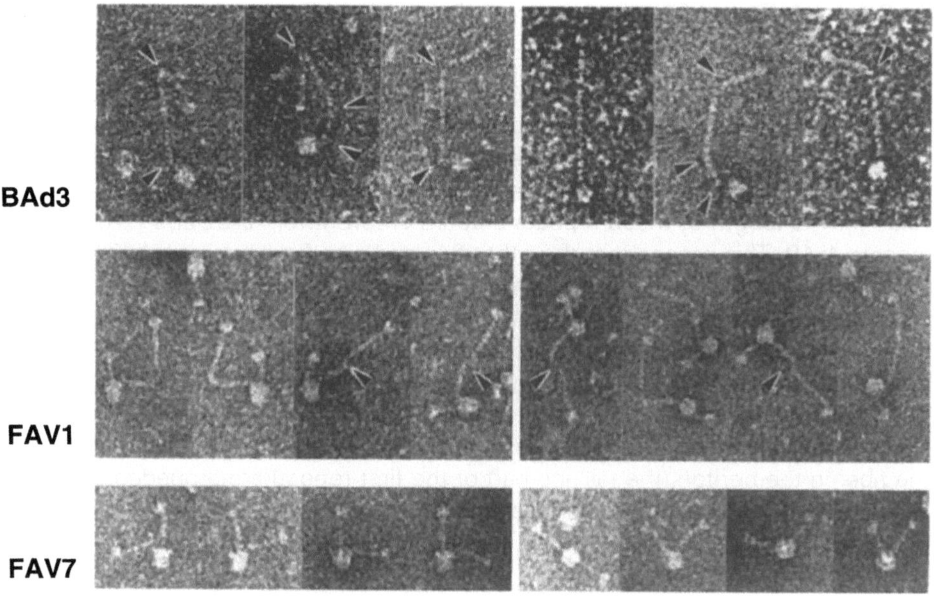


**Fig. 6.** Electron micrographs of negatively stained fibers and pentons of various adenovirus subtypes. Human adenovirus type 2 (*HAd2*): *Top picture*, purified fiber isolated from infected cells. Some trimeric views of the head are *encircled*. Some shafts show a beaded structure, indicated with the *arrowheads*. The fibers in this picture are mainly of the short form 307 Å in length, with the N-terminal tail rearranged into a *white spot* at the end of the shaft opposite the head. *Bottom picture*, purified penton. Two views of the penton can be seen: one with the penton base lying on its side (side view), with some clear examples indicated by *arrows*; another with the base lying end-on, showing a rounder or pentagonal outline, indicated by *arrowheads*. Human adenovirus type 3 (*HAd3*): *Main field*, purified pentons in side view (*arrows*) and end-on view (*arrowheads*.) *Lower galleries* show pentons in side and end-on views as indicated, and the side gallery shows isolated fiber molecules that are stuck to the carbon support film by their heads, showing clearly the trimeric structure of the fiber. Human adenovirus type 40 (*HAd40*): Virus particle showing five fibers, two long ones indicated with *arrowheads* and three short ones. Bovine adenovirus type 3 (*BAd3*): Pentons observed in purified virus preparation. The *arrowheads* indicate where the fiber is bent. Fowl adenovirus types 1 and 7 (*FAV1* and *FAV7*): Pentons observed in purified virus preparation. The *arrowheads* indicate zig-zags in the long FAV1 fiber, probably indicating the presence of two bends. Note that in the side views one fiber always comes straight out of the base (the short one for FAV1) and the other always at an angle of about 90°. Sample were adsorbed onto carbon film and negatively stained with 1% sodium silicotungstate, as described by RUIGROK et al. (1990). The magnification was calibrated with negatively stained crystals of catalase. All figures have the same magnification, indicated by the *bar* under HAd2, apart from the HAd40 virion, which has its own *bar* representing 500 Å. All images by A. BARGE and R. RUIGROK

we can tell by electron microscopy, there does not seem to be a specific number of long or short fibers on each virion. This is consistent with the fact that there is only one penton base polypeptide encoded in the genome of HAd40 and that the N-terminal tails of both fibers are very similar; thus there should be no preference for the incorporation of either long or short pentons in the capsid structure.

### 3.4.2 Correlation of Morphology with Amino Acid Sequence

The sequences of the N-terminal tail and the C-terminal head parts of the HAd2 and HAd3 fiber polypeptides are similar in length. The main difference between the two lies in the number of 15-residue repeats in the fiber shaft. By comparing the length of both fibers we were able to calculate the length per 15-residue repeat, i.e., 13.2 Å per repeat (RUIGROK et al. 1990). However, note that this is the value for the length of those repeats that are different for HAd2 and HAd3, i.e., repeats 4–19. This means that the other five and a half repeats of HAd2 have a total length of 79 Å, corresponding to 14.3 Å per repeat. The values in Table 2 were calculated from the total shaft length divided by the number of repeats (Table 1) without taking into account this difference between the middle 13.2-Å repeats of HAd2 and the external 14.3-Å repeats.

Knowing the length of the shaft, we were able to calculate the length of the N-terminal tail in free fibers, 34 Å, and to deduce that the first 49 amino acids of the fiber in the penton (the tail and part of the first repeat) are embedded inside or associated with the base. These residues are the most conserved in the

**Table 2.** Length measurements from electron micrographs of adenovirus fiber and penton from various strains

| Strain | | Fiber Length (Å) | Penton Length (Å) | | Shaft Length (Å) per repeat[c] |
|---|---|---|---|---|---|
| | | | End-on-view[a] | Side view[b] | |
| HAd2 | | 373±7 (111) | 400±9 (67) | 431±6 (35) | 13.5 |
| HAd5 | | 372±12 (68) | – | – | 13.4 |
| HAd3 | | 160±14 (26) | 181±8 (42) | 227±11 (72) | 14 |
| HAd40 | Long | – | 390±13 (31) | 432±9 (43) | 13.9 |
| | Short | – | 263±18 (47) | 319±11 (20) | 14.3 |
| BAd3 | | – | 795±24 (60) | 835±32 (22) | 14.8 |
| FAV7 | | – | 286±12 (78) | 331±4 (8) | N.K. |
| FAV1 (CELO) | | | | | |
| Long | | – | 583±16 (57) | – | N.K. |
| Short | | – | 232±9 (91) | 268±6 (34) | N.K. |

Whereas FAV1 clearly has two fibers with different lengths, the two fibers of FAV7 are too similar in length to distinguish in end-on views of the penton. For the FAV7 penton in the side view, only the fiber coming out straight was measured. The data are presented as average plus or minus the standard deviation of the population, the number of measurements being indicated in parenthesis. The fiber sequences of FAV1 and 7 have not been determined yet. Data for HAd2 and 3 from Ruigrok et al. (1990), HAd40 from Kidd et al. (1993), BAd3 from Ruigrok et al. (1994), HAd5, FAV1 and FAV7 from R. Ruigrok (unpublished results).

HAd, human adenovirus; BAd, bovine adenovirus; FAV, avian adenovirus; CELO, chick embryo lethal orphan virus; N.K., not known; –, not determined.

[a] Measured from the far side of the base, lying end-on to the tip of the fiber head.

[b] Measured from the bottom of the base to the tip of the fiber head.

[c] The shaft length can be estimated from the fiber length by substracting 34 Å for the amino terminus and 49 Å for the head (RUIGROK et al. 1990). The fiber length can be derived from the penton length by substracting 27 Å from the penton in the end-on view and 58 Å from the penton in the side view (RUIGROK et al. 1990). Where the fiber lengths derived from end-on and side-view pentons was not the same, the average was taken in order to calculate the shaft length. The number of 15-residue repeats in the shaft is given in Fig. 1.

N-terminal half of the fiber (Fig. 2) because they are probably required for specific protein–protein contacts and/or are protected from antigenic pressure. This number of 49 amino acids is an extension of the minimum of 30 protected residues proposed by WEBER et al. (1989) based on the reactivity of an anti-tail antibody with fiber in the presence and absence of penton base.

Using the length per repeat and the length of the tail, the bend in the shaft can be correlated with the third repeat in the shaft. This abberant repeat is conserved in most fiber sequences known so far (Fig. 1; Sect. 3.1.2), with the exception of the subgroup D viruses, which probably means that the bend has some biological purpose. We propose that bending of the fiber is essential to permit interaction of the penton base RGD sequence with the host cell integrins necessary for virus internalization (WICKHAM et al. 1993).

### 3.4.3 Bovine Adenovirus Type 3

The fiber of bovine adenovirus type 3 is very long. The sequence shows the presence of 46.5 repeats in the shaft (MITTAL et al. 1992, 1993), which agrees with the length of the fiber observed in intact pentons (Table 2). The length of the pentons would corresponds to a fiber length of 772 Å and a shaft length per repeat of 14.8 Å. The fiber can be straight, but often shows bends at more or less specific sites (indicated with arrowheads in Fig. 6, BAd3); they appear at 685, 526, and 249 Å from the tip of the head (RUIGROK et al. 1994). The bends possibly correlate with repeats 3–4, 14–15, and 33–34, respectively, which are among the most irregular repeats in the otherwise very regular shaft.

### 3.4.4 Avian Adenoviruses FAV1 and FAV7

Avian adenoviruses have two fibers per penton (LAVER et al. 1971; GELDERBLOM and MAICHLE-LAUPPE 1982), apart from the EDS (egg drop syndrome) adenovirus which also differs considerably in other aspects from the other avian viruses (BRUGH et al. 1984). Avian adenovirus (FAV) serotypes 2–11 all have two fibers of similar length, ranging from 160 to 280 Å, only FAV1 (or CELO, chick embryo lethal orphan virus) has two fibers with different lengths (GELDERBLOM and MAICHLE-LAUPPE 1982). Figure 6 shows pentons of FAV1 and FAV7 in side view and end-on view. The penton lengths are given in Table 2. We have assumed that the penton lengths for the two FAV7 fibers (seen on pentons in the end-on-view) are identical. When using the same calculations as for the human adenoviruses, we can derive fiber lengths of 207 Å and 556 Å for the short and long fibers of FAV1, corresponding to predicted numbers of shaft repeats of 9 and 33, respectively, and 266 Å, corresponding to 13–14 repeats for the FAV7 fiber. The long fiber of FAV1 is bent in two places lying close together, resulting either in a rather shallow curve or in a zig-zag (arrowheads in Fig. 6; FAV1).

For the human viruses, the symmetry mismatch between the trimeric fiber and the pentameric base into which it fits presents an intriguing problem (see

Fig. 13 in VAN OOSTRUM et al. 1987). A possible solution to this problem has been proposed by VAN OOSTRUM and BURNETT (1985), who suggested that there was a structural difference between those penton base monomers which interact with fiber monomers and those which do not. Both fibers in the penton of the avian viruses seem to have the same morphology as the HAd fibers, i.e., beaded character of the shaft, box-like head, and putative trimeric views of the heads of both the long and the short FAV1 fibers, suggesting that they are also both trimeric. Assuming this, it is unlikely that the interaction of both trimeric fibers with the base is the same as described above for HAd2. Looking at the FAV1 pentons in the side view (Fig. 6), it can be seen that the short fiber always comes out of the base in a straight fashion, whereas the long fiber always comes out parallel to the top surface of the base at a 90° angle to the short fiber. The same arrangement can be seen in the side-view images of FAV7. Close inspection of Figs. 2 and 3 in GELDERBLOM and MAICHLE-LAUPPE (1982) also shows side view-oriented pentons with this arrangement of the fibers.This could mean that the two fibers in the FAV penton are not associated in an equivalent manner. Maybe only one avian fiber is associated directly with the penton base, whereas the other could be attached to the first fiber instead. Sequence analysis of the N-terminal parts of the avian fibers is required to help to elucidate this unique structure.

### 3.4.5 Fiber Length as a Characteristic for Subgroup

The length of the fiber seems to be specific for the virus subgroup. Subgroup A was reported to have fiber lengths of 280–310 Å, subgroup B (HAd3 among others) 90–110 Å, subgroup C (HAd2 and HAd5) 230–310 Å, subgroup D 120–130 Å, subgroup E 170 Å, and subgroup F(HAd40 and HAd41) 280–330 Å (PETTERSSON and WADELL 1985). For the cases in which we have determined the fiber lengths (Table 2), we have obtained substantially higher values than published before; this is also true for the fiber lengths of the avian viruses (compare the values in Table 2 with those in GELDERBLOM and MAICHLE-LAUPPE 1982). There may be a number of reasons for this. First, in many cases the length of fiber extending from the penton base was measured and about 40 Å of the fiber sits inside the base, which therefore has to be added to the length. Second, the reported measurements were sometimes done on virus in which the six fibers that are usually observed are not lying flat on the support film. Third, for those cases in which the fiber length was measured from pentons, the two different penton orientations were not recognized. Fourth, in the cases in which isolated fiber was measured it was usually the short form that was observed. The HAd2 fiber length measured by Wrigley and reported in GREEN et al. (1983) was given as 308±27 Å, the same as we found for the short form.

## 3.5 Crystals of the Fiber

To obtain higher-resolution structural information than that achievable by electron microscopy study of single particles, ordered crystals of fiber amenable to X-ray structural analysis are required. The first fiber crystals to be obtained were microcrystals of HAd5 fiber (MAUTNER and PEREIRA 1971, PEREIRA et al. 1975), which were examined further by GREEN et al. (1983). Crystals of HAd2 fiber were first described by DEVAUX et al. (1984, 1987). The most extensive work on fiber crystals is that by DEVAUX et al. (1990), in which two crystal forms each from HAd2 and HAd5 were analyzed by cryoelectron microscopy, image processing, and X-ray diffraction. Fiber from HAd2 crystallizes in two forms depending on whether or not it is cleaved after Tyr-17; both crystal forms of HAd5 contain fiber cleaved after Tyr-17. Cleaved HAd2 fiber crystals can grow to reasonable macroscopic dimensions, but are thin plates and invariably twinned. This, coupled with a very long cell axis (540 Å), makes them unsuitable for full structural analysis, although X-ray diffraction studies have yielded important information useful in assessing a new model of the fiber shaft (see Sect. 3.6). One crystal form of HAd5 also grows to macroscopic size (also thin plates), but these crystals are again systematically disordered. Contributary reasons for the poor diffraction quality of all these crystals could be the apparent heterogeneity of the material as observed by isoelectric focusing or due to partial cleavage of the fiber at a second site in the head following Met-448 in HAd2 fiber (DEVAUX et al. 1990).

Using the results of image analysis of cryoelectron micrographs and arguments based on symmetry, packing models have been devised for each of the four crystal forms (DEVAUX et al. 1990). These confirmed that the fiber is a trimer and has an overall length of 350–370 Å, which is consistent with the results of electron microscopic studies of single particles. Some interesting additional features were observed in the X-ray diffraction patterns of cleaved HAd2 fiber crystals. These show very strong intensity at a Bragg spacing of 4.4 Å (also observed in the crystals of HAd5 fiber) and weaker diffuse intensity corresponding to layer lines of spacing 26.4 Å in a direction parallel to the fiber axis. These features, typical of an axially periodic structure, have been interpreted to arise from the quasi-periodicity of the structure of the fiber shaft (see Sect. 3.6).

Recently, crystals of the C-terminal head domain of HAd2 fiber expressed in the baculovirus system have been obtained (LOUIS et al. 1994). These diffract strongly to beyond 3-Å resolution, but are systematically disordered, making them unsuited for structural analysis. This again may be due to microheterogeneity.

Much more satisfactory crystals have been obtained of the head domain of HAd5 expressed in *E. coli* (HENRY et al. 1994). These have permitted the crystal structure of the receptor domain of HAd5 to be determined to 1.7 Å resolution (XIA et al. 1994, this volume).

## 3.6 Models of the Adenovirus Fiber Shaft

In the paper by GREEN et al. (1983), a cross-β model was proposed for the structure of the adenovirus shaft, a fold that had previously been proposed for Chrysopa silk by GEDDES et al. (1968) and for phage T4 tail fiber by EARNSHAW et al. (1979). In this model the fiber polypeptide would, in the shaft region, be folded into a long, narrow, antiparallel β-sheet, with each 15-residue motif making up two consecutive three-residue antiparallel strands (Sa and Sb, Fig. 2a, b) perpendicular to the fiber length and connected by turns (Ta and Tb). That one of these turns is a five-residue turn (Ta) was argued to be necessary for geometrical reasons. This model was attractive, since it would place all the hydrophobic positions 1, 3, 9, and 11 on one side of the β-sheet and permit the formation of a stable dimer with the two hydrophobic faces in contact. This was consistent with some results at the time that suggested, incorrectly, that the fiber was a dimer (DEVAUX et al. 1984). Furthermore, the model was in rough agreement with the then reported length of the HAd5 fiber, since the two β-strands of each motif would contribute $2 \times 4.7$ Å = 9.4 Å to the length, giving a maximum total shaft length of 208 Å. Additional electron diffraction and circular dichroism data were used to support the model.

The confirmation that the fiber was a trimer and that the axial rise per motif was 13.2 Å, and not the 9.4 Å predicted by the cross-β model, showed that a revised three-dimensional model was required for the fiber shaft. This should maintain the conserved secondary structure predicted by GREEN et al. (1983), but needed to be more extended in length. The key to the new triple-helical model proposed by STOUTEN et al. (1992) was the replacement of the *intra*chain hydrogen-bonding pattern of the cross-β model by *inter*chain hydrogen bonding between the three polypeptides of the molecule. This permits β-strands to be tilted with respect to the transverse direction (i.e., not perpendicular to the fiber axis, as in the cross-β model) and hence permits a longer fiber. With this idea, and using all the known experimental constraints and some derived from general protein structure analysis (e.g., the preferred right-handed twist of β-sheets), a number of atomic models were constructed of a regular four-motif segment of HAd5 fiber (residues 94–5155, boxed in Fig. 2a) and refined by molecular dynamics and energy minimization methods. In the most satisfactory model, the three monomers form a left-handed triple-helical structure with threefold symmetry. Successive 15-residue repeats on the same chain are related by an axial rise of 13.1 Å and left-handed azimuthal rotation of close to 300°. Three threefold-related β-sheets with short strands tilted at an angle of 26° to the transverse direction (compared to 0° in the cross-β model) are formed by intermonomer mainchain hydrogen bonds and give rise to superhelical ribbons covering the surface of the shaft (Fig. 7). The model satisfies criteria of extensive hydrogen bonding, reasonable backbone torsion angles, and burial of the conserved hydrophobic residues (in particular the positions 1, 3, 9, and 11) in the central core of the shaft. Most importantly, the model is consistent with the observed fiber length and diameter. Furthermore, its calculated diffraction pattern shows the characteristic features found in the

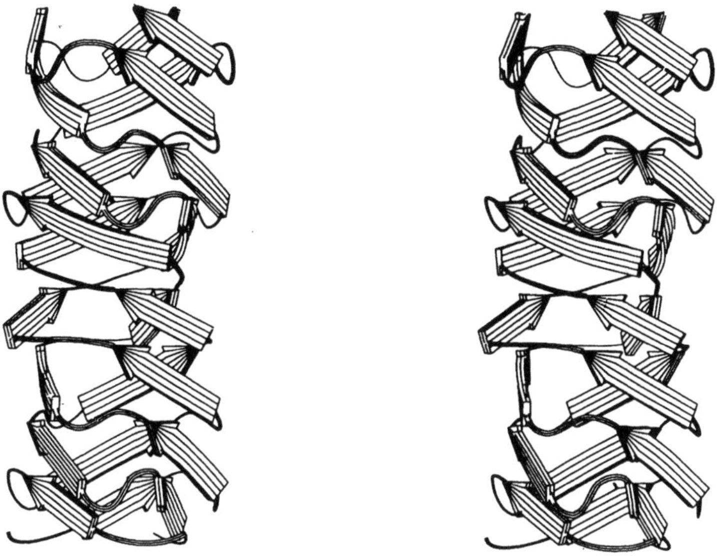

**Fig. 7.** Stereo pair showing the STOUTEN et al. (1992) triple-helical model of the fiber shaft. Broad bands with *arrowheads* represent the β-strand regions (Sa and Sb) and *thin ribbons* represent the turns (Ta and Tb as depicted in Fig. 2b). The β-strands do not have equal lengths, due to sequence variability of the individual motifs. The surface of the fiber can be considered to be made up of three left-handed helical β-sheets with staggered β-strands (follow up in a left-handed helix from the *arrow* at the *bottom right* to that at the *top left*)

observed diffraction pattern from Ad2 fiber crystals, notably layer lines with a spacing of about 1/26.4 Å$^{-1}$ and strong meridional intensity at 1/4.4 Å$^{-1}$ (i.e., at the sixth order). It may be asked how the model explains the fact that the true axial repeat (26.4 Å) corresponds to the axial rise of almost exactly *two* motifs. The answer follows from the helical parameters of the model mentioned above (i.e., one motif corresponds to an axial rise of 13.1 Å and a left-handed azimuthal rotation of −60°) coupled with the threefold symmetry. Starting from a fixed point on one chain, say the conserved proline (glycine), two motifs further along the same chain will give us another proline (glycine) after an axial rise of 26.4 Å and a rotation of −120°; however, since 120° is the threefold symmetry element, if we now apply this symmetry we will find another equivalent proline, but from one of the other chains directly above the original proline (glycine). Thus the whole structure has an axial repeat of 26.4 Å. It is important to note that this feature was not put into the initial models, but emerged as a property of the final model.

The one unsatisfactory feature of the model is that it gives no explanation for the alternating four- and five-residue turns, which, however, are a strongly conserved feature of the primary sequence. Also, since the model was based on the most regular section of the Ad5 sequence, it does not address the effect on the structure of extra long turns or other anomalies. Perhaps related to this problem, the model also does not explain the beaded appearance of the fiber sometimes observed in electron mocroscopy (Sect. 3.4.1). Finally, the model may have to be modified to accommodate the 36-residue "double" repeats observed in the murine and canine adenovirus fiber sequences with the anomalies noted above (Sect. 3.1.2). In this respect we note that 40-residue quasi-repeats have been detected in the sequence of the short-tail fiber of bacteriophage T4, which is composed of the product of gene 12 (Makhov et al. 1993). This molecule has a morphology similar to that of the adenovirus fiber (i.e., N-terminal tail, shaft, and C-terminal head), although the shaft is thicker (38-Å diameter compared to 28 Å for the adenovirus fiber shaft). It is proposed that the quasi-repeats contain four β-sheet segments which may form a triple helical structure related to that proposed for the adenovirus fiber (Makhov et al. 1993). Interestingly, the tail fibers of T-even-type phages composed of the product of gene 36 are apparently dimers and have been proposed to posses the cross-β fold (Earnshaw et al. 1979; Riede et al. 1987).

# 4 Biosynthesis

## 4.1 Transcription

The fiber polypeptide is a late product, as has been shown for subgroup C (serotypes 2 and 5), subgroup B (serotypes 3 and 7), and subgroup F viruses (serotypes 40 and 41), which is encoded in transcription unit L5 of the adenovirus genome (Chow et al. 1977; Kilpatrick et al. 1979; Cladaras and Wold 1985; Signäs et al. 1985; Chroboczek and Jacrot 1987; Hong et al. 1988; Kidd and Erasmus 1989; Kidd et al. 1990). For human adenoviruses, with the exception of subgroup F, one single fiber gene was localized in the right-hand part of the virus genome (Herrisé and Galibert 1981; Herrisé et al. 1981; Cladaras and Wold 1985; Signäs et al. 1985; Chroboczek and Jacrot 1987). In the case of enteric viruses HAd40 and HAd41, two adjacent fiber genes were found, separated by a polyadenylation signal and about 40 bases (Pieniazek et al. 1990; Kidd et al. 1990, 1993). Like most other late products, the fiber polypeptide is translated from a mRNA containing a three-partite leader originating from three different genome regions (Klessig 1977; Zain et al. 1979; Chow and Broker 1978). The HAd2 fiber mRNA may also contain additional leader sequences (Chow and Broker 1978; Uhlén et al. 1982). These x-, y-, and z-leader sequences overlap the E3 region, which is upstream of the fiber gene in the HAd2 genome. Similar sequences, with 54%–65% homology over approximately 100 bases, were also found upstream of the fiber gene in murine adenovirus type 1 DNA (Raviprakash et al. 1989). The significance of these additional leader sequences is unknown; they do not seem to affect the translation of these mRNAs (Dunn et al. 1978), and it is possible that they are processing intermediates. On the other hand, a correct splicing of the fiber mRNA may determine the host range of the virus, since monkey cells which are semi-permissive for replication of adenoviruses appear to splice fiber mRNA in an aberrant fashion (Klessig and Chow 1980). All leader-intervening sequence junctions contain the nucleotides GT and AG at the 5' and 3' ends, respectively, which is also typical for other eukaryotic systems.

## 4.2 Translation

Fiber protein is produced in the viral cycle beginning at 9–11 h after infection of tissue culture cells (White et al. 1969; Hayes et al. 1990). Sequence analysis of radioactive proteins synthesized in vitro in rabbit reticulocytes showed that the synthesis of fiber protein (similar to that of hexon and protein IX) is initiated at the AUG codon closest to the 5' end of its mRNA and that the initiating methionine is retained by fiber and protein IX, while it is removed from hexon (Anderson and Lewis 1980). In contrast with some adenovirus structural proteins, fiber does not undergo proteolytic cleavage during virus maturation.

It was noticed quite early for the subgroup C viruses that various stable structural polypeptides are synthesized at different rates which do not reflect the ratio at which they are present in virions (Klemperer and Pereira 1959; Philipson 1960; Wilcox and Ginsberg 1963). Capsid proteins are made in considerable excess, whereas core polypeptides are in relatively short supply (White et al. 1969). It was estimated that, in the case of HAd5, only about 10% of the structural proteins synthesized are assembled into virions (Wilcox and Ginsberg 1963). This permits the isolation of large amounts of some structural proteins, fiber among others, from infected cell extract (Boulanger and Puvion 1973; Pettersson et al. 1968). However, no systematic studies have been done to compare, from structural and functional points of view, the fiber protein isolated from infected cells with that in virions. The reason for this overproduction is not understood and the situation could be somewhat different for other adenovirus subgroups; a large excess of structural proteins is not observed in the extracts of HeLa of KB cells infected with HAd3 (C. Albiges-Rizo and J. Chroboczek, unpublished results).

Fiber polypeptide synthesis seems to be very fast. By following the fate of radioactive polypeptides synthesized in infected cells, it was found that fiber, like penton base, is synthesized and released from the ribosomal site of synthesis in approximately 2 min. On the other hand, assembly of penton capsomere (penton base and fiber) is rather slow; only 24% is assembled by 18.5 min, and the rest is assembled over a period of 10–11 h (Horwitz et al. 1969).

As for other late messengers, continued initiation of translation of fiber mRNA requires the production of the small viral RNA virus-associated (VA) RNA I (Thimmappaya et al. 1982). The primary role of VA RNA I is to overcome inhibitory consequences of the activation of a cellular protein kinase (named Da1 or P1/eIF-2a), induced by the adenovirus infection, which phosphorylates the α-subunit of the translation initiation factor eIF-2 (Schneider and Shenk 1987; Mathews and Shenk 1991). It was found that the extent of phosphorylation of eIF-2α during the late phase of infection is high enough to completely inhibit protein synthesis under other circumstances (O'Malley et al. 1989). To account for these findings it was proposed that the adenovirus-infected cell has two compartments, one containing viral mRNA and VA RNA I and the other cellular mRNA species but no VA RNA I. Only the translation of viral mRNA would then be protected by virtue of the presence of VA RNA I (O'MALLEY et al. 1989).

Such compartmentalization could explain the reduced level of newly synthesized viral late proteins observed in 293 cells infected with a mutant of HAd5, H5*ts*1 (HAYES et al. 1990). The mutation was mapped to the L4 100-kDa nonstructural protein and identified as a replacement of Ser-446 with Pro. The mutation did not affect either early product synthesis or the stability of late proteins, but it resulted in the impairment of the initiation of translation of late proteins. It seems that the 100-kDa protein is necessary for the efficient translation of viral late mRNA species. The authors hypothesize that late mRNA species are transcribed from replicated rather than parental viral DNA molecules and that 100-kDa protein associating with such late phase-specific messengers will direct them to, or maintain them in, a physical or biochemical compartment in the cytoplasm in which their translation could be facilitated. It should be noted that the L4 100-kDa protein is implicated in hexon trimerization and transport (CEPKO and SHARP 1983) and if its association with late messengers can be proved, it will be an example of a multifunctional viral chaperone protein.

## 4.3 Glycosylation

Fibers of some human serotypes are glycosylated: HAd2 fiber has three to four and HAd5 fiber has 1.7–2.5 residues of *N*-acetylglucosamine per protein molecule linked to the polypeptide chain by an alkali sensitive *O*-glycosidic bond (ISHIBASHI and MAIZEL 1974; CAILLET-BOUDIN et al. 1989; MULLIS et al. 1990). Similarly, HAd12 (subgroup A) fiber is a glycoprotein (BRÜGGEMANN et al. 1985), whereas fibers of serotypes 3 and 7 (subgroup B), 9 (subgroup D), and 4 (subgroup E) are not glycosylated (CAILLET-BOUDIN et al. 1989; MULLIS et al. 1990). It was found for HAd2 and HAd5 (subgroup C) fibers that both cytoplasmic and nuclear fibers are glycosylated and that glycosylation occurs in the cytoplasm and is not affected by known inhibitors of O- or N-glycosylation (CAILLET-BOUDIN et al. 1989). These results are in agreement with recent data showing that unlike N-linked and most O-linked glycosylation events, which occur within the lumen of the endoplasmic reticulum and the Golgi apparatus, O-linked *N*-acetylglucosamine is added to the proteins in the cytoplasm after they emerge from free polysomes (HALTIWANGER et al. 1990). The glycosylation site is probably located in the shaft part of the fiber, since when [$^{14}$C] glucosamine-labeled HAd2 fiber was hydrolyzed under acidic conditions yielding N-terminal and C-terminal fragments, the N-terminal 44-kDa fragment contained the label (CAILLET-BOUDIN et al. 1989). In accordance with this result, HAd2 shafts devoid of heads that were produced in the baculovirus expression system were glycosylated (NOVELLI and BOULANGER 1991) whereas the head domain of HAd2 produced in the baculovirus system was not (LOUIS et al. 1994).

Native HAd2 fiber could not be efficiently labeled with galactosyl-transferase, but SDS-denatured and boiled fibers could, suggesting that GlcNAc residues are probably inaccessible within the fiber structure and that they may play a role in the assembly or stabilization of HAd2 or HAd5 fiber trimers (MULLIS et al. 1990). There have been a limited number of glycoprotein-folding studies; recently, KERN et al. (1992), comparing stability and folding of nonglycosylated *versus* glycosylated forms of yeast invertase, found that glycosylation increases the overall stability of the native form against thermal and solvent denaturation. In addition, there are some data from other systems which suggest that this kind of O-linked glycosylation is important for the proper assembly of multimeric complexes (HOLT et al. 1987). Indeed, a mutant of HAd5 which fails to glycosylate its fibers at the nonpermissive temperature produces trimeric fibers with changed immunoreactivity (CHENG CHEE-SEUNG and GINSBERG 1982), implying some change in structure.

The functional significance of the carbohydrate moiety on the fiber remains unknown. Decisive experiments to show that the carbohydrate is necessary for proper folding are lacking. Furthermore, since all fibers seem to have a highly conserved overall organization, it seems unlikely that the sugar plays a crucial role in the structure since the majority of known fibers are not glycosylated.

## 4.4 Trimerization

The deletion of the nuclear targeting signal resulted in the accumulation of trimeric HAd2 fiber in the cytoplasm during expression in the vaccinia system, which shows that fiber trimerization occurs in the cytoplasm (HONG and ENGLER 1991). In this system HAd2 fiber was expressed without other adenoviral proteins and was correctly trimerized. HAd3 fiber protein produced in the *E. coli* expression system was not soluble and was recovered in inclusion bodies (ALBIGES-RIZO et al. 1991). The protein was solubilized in urea, and upon slow removal of urea it reassociated into trimers, as shown by nondenaturing gel electrophoresis and cross-link analysis. Recently, the HAd3 fiber was expressed in the eukaryotic baculovirus system. This protein is soluble, trimeric, and resistant to proteolysis. It is functional since it inhibits the attachment of the virus (A.M. DI GUILMI and J. CHROBOCZEK, unpublished results). The head domain of HAd2 expressed in the baculovirus system was also found to associate into trimers, suggesting that the head alone is sufficient for trimerization (LOUIS et al. 1994). This series of experiments shows that folding and trimerization of fiber does not depend on the action of other viral proteins. This is different from the situation encountered for the hexon protein, which needs the viral 100-kDa protein for both trimerization and translocation to the nucleus (CEPKO and SHARP 1983).

These results raise the question of the possible involvement of eukaryotic cell factor(s) such as chaperones in the proper folding and trimerization of the fiber. Indeed, it was found that the heat shock protein HSP70 can be coimmunoprecipitated with fiber protein from the extracts of cells infected with HAd5 (MACEJAK and LUFTIG 1991). However, the significance of this observation is not clear, since only about 5% of newly synthesized protein was contained in the complex. This complex may represent a dead-end for misfolded fibers. The explanation for the possible role of chaperones in folding and trimerization of fiber protein awaits further experiments.

### 4.5 Transport

To complete the viral cycle the adenovirus structural proteins have to be translocated to the cell nucleus, which is the site of virus assembly. Using a recombinant vaccinia virus vector expressing HAd2 fiber, it was shown that the trimeric glycosylated protein was transported to the nucleus and that the nuclear localization signal seems to be encoded by the sequence **KRAR** (amino acids 2–5), since deletion of this sequence resulted in the accumulation of trimeric fiber in the cytoplasm, and addition of the sequence **TKRVRL** (found at the beginning of HAd7 fiber) to the N terminus of this mutant restored correct targeting (HONG and ENGLER 1991). In this work the fiber protein expressed without any other viral proteins was correctly transported to the nucleus, which shows that the fiber contains all the signals required for nuclear targeting. Similar motifs are found at the N terminus of all known fiber sequences with the exception of murine serotype 1, in which a similar sequence, 477-KRVKKR, is found in the C-terminal head domain (RAVIPRAKASH et al. 1989; see Sect. 3.1.1).

## 5 Involvement in Virus Maturation

Studies with mutant viruses producing mutant or no fiber protein have revealed a possible additional function of fiber, connected with virus maturation. HAd5 temperature-sensitive mutants producing viral particles with underprocessed proteins VI, VII, and VIII (EDVARDSSON et al. 1978) have been mapped genetically and physically to the right end of the adenovirus genome (SAMBROOK et al. 1975), which encompasses the fiber gene, and they are fiber defective (MAUTNER et al. 1975; RUSSEL et al. 1972). Two other fiber-minus mutants—H5*ts* 142, which was mapped to the right end of the adenovirus genetic map (CHEE-SHEUNG and GINSBERG 1982) and H2*ts* 125, with a mutation in the fiber gene leading to the exchange of Ala434 to Val (BOUDIN et al. 1983)—both produce in infected cells empty or intermediate viral particles, and gel analysis of structural proteins revealed that there is no proteolytic maturation of proteins VI, VII, and VIII (CHEE-SHEUNG and GINSBERG 1982; D'HALLUIN et al. 1980). The first mutant at the nonpermissive temperature produces fiber apparently trimeric but with changed antigenicity, and the second mutant synthetizes a temperature-sensitive fiber polypeptide with aberrant gel mobility (MARTIN et al. 1975), both implying changes in the fiber protein shape. Another class of fiber-minus HAd2 and HAd5 deletion mutants has been described which lack the DNA region comprising the fiber gene and a large part of the E3 region (FALGOUT and KENTNER 1988). They make substantial amounts of virus particles with subnormal amounts of fiber (which comes from helper virus contaminants) and with precursor proteins pVI, pVII, pVIII that are not proteolytically matured. CHATTERJEE and FLINT (1987) have suggested that adenovirus-processing protease should be self-cleaved for full activity. Subsequently, FALGOUT and KETNER (1988) determined the degree of protease processing in one of the mutants, H5*dl*1021, and found that the protease is underprocessed in mutant virions. This suggested that in the absence of fiber the virion protease is not cleaved and remains inactive. Recent experiments have not confirmed that autocatalysis of the endoprotease is necessary for its function (J. WEBER, personal communication). It could be that during virion assembly addition of fiber triggers subsequent virion maturation, either through the processing of the viral protease or via another mechanism. Further studies of fiber mutants are needed to clarify the involvement of fiber protein in the last steps of adenovirus assembly.

## 6 Perspectives

The adenovirus fiber has been known as a distinct protein for 35 years and has been the subject of intense research for the last 10 years. While many aspects of its structure and function have been elucidated, several important questions remain unanswered: e.g., how does the fiber solve the symmetry mismatch problem in its interaction with the penton base? What is the nature of the receptor-binding site on the head? How does the fiber trimerize in vivo? Many of these questions could be addressed if the atomic structure of the fiber (and indeed of the whole penton) were known. This should be a major future goal.

Another essential element that is currently unknown is the identity of the primary cellular receptor that interacts with the fiber. Knowledge of the primary receptors and their position on the cell surface may help to understand why some fibers are long, some are short, some are straight, and some are bent as well as to extend our currently meagre understanding of the details of virus–receptor interactions.

Adenoviruses infect nondividing cells, and recombinant adenoviruses are now being used for gene transfer. Structural knowledge of the initial cell attachment mediated by fiber may help in the design of new subviral vehicles for drug and gene transfer.

*Acknowledgments.* The authors wish to express profound thanks to Bernard Jacrot, who initiated the work on adenovirus at the European Molecular Biology Laboratory (EMBL) in Grenoble and who ever since, through his enthusiasm and scientific insight, has been a major force behind the continuing work on the fiber. We would also like to thank our numerous collaborators who have contributed to the research on adenovirus fiber and whose results, sometimes unpublished, have been cited in this review.

## References


Albiges-Rizo C, Barge A, Ruigrok RW, Timmins PA, Chroboczek J (1991) Human adenovirus serotype 3 fiber protein. Comparison of native and recombinant proteins. J Biol Chem 266: 3961–3967

Anderson CW, Lewis JB (1980) Amino-terminal sequence of adenovirus type 2 proteins: hexon, fibre component IX and early protein 1B-15K. Virology 104: 27–41

Boudin M-L, Boulanger P (1982) Assembly of adenovirus penton base and fiber. Virology 116: 589–604

Boudin M-L, Rigolet M, Lemay P, Galibert F, Boulanger P (1983) Biochemical and genetic characterization of a fiber-defective temperature-sensitive mutant of type 2 adenovirus. EMBO J 2: 1921–1927

Boulanger PA, Puvion F (1973) Large-scale preparation of soluble adenovirus hexon, penton and fiber antigens in highly purified form. Eur J Biochem 39: 37–42

Brüggemann U, Klenk HD, Doerfler W (1985) Increased infectivity of extracellular adenovirus type 12. J Virol 55: 117–125

Brugh M, Beard CW, Villegas P (1984) Experimental infection of laying chickens with adenovirus 127 and with a related virus isolated from ducks. Avian Dis 28: 168–178

Caillet-Boudin M-L (1989) Complementary peptide sequences in partner proteins of the adenovirus capsid. J Mol Biol 208: 195–198

Caillet-Boudin M-L, Strecker G, Michalski J-C. (1989) O-Linked GlcNAc in serotype-2 adenovirus fiber. Eur J Biochem 184: 205–211

Cepko CL, Sharp PA (1983) Analysis of Ad5 hexon and 100k ts mutants using conformation-specific monoclonal antibodies. Virology 129: 137–154

Chardonnet Y, Dales S (1970a) Early events in the interaction of adenovirus with HeLa Cells. I. Penetration of type 5 and intracellular release of the DNA genome. Virology 40: 462–477

Chardonnet Y, Dales S (1970b) Early events in the interaction of adenovirus with HeLa cells. II. Comparative observations on the penetration of types 1, 5, 7 and 12. Virology 40: 478–485

Chatellard C. Chroboczek J (1989) Synthesis of human adenovirus type 2 fiber protein in *Escherichia coli* cells. Gene 81: 267–274

Chaterjee PK, Flint S (1987) Adenovirus type 2 endopeptidase: an unusual phosphoprotein enzyme matured by the autocatalysis. Proc Natl Acad Sci USA 84: 714–718

Cheng Chee-Seung C, Ginsberg HS (1982) Characterization of a temperature-sensitive fiber mutant of type 5 adenovirus and effect of the mutation on virion assembly. J Virol 42: 932–950

Chow LT, Broker TR (1978) The spliced structures of adenovirus 2 fiber message and the other late mRNAs. Cell 15: 497–510

Chow LT, Roberts JM, Lewis JB, Broker TR (1977) A map of cytoplasmic RNA transcripts from lytic adenovirus type 2, determined by electron microscopy of RNA:DNA hybrids. Cell 11: 819–836

Chroboczek J, Jacrot B (1987) The sequence of adenovirus fiber: similarities and differences between serotypes 2 and 5. Virology 161: 549–554

Cladaras C, Wold WSM (1985) DNA sequence of the early E3 transcription unit of Ad5. Virology 140: 28–43.

Cuzange A, Chroboczek J, Jacrot B (1994) The penton base of human adenovirus type 3 has the RGD motif. Gene 146: 257–259

Davison AJ, Telford EAR, Watson MS, McBride K, Mautner V (1993) The DNA sequence of adenovirus type 40. J Mol Biol 234: 1308–1316

Defer C, Belin M-T, Caillet-Boudin M-L, Boulanger P (1990) Human adenovirus host-cell interaction: a comparative study with members of subgroups B and C. J Virol 64: 3661–3673

Devaux C, Berthet-Colominas C, Timmins PA, Boulanger P, Jacrot B (1984) Crystal packing and stoichiometry of the fiber protein of adenovirus type 2. J Mol Biol 174: 727–737

Devaux C, Caillet-Boudin M-L, Jacrot B, Boulanger P (1987) Crystallization, enzymatic cleavage, and the polarity of the adenovirus type 2 fiber. Virology 161: 121–128

Devaux C, Adrian M, Berthet-Colominas C, Cusack S, Jacrot B (1990) Structure of adenovirus fibre. I. Analysis of crystals of fiber from adenovirus serotype 2 and 5 by electron microscopy and X-ray crystallography. J Mol Biol 215: 567–588

D'Halluin J-C, Milleville M, Martin GR, Boulanger P (1980) Morphogenesis of human adenovirus type 2 studied with fiber- and fiber and penton-base defective temperature-sensitive mutants. J. Virol 33: 88–99

Dragulev BP, Sira S, Abouhaider MG, Campbell JB (1991) Sequence analysis of putative E3 and fiber genomic regions of two strains of canine adenovirus type 1. Virology 183: 298–305

Dunn AR, Matthews MB, Chow LT, Sambrook J, Keller W (1978) A supplementary adenovirus leader sequence and its role in messenger translation. Cell 15: 511–526

Earnshaw WC, Goldberg EB, Crowther RA (1979) The distal half of the tail fibre of bacteriophage T4. J Mol Biol 132: 101–131
Edvardsson B, Ustacelebi S, Williams J, Philipson L (1978) Assembly intermediates among adenovirus type 5 temperature-sensitive mutants. J Virol 25: 641–651
Everett S, Ginsberg HS (1958) A toxin-like material separable from type 5 adenovirus particles. Virology 6: 770–771
Falgout B, Ketner G (1988) Characterization of adenovirus particles made by deletion mutants lacking the fiber gene. J Virol 62: 622–625
Fitzgerald DJP, Padmanabhan R, Pastan I, Wilingham MC (1983) Adenovirus-induced release of epidermal growth factor and *Pseudomonas* toxin into the cytosol of KB cells during receptor-mediated endocytosis. Cell 32: 607–617
Geddes AJ, Parker KD, Atkins EDT, Beighton E (1968) Cross-β conformation in proteins. J Mol Biol 32: 343–358
Gelderblom H, Maichle-Lauppe I (1982) The fibers of fowl adenoviruses. Arch Virol 72: 289–298
Ginsberg HS (1979) Adenovirus structural proteins. In: Fraenkel-Conrat H, Wagner RR (eds) Comparative virology, vol B. Plenum, New York, pp 409–427
Ginsberg HS, Pereira HG, Valentine RC, Wilcox WC (1966) A proposed terminology for the adenovirus antigens and virions morphological subunits. Virology 28: 782–783
Green NM, Wrigley NG, Russel WC, Martin SR, McLachlan A (1983) Evidence for a repeating cross β sheet structure in the adenovirus fibre. EMBO J 2: 1357–1365
Gruber WC, Russell DJ, Tibbetts C (1993) Fiber gene and genomic origin of human adenovirus type 4. Virology 196: 603–611
Haltiwanger RS, Holt GD, Hart GW (1990) Enzymatic addition of O-GlcNAc to nuclear and cytoplasmic proteins. J Biol Chem 265: 2563–2568
Hayes BW, Telling GC, Myat MM, Williams JF, Flint SJ (1990) The adenovirus L4 100-kilodalton protein is necessary for efficient translation of viral late mRNA species. J Virol 64: 2732–2742
Hennaché B, Boulanger P (1977) Biochemical study of the KB cell receptor for adenovirus. Biochem J 166: 237–247
Henry L, Xia D, Wilke M, Deisenhofer J, Gerard RD (1994) Characterization of the knob domain of the adenovirus type 5 fiber protein expressed in *Escherichia coli*. J Virol 68: 5239–5246
Herissé J, Galibert F (1981) Nucleotide sequence of the EcoRI fragment of the adenovirus 2 genome. Nucleic Acids Res. 9: 1229–1240
Herissé J, Rigolet M, Dupont de Dinechin S, Galibert F (1981) Nucleotide sequence of adenovirus 2 DNA fragment encoding for the carboxylic region of the fiber protein and the entire E4 region. Nucleic Acids Res 9: 4023–4042
Higgins DG, Bleasby AJ, Fuchs R (1992) Improved software for multiple sequence alignment. CABIOS 8: 189–191
Holt GD, Snow CM, Senior A, Haltiwanger RS, Gerace L, Hart GW (1987) Nuclear pore complex glycoproteins contain cytoplasmically disposed O-linked N-acetyl-glucosamine residues. J Cell Biol 104: 1157–1164
Hong JS, Engler JA (1991) The amino terminus of the adenovirus fiber protein encodes the nuclear localization signal. Virology 185: 758–767
Hong JS, Mullis KG, Engler JA (1988) Characterization of the early region 3 and fiber genes of Ad7. Virology 167: 545–553
Hopp TP, Woods KR (1981) Prediction of protein antigenic determinants from amino acid sequences. Proc Natl Acad Sci US 78: 3824–3828
Horwitz MS, Scharff MD, Maizel JV (1969) Synthesis and assembly of adenovirus 2. I. Polypeptide synthesis, assembly of capsomeres and morphogenesis of the virion. Virology 39: 682–694.
Hynes RO (1992) Integrins: versatility, modulation and signaling in cell adhesion. Cell 69: 11–25
Ishibashi M, Maizel JV (1974) The polypeptides of adenovirus. VI. The early and late glycopeptides. Virology 58: 345–361
Jones N, Shenk T (1979) Isolation of adenovirus type 5 host range deletion mutants defective for transformation of rat embryo cells. Cell 17: 683–689
Kern G, Schulke N, Schmidt FX, Jaenicke R (1992) Stability, quaternary structure, and folding of internal, external and core-glycosylated invertase from yeast. Protein Sci 1: 120–131
Kidd AH, Erasmus MJ (1989) Sequence characterization of the adenovirus 40 fiber gene. Virology 172: 134–144
Kidd AH, Erasmus MJ, Tiemessen CT (1990) Fibre sequence heterogeneity in subgroup F adenoviruses. Virology 179: 139–150
Kidd AH, Chroboczek J, Cusack S, Ruigrok RWH (1993) Adenovirus type 40 virions contain two distinct fibres. Virology 192: 73–84

Kilpatrick BA, Gelinas RE, Broker TR, Chow LT (1979) Comparison of late mRNA splicing among class B and class C adenoviruses. J Virol 30: 899–912

Klemperer HG, Pereira HG (1959) Study of adenovirus antigens fractionated by chromatography on DEAE cellulose. Virology 9: 536–545

Klessig DF (1977) Two adenovirus mRNAs have a common 5' terminal leader sequence encoded at least 10kb upstream from their main coding regions. Cell 12: 9–21

Klessig DF, Chow LT (1980) Incomplete splicing and deficient accumulation of the fiber messenger RNA in monkey cells infected by human adenovirus type 2. J Mol Biol 139: 221–242

Laver WG, Banfield Younghusband H, Wrigley NG (1971) Purification and properties of chick embryo lethal orphan virus. Virology 45: 598–614

Lonberg-Holm K, Crowell RL, Philipson L (1976) Unrelated animal viruses share receptors. Nature 259: 679–681

Louis N, Fender P, Barge A, Kitts P, Chroboczek J (1994) Cell-binding domain of adenovirus serotype 2 fiber. J Virol 66: 4104–4106

Macejak DG, Luftig LB (1991) Association of HSP70 with the adenovirus type 5 fiber protein in infected HEp-2 cells. Virology 180: 120–125

Makhov AM, Trus BL, Conway JF, Simon MN, Zurabishvili TG, Mesyanzhinov VV, Steven AC (1993) The short tail-fiber of bacteriophage T4: molecular structure and a mechanism for its conformational transition. Virology 194: 117–127

Martin GR, Warocquier R, Boulanger P (1975) Quantitation of adenovirus soluble antigens by crossed immuno-electrophoresis: application to serological characterization of mutants. Intervirology 5: 162–172

Mathews MB, Shenk T (1991) Adenovirus virus-associated RNA and translation control. J Virol 65: 5657–5662

Mautner V, Pereira HG (1971) Crystallisation of a second adenovirus protein (the fiber). Nature 230: 456–457

Mautner V, Williams J, Sambrook J, Sharp PA, Grodzicker T (1975) The location of the genes coding for hexon and fiber proteins in adenovirus DNA. Cell 5: 93–99

Meager A, Butters TD, Mautner V, Hughes RC (1976) Interaction of KB-cell glycoproteins with an adenovirus capsid protein. Eur J Biochem 61: 345–353

Mei Y-F, Wadell G (1993) Hemagglutination properties and nucleotide sequence of the fiber gene of adenovirus genome types 11p and 11a. Virology 194: 453–462

Mittal SK, Prevec L, Babiuk LA, Graham FL (1992) Sequence analysis of bovine adenovirus type 3 early region 3 and fibre protein genes. J Gen Virol 73: 3295–3300

Mittal SK, Prevec L, Babiuk LA, Graham FL (1993) Sequence analysis of bovine adenovirus type 3 early region 3 and fiber protein genes (Corrigendum). J Gen Virol 74: 2825

Mullis KG, Haltiwanger RS, Hart GW, Marchase RB, Engler JA (1990) Relative accessibility of N-acetylglucosamine in trimers of the adenovirus types 2 and 5 fiber proteins J Virol 64: 5317–5323

Neumann R, Chroboczek J, Jacrot B (1988) Determination of the nucleotide sequence for the penton-base gene of human adenovirus type 5. Gene 69: 153–157

Norrby E (1966) The relationship between the soluble antigens and the virion of adenovirus type 3. I. Morphological characteristics. Virology 28: 236–248

Norrby E (1969) The relationship between the soluble antigens and the virion of adenovirus type 3. IV. Immunological complexity of soluble components. Virology 37: 565–576

Novelli A, Boulanger P (1991) Deletion analysis of functional domains in baculovirus-expressed adenovirus type 2 fiber. Virology 185: 365–376

O'Malley RP, Duncan RF, Hershey JWB, Mathews MB (1989) Modification of protein synthesis initiation factors and the shut-off of host protein synthesis in adenovirus-infected cell. Virology 168: 112–118

Pereira HG (1958) A protein factor responsible for the early cytopathic effect of adenoviruses. Virology 6: 601–611

Pereira HG, Wrigley N, Mautner V (1975) Polymerization of adenovirus proteins. In: Mano E (ed) Proceedings of the International Symposium on Macromolecules. Elsevier, Amsterdam, pp 451–463

Persson R, Wohlfart C, Svensson E, Everitt E (1983) Virus-receptor interaction in the adenovirus system: characterization of the positive co-operative binding of virions on HeLa cells. J Virol 54: 92–97

Pettersson U, Wadell G (1985) Antigenic structure of adenoviruses. In: van Rgenmortel MHV, Neurath AR (eds) Immunochemistry of viruses. The basis for serodiagnosis and vaccines. Elsevier, Amsterdam

Pettersson U, Philipson L, Höglun S (1968) Structural proteins of adenoviruses. II. Purification and characterization of the adenovirus type 2 fiber antigen. Virology 35: 204, 215

Philipson L (1960) Fractionation on DEAE cellulose of components associated with adenovirus reproduction. Virology 10: 459–465

Philipson L, Lonberg-Holm K, Pettersson U (1968) Virus-receptor interaction in an adenovirus system. J Virol 2: 1064–1075
Pieniazek NJ, Slemenda SB, Pieniazek D, Velarde J, Luftig RB (1989) Sequence of human enteric adenovirus type 41 (Tak) fiber protein gene. Nucleic Acids Res 17: 9474
Pieniazek NJ, Slemenda SB, Pieniazek D, Velarde J, Luftig RB (1990) Human enteric adenovirus type 41 (Tak) contains a second fiber protein gene. Nucleic Acids Res 18: 1901
Praszkier J, Ginsberg HS (1987) Isolation and characterization of temperature-sensitive mutants of adenovirus type 7. J Virol 61: 3089–3095
Raviprakash KS, Grunhaus A, El Kholy MA, Horwitz MS (1989) The mouse adenovirus type 1 contains an unusual E3 region. J Virol 63: 5455–5458
Riede I, Schwarz H, Jähnig F (1987) Predicted structure of tail-fiber proteins of T-even type phages. FEBS Lett 215: 145–150
Rowe WP, Hartley JW, Roizman B, Levy HB (1958) Characterisation of a factor formed in the course of adenovirus infection of tissue cultures causing detachment of cells from glass. J Exp Med 108: 713–729
Ruigrok RWH, Barge A, Albiges-Rizo C, Dayan S (1990) Structure of adenovirus fibre. II. Morphology of single fibers. J Mol Biol 215: 589–596
Ruigrok RWH, Barge A, Mittal SK, Jacrot B (1994) The fibre of bovine adenovirus is very long but bent. J Gen Virol 75: 2069–2073
Russell WC, Newman C, Williams J (1972) Characterization of temperature-sensitive mutants of adenovirus type 5—serology. J Gen Virol 17: 265–279
Sambrook J, Williams JF, Sharp P, Grodzicker T (1975) Physical mapping of temperature sensitive mutations of adenoviruses. J Mol Biol 97: 369–390
Schneider RJ, Shenk T (1987) Impact of virus infection on host cell protein synthesis. Ann Rev Biochem 56: 317–332
Sheppard M, Trist H (1992) Characterization of the avian adenovirus penton base. Virology 188: 881–886
Shieh W-J. (1992) Variations and phylogenetic analyses of adenovirus fiber gene. Dissertation, Vanderbildt University
Signäs C, Akusjärvi G, Pettersson U (1985) Adenovirus 3 fiber polypeptide gene: implications for the structure of the fiber protein. J Virol 53: 672–678
Silver L, Anderson CV (1988) Interaction of human adenovirus serotype 2 with human lymphoid cells. Virology 165: 377–387
Sprengel J, Schmitz B, Heuss-Neitzel D, Zock C, Doefler W (1994) The nucleotide sequence of human adenovirus type 12 DNA: a comparative functional analysis. J Virol 68: 379–389
Stewart PL, Burnett RM, Cyrklaff M, Fuller SD (1991) Image reconstruction reveals the complex molecular organization of adenovirus. Cell 67: 145–154
Stouten PFW, Sander C, Ruigrok RWH, Cusack S (1992) A new triple-helical model for the adenovirus fiber shaft. J Mol Biol 226: 1073–1084
Strassheim ML, Gruenberg A, Veijalainen P, Sgro J-P, Parrish CR (1994) Two dominant neutralizing antigenic determinants of canine parvovirus are found on the threefold spike of the virus capsid. Virology 198: 175–184
Sundquist B, Pettersson U, Thelander R, Philipson L (1973) Structural proteins of adenoviruses. IX. Molecular weight and subunit composition of adenovirus type 2 fiber. Virology 51: 252–256
Svensson U, Persson R, Everitt E (1981) Virus-receptor interaction in the adenovirus system. I. Identification of virion attachment proteins of the HeLa cell plasma membrane. J. Virol 38: 70–81
Thimmappaya B, Weinberger C, Schneider RJ, Shenk T (1982) Adenovirus VA I RNA is required for efficient translation of viral mRNAs at late times after infection. Cell 31: 543–551
Uhlén M, Svensson C, Josephson S, Aleström P, Chatapadhyaya JB, Pettersson U, Philipson L. Leader arrangement in the adenovirus fiber mRNA. (1982) EMBO J 1: 249–254
Valentine RC, Pereira HG (1965) Antigens and structure of the adenovirus. J Mol Biol 13: 13–20
van Oostrum J, Burnett RM (1985) The molecular composition of the adenovirus type 2 virion. J Virol 56: 439–448
van Oostrum J, Smith PR, Mohraz M, Burnett RM (1987) The structure of the adenovirus capsid III. Hexon packing determined from electron micrographs of capsid fragments. J Mol Biol 198: 73–89
Varga MJ, Weibull C, Everitt E (1991) Infectious entry pathway of adenovirus type 2. J Virol 65: 6061–6070
Wadell G, Norrby E (1969) Immunological and other biological characteristics of penton of human adenoviruses. J Virol 4: 671–680
Watson G, Burdon MG, Russell WC (1988) An antigenic analysis of the adenovirus type 2 fibre polypeptide. J Gen Viron 69: 525–535

Weber JM, Talbot BG, Delorme L (1989) The orientation of the adenovirus fiber and its anchor domain identified through molecular mimicry. Virology 168: 180–182
White DO, Scharff MD, Maizel JV (1969) The polypeptides of adenovirus III. Synthesis in infected cells. Virology 3: 395–406
Wickham T, Mathias P, Cheresh DA, Nemerow GR (1993) Integrins $a_vb_3$ and $a_vb_5$ promote adenovirus internalization but not virus attachement. Cell 73: 309–319
Wilcox WC, Ginsberg HS (1963) Structure of type 5 adenovirus. I. Antigenic relationship of virus structural proteins to virus specific soluble antigens from infected cells. J Exp Med 118: 295–306
Wohlfart CEG, Svensson UK, Everitt E (1985) Interaction between HeLa cells and adenovirus type 2 virions neutralized by different sera. J Virol 56: 896–903
Xia D, Henry LJ, Gerard RD, Deisenhofer J (1994) Crystal structure of the receptor-binding domain of adenovirus type 5 fibre protein at 1.7 Å resolution. Structure 2: 1259–1270.
Yeh H-Y, Luftig RB (1991) Human enteric adenovirus type 41 binding to HEp-2 cells is specific. Virology 183: 410–414
Zain S, Sambrook J, roberts RJ, Keller W, Fried M, Dunn AR (1979) Nucleotide sequence analysis of the leader segments in a cloned copy of adenovirus 2 fiber mRNA. Cell 16: 851–861

# Transcription of Adenovirus RNA Polymerase III Genes

R. PRUZAN[1,2] and S.J. FLINT[1]



## 1 Introduction

Eukaryotic RNA polymerase (RNAp) III is responsible for synthesis of abundant, low molecular mass RNA species, including tRNA, 5S rRNA, and U6 snRNA (see ROEDER 1976; GEIDUSCHEK and TOCCHINI-VALENTINI 1988). Although they are non-coding, RNAp III transcripts participate in such essential processes as translation, RNA processing, and protein targeting (see RICH and RAJ BHANDARY 1976; GUTHRIE and PATTERSON 1988; DREYFUSS et al. 1988; STEITZ et al. 1988; WOLIN and WALTER 1991; YOUNG 1991). Transcription of a viral genome by RNAp III was first recognized 20 years ago, when it was established that a small RNA specific to adenovirus-infected cells, virus-associated (VA) RNA (REICH et al. 1966), is synthesized by this enzyme. All adenovirus genomes examined to date encode at least one VA RNA species. Investigation of the organization of the transcriptional control regions of these viral RNAp III genes made an important contribution to the unexpected discovery of the intragenic location of elements of RNAp III promoters. As discussed in this review, the VA RNA I promoter of subgroup C adenoviruses, which is exceptionally active in in vitro systems, has provided an important model for investigation of the mechanism of initiation of transcription by this RNA polymerase. Elucidation of the part played by VA RNA I during

[1] Department of Molecular Biology, Princeton University, Princeton, NJ 08544, USA
[2] Present address: Cold Spring Harbor Laboratory, Cold Spring Harbor, NY, USA

adenovirus infection has, moreover, identified a previously unknown mechanism by which viruses can evade cellular defence systems. Indeed, the properties of adenovirus VA RNA I have established the paradigm for small RNAs encoded by other viruses (see MATHEWS 1995).

Much more recently, investigation of RNAp III transcription of the adenoviral genome has yielded another surprise: an RNAp III promoter, first identified in vitro but operating in subgroup C adenovirus-infected cells, is superimposed on the RNAp II promoter of the E2 early (E2E) transcription unit. The unusual organization of this third, adenoviral RNAp III transcription unit raises new questions about both the mechanisms that specify promoter recognition by RNAp II or by RNAp III and the roles of RNAp III transcription.

## 2 Virus-Associated RNAs and Their Transcription

### 2.1 Virus-Associated RNA Genes

By virtue of its abundance in subgroup C adenovirus-infected cells, the small RNA species that we now know as VA RNA I was among the first adenoviral gene products to be characterized in detail. Adenovirus types 2 (Ad2) and 5 (Ad5) encode two VA RNAs VA RNA I and VA RNA II (MATHEWS 1975; SÖDERLUND et al. 1976), which are transcribed by RNAp III: their synthesis is nuclei isolated from infected cells is resistant to concentrations of α–amanitine that inhibit RNAp II, but sensitive to the higher concentrations of the drug at which RNAp III, is inhibited (PRICE and PENMAN 1972; WEINMANN et al. 1974, 1976; SÖDERLUND et al. 1976; VENSTRÖM et al. 1978a). VA RNA I and VA RNA II are encoded by separate transcription units, located near map coordinate 30 in the r-strand of the Ad2 or Ad5 genomes (MATHEWS 1975; PETTERSSON and PHILIPSON 1975; SÖDERLUND et al. 1976; PAN et al 1977; MATHEWS and PETTERSSON 1978; AKUSJÄRVI et al. 1980). These RNAp III transcription units are thus included within the major late transcription unit transcribed by RNAp II. A similar arrangement of two VA RNA genes is observed in the genome of Ad7, a representative of human subgroup B serotypes (ENGLER et al. 1983), but more distantly related adenoviruses differ in the number or location of their VA-RNA genes. The genomes of the subgroup A human virus Ad12 and of simian Ad7 appear to contain a single VA RNA gene, at a location analogous to the Ad2, Ad5, or Ad7 genes (FÖHRING et al. 1979; LARSSON et al. 1986a; SHU et al.1986). The single VA RNA detected in permissive cells infected by chicken embryo lethal orphan (CELO) virus is, by contrast, encoded by a gene transcribed in a leftward direction and located near map coordinate 90 (LARSSON et al. 1986b). CELO virus VA RNA, some 90 nucleotides (n) long, is, moreover considerably shorter than human or simian adenoviral VA RNA, which are approximately 160 bases in length. Despite such variations, each human or simian serotype examined encodes a VA RNA species that can functionally substitute for VA RNA I of Ad2 or Ad5 (see MATHEWS and SHENK 1991; MA and MATHEWS 1993). As it is only the VA RNAs of the subgroup C serotypes that have been studied in any detail, subsequent sections will consider the properties of these RNA and their expression.

## 2.2 Virus-Associated VA RNA I Promoter

An in vitro system in which RNAp III can transcribe VA RNA I was first described in 1978 (WU 1978). In stark contrast to purified RNAp III, which nonspecifically transcribed both strands of Ad2 DNA (JAEHNING et al. 1977), HeLa cells extracts were shown to support synthesis of discrete RNA species that specifically hybridized to the Ad2 VA RNA I gene (WU 1978). Thus, this pioneering study provided important evidence that accessory factors are essential for specific initiation of transcription by RNAp III. The high efficiency of VA RNA I transcription of such unfractionated extracts has facilitated analysis of the organization of the VA RNA I promoter, permitting development of a detailed description of its architecture.

Sequencing of the VA RNA genes (OHE and WEISSMANN 1970; PAN et al. 1977; THIMMAPPAYA et al. 1979; AKUSJÄRVI et al. 1980), in conjunction with mapping the termini of VA RNA I synthesized in infected cells (CELMA et al. 1977a, b; VENNSTRÖM et al. 1978b; THIMMAPPAYA et al. 1979), established that VA RNA I transcription initiates at two sites in vivo, with a G here designated position +1 and less frequently with the A at position –3 (see Fig. 1). Initiation from the latter site is not essential, as a virus that does not produce this form of VA RNA I, because of a 2 bp deletion some 20 bp upstream, grows as well as wild-type Ad5 in HeLa cells (THIMMAPPAYA et al. 1979). Indeed this virus, *dl* 309, has been used as the wild type in many subsequent analyses. RNAp III specifically initiates transcription from both the +1 and –3 sites in the unfractionated in vitro transcription systems used to identify elements of the VA RNA I promoter (for example, FOWLKES and SHENK 1980). Analysis of the effects of deletions encompassing various segments of the Ad2 or Ad5 genes upon VA RNA I transcription initially localized sequences essential for promoter activity to the internal sequence +7 to +72 (FOWLKES and SHENK 1980; GUILFOYLE and WEINMANN 1981). This internal control region of the VA RNA I gene exhibits little homology to that of the cellular 5S rRNA gene, which was identified just previously (BOGENHAGEN et al. 1980; ENGELKE et al. 1980; SAKONJU et al. 1980), but was immediately recognized as similar to internal sequences of tRNA genes implicated in promoter activity (FOWLKES and SHENK 1980). Subsequent experiments confirmed that the VA RNA I internal control regions are very similar to those of tRNA genes, which comprise two internal elements, termed the A and B boxes (see GEIDUSCHEK and TOCCHINI-VALENTINI 1988).

Analysis of the efficiencies with which a large number of VA RNA I promoters carrying internal deletions, insertions, or substitutions were transcribed in vitro

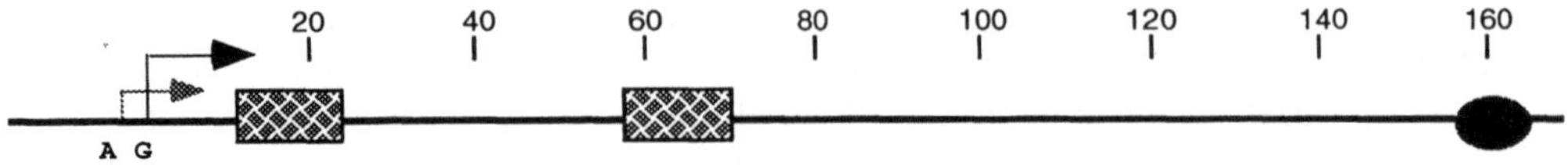


| | A box | B box | Termination |
|---|---|---|---|
| VA-RNAI | GTGGTCTGGTGG | GGTTCGAACCC | CCTTTTGGC |
| VA-RNAII | GTAGCCGGAGGG | GGTTCGAGTCT | CCTTTTTTGCTTTTCCC |
| tRNA | GTRGYYYGRYGG | GGTTCGANTCC | |

**Fig. 1.** Organization and promoter sequences of the subgroup C adenovirus virus-associated (VA) RNA I gene. The VA RNA I gene, shown to scale, is represented by the *horizontal line* at the top of figure. Major and minor sites of initiation of transcription by RNAp III are indicated by the *solid* and *dashed arrows*, respectively, drawn in the direction of transcription. The locations of the internal A and B box sequences discussed in the text and of the termination region are indicated by *boxes* and an *oval*, respectively. The sequences of these promoter elements of the VA RNA I gene are shown below, as are the corresponding VA RNA II gene sequences and consensus A and B box sequences described in the text. *R*, purine; *Y*, pyrimidine; *N*, any nucleotide

idetified promoter-proximal and promoter-distal control elements, analogous to these A and B boxes (Bhat et al. 1983; Cannon et al. 1986; Rohan and Ketner 1987; Snouwaert et al. 1987; Wu et al.1987; Railey and Wu 1988). These elements, which have been localized to +12/+14 to +24 and +58 to +69 (see Fig. 1), occupy positions in the VA RNA I gene similar to those of the A and B box elements of cellular tRNA genes (see Geiduschek and Tocchini-Valentini 1988). Moreover, they match very well A and B box consensus sequences derived from tRNA genes of a variety of eukaryotes (see Fig. 1). The close relationship of the VA RNA I promoter to these cellular RNAp III promoters is further emphasized by the ability of hybrid promoters, in which A or B box sequences of the VA RNA I gene were replaced by the corresponding elements of a *Xenopus* $tRNA^{met}$ gene, to support accurate and efficient transcription by RNAp III (Bhat et al. 1983). Both the A and B box elements of the VA RNA I gene are essential for its efficient transcription in adenovirus-infected cells (Bhat and Thimmappaya 1984). By contrast to that of certain tRNA genes (see Geiduschek and Tocchini-Valentini 1988), the A box of the VA RNA I promoter appears less important than the B box in in vitro systems, for mutations that alter most of the A box sequence but do not eliminate promoter activity have been described (e.g., Snouwaert et al. 1987; Railey and Wu 1988). It has not been established whether these mutations elicit a similar phenotype in infected cells.

The distance between the A and B box sequences of the VA RNA I promoter is important for optimal activity: promoters with deletions of only 4–6 bp, within a region whose substitution induced only modest reductions in promoter activity (e.g., Cannon et al. 1986; Snouwaert et al. 1987; Railey and Wu 1988), were transcribed at 1%–2% the efficiency of the wild type (BHAT and THIMMAPPAYA 1983; CANNON et al.1986). By contrast, insertions of up to 46 bp had little effect (BHAT and THIMMAPPAYA 1983; CANNON et al. 1986) and even expansion of the spacer region by 70–120 bp was less deleterious than removal of 4–6 bp (CANNON et al. 1986). These observations suggest that a minimum separation of the A and B boxes is required for optimal interaction among components of the RNAp III transcriptional machinery during initiation of VA RNA I transcription.

Although the elements that comprise the internal region of the VA RNA I promoter are the most important, sequences immediately upstream of and surrounding the sites of initiation also govern the efficiency and specificity of transcription. Although not essential for promoter activity, the 5' flanking region modulates the activity of the VA RNA I promoter. Substitution of sequences between positions –17 to +2, for example reduced VA RNA I synthesis by up to a factor of 2 (RAILEY and WU 1988). Mutation of 5' flanking sequences of tRNA genes can have similar effects (see GEIDUSCHEK and TOCCHINI-VALENTINI 1988). Deletion of 5' sequences to various positions between –33 and –7 specifically inhibited initiation from the –3(A) site, whereas more extensive deletion into coding sequences induced initiation from novel purine sites (FOWLKES and SHENK 1980; GUILFOYLE and WEINMANN 1981; WU et al. 1987). Similarly, substitutions within the region from –11 to –1 elicited the use of alternative initiation sites 5–10 bp upstream or downstream of the major wild-type site (RAILEY and WU 1988). Thus, the 5'-flanking sequence of the VA RAN I gene clearly plays a role in determining initiation specificity. A second important determinant of the site of initiation is the A box, for mutations within its 5' segment also elicit novel initiation sites (RAILEY and WU 1988). Furthermore, when the distance between the A and B boxes was expanded by up to 70 bp, initiation a fixed distance upstream of the A box was observed (CANNON et al. 1986). Thus, interactions among RNAp III factors and the 5'-flanking sequence and internal elements of the VA RNA I promoter that (above a minimum distance) are relatively indifferent to the length or helical register of the DNA sequence lying between the A and B boxes appear to position RNAp III for initiation at specific sites.

Little attention has been paid to termination of VA RNA I transcription. The 3' ends of VA RNA I lie within a sequence that comprises a block of T in the nontranscribed stran (see Fig. 1). Such sequences of four or more T flanked by GC-rich sequences in the nontranscribed strand are typical of many RNAp III termination sites (BOGENHAGEN and BROWN 1981; see GEIDUSCHEK and TOCCHINI-VALENTINI 1988). Presumably, this VA RNA I sequence specifies termination of transcription, but this function has not been demonstrated directly. Certain substitutions in both the A and B box sequences of the VA RNA I have been reported to induce termination at alternative sites 10–15 bp upstream of the wild-type sites (RAILEY and WU 1988). The mechanism by which these sequences might govern termination at sites a considerable distance downstream has not been investigated.

Although examined in considerably less detail, the VA RNA II promoters of Ad2 and Ad5 appear to possess an organization similar to the VA RNA I promoters: the B box sequence that occupies positions +59 to +65 of the Ad2 VA RNA I gene is identical to the sequence +57 to +63 of the Ad2 VA RNA II gene (Fig. 1): the VA RNA I promoter is a strong competitor of VA RNA II transcription both in vitro and in vivo (FOWLKES and SHENK 1980; BHAT and THIMMAPPAYA 1984); mutations that impair VA RNA I transcription stimulate VA RNA II transcription in in vitro systems (FOWLKES and SHENK 1980; GUILFOYLE and WEINMANN 1981) and in infected cells (BHAT and THIMMAPPAYA 1984), and human cell extracts yield similar footprints over the B boxes of the VA RNA I and VA RNA II promoters (VAN DYKE and ROEDER 1987).

By virtue of the strong similarities of the organization and sequences of their intragenic elements, the VA RNA I and cellular tRNA gene promoters are considered members of the same class of RNAp III internal control regions, the type 2 promoters (see GEIDUSCHEK and TOCCHINI-VALENTINI 1988). Consistent with this classification, specific initiation of transcription by RNAp III from all type 2 promoters requires the same transcription factors (TF), TFIIIB and TFIIIC. These factors were initially identified as the fractions recovered following chromatography of cytoplasmic or nuclear extracts on phosphocellulose required for accurate and efficient initiation of transcription by RNAp III (SEGALL et al. 1980; SHASTRY et al. 1982). Competition transcription assays demonstrated that the same factors that are present in such partially purified preparations of TFIIIB and TFIIIC mediate transcription by RNAp III from the viral VA RNA I and the cellular tRNA promoters (LASSAR et al. 1983). Furthermore, elegant template commitment assays identified binding of TFIIIC to the internal control region as the first step in initiation of transcription of both tRNA or VA RNA I genes (LASSAR et al. 1983; see GEIDUSCHEK and TOCCHINI-VALENTINI 1988; GABRIELSEN and SENTENAC 1991). The interaction of TFIIIC with the VA RNA I and at least some tRNA gene promoters is of sufficient stability to sequester the factor and thus render transcription from these promoters resistant to challenge from a second template, added once TFIIIC has bound to the first (LASSAR et al. 1983; CAREY et al. 1986; DEAN and BERK 1988; see GEIDUSCHEK and TOCCHINI-VALENTINI 1988). In the case of the VA RNA I promoter, the promoter-distal B box sequence is essential for transcription and formation of such stable preinitiation complexes via TFIIIC binding (FOWLKES and SHENK 1980; GUILFOYLE and WEINMANN 1981; BHAT et al. 1983; LASSAR et al. 1983; ROHAN and KETNER 1987; SNOUWAERT et al. 1987). Indeed, partially or highly purified human TFIIIC specifically protects the VA RNA I B box, but not the A box, sequence from nuclease digestion or chemical cleavage (FUHRMAN et al.1984; CAREY et al. 1986; VAN DYKE and ROEDER 1987; HOEFFLER et al. 1988; YOSHINAGA et al. 1987, 1989; DEAN and BERK 1988; SCHNEIDER et al. 1989). Such sequence-specific interaction of TFIIIC with the B box must occur before TFIIIB canbe recruited to the VA RNA I (or tRNA) promoters, and only after both factors are assembled can RNAp III enter the preinitiation complex (LASSAR et al. 1983; CAREY et al. 1986; see GEIDUSCHEK and TOCCHINI-VALENTINI 1988; GABRIELSEN and SENTENAC 1991).

Although this assembly pathway has been recognized for some time, a detailed, molecular description of initiation of RNAp III transcription from the VA RNA I promoter that accounts for the contributions of all promoter elements described in previous paragraphs is not yet possible. A significant impediment to progress in this area has been the difficulty of, and often apparently contradictory results obtained during, purification and characterization of the mammalian factors necessary for initiation of transcription by RNAp III. However, recent progress in the purification and analysis of components of both TFIIIB and TFIIIC hold out hope that this situation will be remedied in the foreseeable future.

TFIIIC, whose specific interaction with the B box sequence initiates assembly of transcription complexes of the VA RNA I promoter, can be monitored during purification by virtue of its sequence-specific DNA-binding or transcription activities. Surprisingly, however, TFIIIC B box complexes examined by footprinting methods or using transcription assays were reported to exhibit different properties (CAREY et al. 1986). Furthermore, in some experiments, B box-binding and TFIIIC transcriptional activity did not exactly cochromatograph (e.g., VAN DYKE and ROEDER 1987). It has since been established that HeLa cells contain both transcriptionally active and inactive forms of TFIIIC, each capable of binding specifically to the B box of the VA RNA I gene (HOEFFLER et al. 1988). The presence of these two forms and the sensitivity of their relative concentrations to cell growth state (HOEFFLER et al. 1988) may account for these puzzling observations and the relatively slow progress in purification of transcriptionally active human TFIIIC.

There is general agreement that one component of human TFIIIC, originally defined as the activity supplied by the 0.5 *M* phosphocellulose fraction, is a high molecular mass (some 500 kDa) complex, comprising five polypeptides of 220–230, 110, 100–102, 80–90, and 60–63 kDa (YOSHINAGA et al. 1989; SCHNEIDER et al. 1989; KOVELMAN and ROEDER 1992). This TFIIIC component, which has been purified by largely different procedures by different investigators, binds specifically to the B boxes of VA RNA I and tRNA genes (YOSHINAGA et al. 1987, 1989; H.R.SCHNEIDER et al. 1989; KOVELMAN and ROEDER 1992). Its largest polypeptide can be specifically cross-linked to the VA RNA I promoter or its B box sequence (YOSHINAGA et al. 1989; KOVELMAN and ROEDER 1992), and specific binding of polypeptides of 110–12 and 68 kDa or less to the VA RNA I promoter has also been detected using blotting assays (SCHNEIDER et al. 1989, 1990). These properties implicate this component of human TFIIIC activity in recognition of the B box of the VA RNA I promoter, and thus initiation of the sequence of reactions that assemble preinitiation complexes on this viral gene. In the yeast *Saccharomyces cerevisiae*, interaction of TFIIIC, quite similar in mass and composition to this component of human TFIIIC activity, with the A and B boxes of a tRNA gene promoter appears necessary and sufficient to recruit TFIIIB (BARTHOLOMEW et al. 1990, 1991). By contrast, the purified, human TFIIIC component is not sufficient to substitute for the TFIIIC activity of the 0.5 *M* phosphocellulose fraction: during purification of TFIIIC, Berk and colleagues separated this component, termed TFIIIC2, from a second factor, TFIIIC1, also essential for transcription of VA RNA I or tRNA genes (YOSHINAGA et al. 1987). Although no factor corresponding to TFIIIC1 was detected by KOVELMAN and ROEDER (1992), the factor equivalent to TFIIIC2, purified some 4000-fold as judged by specific DNA-binding activity, exhibited less than a ten fold purification of specific transcriptional activity. The second essential component, TFIIIC1, identified by Berk and colleagues has not been extensively characterized. Nevertheless, TFIIIC1 and TFIIIC2 can be distinguished by their interactions with different internal elements of type 2 RNAp III promoters, the A and B boxes, respectively (YOSHINAGA et al. 1987). Binding of TFIIIC2 to the B box can be detected in the absence of TFIIIC1, but the latter factor interacts only very weakly with the A box sequence in the absence of TFIIIC2 (YOSHINAGA et al. 1987). As yeast TFIIIC protects not only B box, but also A box, sequences of several yeast tRNA genes in footprinting assays (e.g., KASSAVETIS et al. 1989), whereas purified, human TFIIIC2 protects only the B box of the VA RNA I promoter (YOSHINAGA et al. 1987, 1989; DEAN and BERK 1988; SCHNEIDER et al. 1989), TFIIICI represents an obvious candidate for a factor that either directly recognizes the VA RNA I gene A box or stabilizes interaction between this promoter element and TFIIIC2. However, its dispensability for recruitment of TFIIIB to TFIIIC2-promoter DNA complexes (DEAN and BERK 1988) appears to argue against such a role. Thus, the mechanism of recognition of the A box of the VA RNA I promoter, presumably essential for correctly positioning human TFIIIC components for recruitment of TFIIIB into the assembling complex, is not known. Nor is it known whether human TFIIIB, once bound, interacts with promoter sequences upstream of and spanning initiation sites, as does yeast TFIIIB (e.g., BARTHOLOMEW et al. 1991; KASSAVETIS et al. 1992), perhaps because significant progress in the characterization of human TFIIIB has been made only recently.

The TATA-binding protein (TBP), 38 kDa in humans, was identified by virtue of its specific binding to the TATA elements of typical RNAp II promoters (see GREENBLATT 1991). Although TBP alone supports basal transcription by this enzyme from TATA element-containing promoters, this polypeptide normally operates in association with additional polypeptides (TATA-associated factors, TAFs) in the 750-kDa or larger, multiprotein TFIID complex (see GREENBLATT 1991; SHARP 1992; RIGBY 1993; HERNANDEZ 1993). The TBP was subsequently shown to be a component of both the RNAp I-specific factor SL-1 (COMAI et al. 1992) and factors that could supply TFIIIB activity in reconstituted RNAp III systems (WHITE et al. 1992; WHITE and JACKSON 1992; TAGGART et al. 1992; LOBO et al. 1992; SIMMEN et al. 1992; CHIANG et al.1993; see RIGBY 1993). Genetic experiments in yeast confirmed the essential role of TBP in transcription by all three RNA polymerases (CORMACK and STRUHL 1992; SCHULTZ et al.1992). These important findings have led to a reemphasis of the similarities, rather than the differeces, between the mechanisms by which the three eukaryotic RNA polymers initiate transcription (see, for example, RIGBY 1993; HERNANDEZ 1993).

Neither TBP itself nor TFIID can support RNAp III transcription from type 2 promoters, such as that of the VA RNA I gene (LOBO et al. 1992; SIMMEN et al. 1992; TAGGART et al. 1992; WHITE and JACKSON 1992), indicating that TFIIIB must contain RNA polymerase III-specific TAF. However, the number and nature of the constituents of transcriptionally active TFIIIB have not yet been established, as different investigators have reported separation of TFIIIB activity present in the 0.3 *M* KCl phosphocellulose fraction into at least two components under different experimental conditions. It has been clearly established that such TFIIIB activity includes a component that comprises TBP in association with a number of additional polypeptides (LOBO et al. 1992; TAGGART et al. 1992; SIMMEN et al. 1992; WHITE and JACKSON 1992; WHITE et al. 1992; CHIANG et al. 1993). The relationships among the TBP-associated polypeptides observed in factors that were necessary, alone or in conjunction with others, to substitute for the TFIIIB activity of the 0.3 *M* KCl phosphocellulose fraction have not yet been established, nor has it been demonstrated that each is required for activity. Further complicating the picture is the presence, in this same phosphocellulose fraction, of a TBP-containing complex capable of supporting basal, but not activated, transcription by RNAp II (TIMMERS and SHARP 1991). Whether this complex, termed B-TFIID (TIMMERS and SHARP 1991), supports RNAp III transcription is an open question: in some experiments, TFIIIB and B-TFIID activities exhibited different chromatographic properties (TIMMERS and SHARP 1991; TIMMERS et al. 1992; WHITE and JACKSON 1992), whereas others fractionation schemes have led to the conclusion that B-TFIID is likely to be equivalent to, or at least part of, the TBP-containing component of TFIIIB (LOBO et al. 1992 SIMMEN et al. 1992). It is, thus, not yet possible to decide whether B-TFIID is one and the same as (or at least part of) the TBP-containing TFIIIB component, in which case its ability to support basal transcription by RNAp II would be ascribed to the presence of some free TBP in B-TFIID preparations, or whether B-TFIID and the TBP-containing component of TFIIIB activity are biochemically and functionally distinguishable TBP-containing factors.

Another issue that has yet to be settled is the number, and nature, of factors present in the 0.3 *M* KCl phosphocellulose fraction that are essential for transcription by RNAp III, but do not contain TBP: several groups of investigators have each identified one such activity (LOBO et al. 1992; TAGGART et al. 1992; CHIANG et al. 1993), but none has yet been described in detail. It has not, therefore, been possible to compare the properties of these apparently essential components of TFIIIB activity, which were separated from TBP-containing complexes by different procedures by different investigators.

Although much remains to be learnt about the components and molecular properties of human RNAp III transcription factors, it is the recent advances in their characterization that permit the kinds of specific questions raised in previous paragraphs of this section to be framed. The resurgence of interest in the mechanisms by which mammalian RNAp III initiates transcription, spurred in part by the recognition that TBP is an essential player, promises a molecular description of assembly of preinitiation complexes on type 2 RNAp III promoters approaching the sophistication that has been achieved in the yeast system (e.g., BARTHOLOMEW et al. 1990, 1991; KASSAVETIS et al. 1992; see HERNANDEZ 1993, for a review). Such a description should pave the way for not only elucidation of such subsequent steps in transcription as recruitment of RNAp III, initiation of RNA synthesis, and the transition to elongation, but also a better understanding of the mechanisms that regulate transcription of VA RNA genes.

## 2.3 Regulation of Virus-Associated RNA I Transcription

Both the VA RNA I and VA RNA II genes are expressed from the beginning of the subgroup C adenovirus infectious cycle, but production of these RNA is not coordinately regulated: the rate of synthesis of VA RNA II increases modestly between 3 and 8 h after infection but then levels off, whereas the rate of VA RNA I production accelerates continuously during the late phase of infection (Söderlund et al. 1976). The temporal pattern of production of the single VA RNA species of Ad12 has been reported to resemble that of Ad2 or Ad5 VA RNA I (Varrichio and Raska 1976). Although synthesis of viral DNA provides new templates for transcription of VA RNA from the beginning of the late phase, the different temporal profiles of VA RNA I and VA RNA II synthesis indicate that specific mechanisms must stimulate VA RNA I and/or repress VA RNA II transcription. Repression of both VA RNA I and VA RNA II transcription has been reported, but the mechanisms that regulate VA RNA synthesis during productive infection are incompletely understood.

The VA RNA II gene is transcribed much less efficiently in in vitro systems than the VA RNA I gene, with differences of up to 50-fold reported (e.g., Fowlkes and Shenk 1980; Guilfoyle and Weinmann 1981; Wu and Cannon 1986). Thus, the former promoter is intrinsically weaker than the latter. This difference might account for the greater rate of synthesis of VA RNA I during the early phase of infection. However, a cellular factor that could regulate the relative rates of VA RNA I and VA RNA II transcription in infected cells has also been described. This factor, termed VA RNA I-binding protein (VBP), binds specifically to an intragenic sequence of the VA RNA I gene immediately upstream of and overlapping the B box sequence (van Dyke and Roeder 1987). Addition of partially purified VBP to transcription reactions containing both the VA RNAI and VA RNAII promoters and partially purified RNAp III, TFIIIB, and TFIIIC reduced the ratio of VA RNAI to VA RNA II transcription from 10:1 to 2:1 (van Dyke and Roeder 1987). This change was the result of inhibition of VA RNA I synthesis, with a concomitant increase in production of VA RNA II (van Dyke and Roeder 1987). It has, therefore, been suggested that during the early phase of infection VBP and TFIIIC, which bind to overlapping intragenic sites, would compete for the limited number of VA RNA I promoters available. Such competition would impair initiation of VA RNA I transcription, at the same time permitting interaction of TFIIIC with the B box of the VA RNA II promoter and thus initiation of transcription of this gene. During the late phase of infection, the concentration of VA RNA I promoters would no longer be limiting, permitting TFIIIC access to VBP-free VA RNA I genes. Thus, it was postulated that the relative rates of VA RNA I and VA RNA II transcription during the late phase of infection would be determined by the interactions of TFIIIC with the B box sequences of these genes (van Dyke and Roeder 1987). Although features of these promoters other than their B boxes appear to determine their different strengths (see next paragraph), binding of VBP to the VA RNA I promoter would nicely account for the fact that, despite the large differences in strengths of the VA RNA I and VA RNA II promoters observed in vitro, the former is transcribed only slightly more efficiently than the latter during the early phase of infection (SÖDERLUND et al. 1976). It would be of considerable interest to determine whether mutations that prevent or impair binding of VBP to the VA RNA I promoter, but do not alter its interaction with TFIIIC, lead to elevated ratios of VA RNA I to VA RNA II transcription during the early phase of infection.

Transcription from the VA RNA I promoter has, furthermore, been shown to repress VA RNA II synthesis during the late phase of infection: mutations of the A or B box sequences of the Ad5 VA RNA I gene that reduced VA RNA I synthesis to undetectable levels in infected cells resulted in at least a tenfold increase in VA RNA II transcription (BHAT and THIMMAPPAYA, 1984). Moreover, when VA RNA I transcription was inhibited by such mutations, the kinetics of VA RNA II production during the late phase resembled those of VA RNA I synthesis in wild-type virus-infected cells (BHAT and THIMMAPPAYA 1984). These properties clearly establish that active VA RNA I transcription represses VA RNA II transcription during the late phase of infection. The intrinsic differences in the strengths of the two promoters suggest that the VA RNA I promoter out-competes the VA RNA II promoter for a component of the RNAp III transcriptional machinery limiting in infected cells. The ability of VA RNA I to inhibit VA RNA II transcription in vitro (FOWLKES and SHENK 1980) and the stimulation of VA RNA II transcription in infected cells induced by mutation of the A or B box sequences of the VA RNA I promoter (BHAT and THIMMAPPAYA 1984) are consistent with such a competition mechanism. Nevertheless, this mechanism has not been well defined at the molecular level, not least because the interactions of TFIIIC and other components of the RNAp III transcriptional machinery with these two viral promoters have not been directly compared. It has, however, been established that, under conditions in which the VA RNA I gene is transcribed 50-fold more efficiently than VA RNA II, the intact VA RNA II gene or its B box are almost as effective competitors of RNAp III transcription as the VA RNA I internal control region (WU and CANNON 1986). This result is not surprising in view of the close similarity of the VA RNA I and VA RNA II B box sequences discussed previously (see Fig. 1). It therefore appears that TFIIIC-B box sequence interactions are not the primary determinants of the large differences in the strengths of these two promoters. Rather, other distinctive features of the VA RNA II gene must weaken its promoter, rendering it sensitive to competition from VA RNA I transcription in infected cells. Such features might include its A box, whose sequence conforms less well to a tRNA gene A box consensus than does that of the VA RNA I gene A box (Fig. 1). The somewhat shorter (by 2 bp) distance between the A and B box sequences of the VA RNA II promoter might also contribute to its weaker activity, for decreasing the spacing between these elements of the VA RNA I promoter can reduce transcription 50- to 100fold (BHAT and THIMMAPPAYA 1983; CANNON et al. 1986). The most surprising property of the VA RNA II promoter is, however, the presence of an internal termination site immediately downstream of the A box (WU and CANNON 1986). This sequence induces premature termination of some 50% of VA RNA II transcripts made in vitro (WU and CANNON 1986) and could thus contribute to the lower levels of VA RNA II than of VA RNA I produced in infected cells.

Whether such premature termination of VA RNA II transcription occurs in vivo has not been established.

It is, therefore, clear that the relative rates of VA RNA I and VA RNA II transcription change remarkably during the course of adenovirus infection and that VA RNA I transcription inhibits expression of the VA RNA II gene by a competition mechanism. The molecular mechanism of such competition has not, however, been fully elucidated, nor has it been established whether the rate of VA RNA I transcription is specifically regulated during the course of infection: the VBP described above might, for example, repress VA RNA I transcription during the early phase of infection. Whether the increased rate of VA RNA I transcription during the late phase can be explained simply by the increased concentration of templates, and possibly titration of VBP, also remains to be addressed. Such issues have received little attention, perhaps because the RNAp III system is not generally regarded as highly regulated. Be that as it may, the limited data available suggest that the alterations in both the relative and absolute rates of VA RNA I and VA RNA II transcription that occur as the infectious cycle progresses represent the net balance of several regulatory circuits.

One aspect of VA RNA I transcriptional regulation that has been investigated in some detail is its activation by adenovirus E1A proteins. The 289R E1A protein stimulates transcription from a wide variety of RNAp II promoters in infected or transfected cells (see BERK 1986; FLINT and SHENK 1989; SHENK and FLINT 1991). This protein is also an activator of tRNA and VA RNA I gene expression in transfected cells (BERGER and FOLK 1985; GAYNOR et al. 1985). The ability of the E1A protein to activate transcription from both RNAp II and type 2 RNAp III promoters, which share no obvious architectural features, and its lack of sequence-specific DNA-binding activity suggest that activation must be the result of regulation of the activities of cellular transcription factors. Thus, the E1A protein could facilitate the formation of stable preinitiation complexes or increase the rate of initiation from such complexes by direct or indirect modification of the activities of cellular transcription factors (see BERK 1986; FLINT and SHENK 1989). In the case of the VA RNA I promoter, alterations in TFIIIC have been implicated in its E1A protein-dependent transcriptional activation.

Extracts prepared from Ad5-infected cells transcribe the VA RNA I promoter more efficiently than do those prepared from cells infected by a virus, *dl* 312, that can express no E1A gene products (HOEFFLER and ROEDER 1985; YOSHINAGA et al. 1986). The greater activity of the former extracts has been shown, by addition of specific fractions obtained from infected cells to reconstituted transcription systems, to be the result of increased TFIIIC activity (HOEFFLER and ROEDER 1985; YOSHINAGA et al. 1986). Based on the results of template commitment assays, YOSHINAGA et al. (1986) concluded that expression of E1A proteins in infected cells increased the concentration of TFIIIC. As TFIIIC was the limiting component in the S100 extracts used in these experiments, the four to eightfold increases in TFIIIC concentration observed in Ad5-infected cell extracts accounted well for the increase of some five fold in the efficiency of VA RNA I transcription (YOSHINAGA et al. 1986). By contrast, using similar template commitment assays, Roeder and colleagues could detect no appreciable change in the TFIIIC concentration in Ad5-infected cells (HOEFFLER et al. 1988). This discrepancy appears to be the result of several differences in experimental conditions: one group (HOEFFLER et al. 1988) prepared nuclear extracts, which contain a greater proportion of total cellular TFIIIC than the S100 extracts used by the second (YOSHINAGA et al. 1986), and cells were infected under quite different conditions. Indeed, it has been suggested that the increased concentration of TFIIIC observed by YOSHINAGA et al. (1986) following prolonged incubation of infected cells in the presence of a DNA synthesis inhibitor might be a secondary consequence of an E1A protein-dependent modification of TFIIIC that increases its activity (see below) and perhaps stability (HOEFFLER et al. 1988).

When the interactions of TFIIIC with the VA RNA I gene promoter were examined using electrophoretic mobility shift and footprinting assays, expression of E1A proteins in Ad5-infected cells was observed to stimulate formation of a specific TFIIIC–VA RNA I promoter complex in the absence of a corresponding increase in TFIIIC concentration (HOEFFLER et al. 1988). The TFIIIC responsible for formation of this complex cochromatographed with transcriptionally active TFIIIC, whereas a fraction capable of forming a second, more rapidly migrating complex exhibited little transcriptional activity (HOEFFLER et al. 1988). Both complexes, nevertheless, contained TFIIIC, identified by specific protection of the B box of this promoter. In vitro treatment of TFIIIC preparations with phosphatase increased formation of the faster migrating, at the expense of the more slowly migrating, complex (HOEFFLER et al. 1988). These observations therefore identified two forms of TFIIIC in human cells: although both specifically recognized the B box of the VA RNA I promoter, only the form detected as the more slowly migrating TFIIIC–promoter DNA complex exhibited transcriptional activity. Expression of E1A proteins in Ad5-infected cells increased the concentration of the transcriptionally active form of TFIIIC without inducing a significant increase in the absolute TFIIIC concentration (HOEFFLER et al. 1988). As this form of TFIIIC was more highly phosphorylated than the transcriptionally inactive form, the E1A proteins appear to stimulate phosphorylation of TFIIIC, directly or indirectly (HOEFFLER et al. 1988).

More recently, transcriptionally active and inactive forms of human TFIIIC have been compared following their separation and extensive purification (KOVELMAN and ROEDER 1992). Although similar in composition to the transcriptionally active TFIIIC component described in Sect. 2.2, the inactive form lacks the 110-kDa polypeptide present in the former and contains additional polypeptides of 95 and 77 kDa (KOVELMAN and ROEDER 1992). Despite their radically different abilities to support VA RNA I transcription, the two forms of TFIIIC bind the VA RNA I promoter with similar affinities (KOVELMAN and ROEDER 1992). Thus, the E1A protein-dependent modification(s) that convert transcriptionally inactive to active TFIIIC (and vice versa) must alter some essential reaction in the pathway of initiation of RNAp III transcription that occurs after specific binding of TFIIIC to the B box.

Although the notion that E1A proteins regulate the modification state, and hence transcriptional activity, of human TFIIIC is attractive, it should be noted that the Ad5-infected cell-specific alterations in TFIIIC observed by HOEFFLER et al. (1988) may not be directly mediated by E1A gene products. Efficient transcription of adenoviral early genes requires production of the 289R E1A protein (see BERK 1986; SHENK and FLINT 1991). Thus, the E1A deletion mutant (*dl* 312)-infected cells used as the source of extracts that lack E1A proteins also contained severely reduced concentrations, compared to Ad5-infected cell extracts, of all other viral gene products. Of potential interest in this regard is a recent report that the E1B 19-kDa protein stimulates production of VA RNA I in transfected cells (SOLLERBRANT et al. 1993).

In view of the complications inherent in interpretation of differences in transcriptional activity exhibited by extracts prepared from wild-type and E1A mutant virus-infected cells, an in vitro system in which stimulation of VA RNA I transcription by the 289R E1A protein could be reproduced would be highly desirable. The 289R E1A protein synthesized in *E. coli* has been reported to stimulate VA RNA I transcription by up to a factor of 5 (DATTA et al. 1991). The E1A protein expressed in insect cells from a baculovirus vector is, however, much more active, stimulating VA RNA I transcription in HeLa cell extracts by as much as 50-fold, probably because it was appropriately modified in insect cells (PATEL and JONES 1990). The dramatic stimulatory activity of exogenous 289R E1A protein demonstrates unequivocally that this protein can activate VA RNA I transcription by a mechanism that does not require synthesis of new proteins, including other viral proteins in infected cells. Moreover, some evidence that bacterially expressed E1A protein itself might alter TFIIIC activity has been reported: when VA RNA I transcription in HeLa whole cell extracts was specifically inhibited by addition of an oligonucleotide containing the B box sequence of the VA RNA I gene, addition of exogenous 289R E1A protein restored VA RNA I synthesis to the level seen in untreated extracts (DATTA et al. 1991). These results suggest that bacterially synthesized 289R E1A protein might directly or indirectly convert inactive TFIIIC not sequestered by the oligonucleotide into a transcriptionally active form of the factor (DATTA et al. 1991). Such a conversion has not, however, been demonstrated. Nor has it been established whether baculovirus-expressed 289R E1A protein, whose considerable stimulatory activity (PATEL and JONES 1990) should greatly facilitate investigation of the molecular mechanism by which it activates VA RNA I transcription, alters TFIIIC activity.

Ironically, the physiological significance of stimulation of VA RNA I transcription by the E1A protein has never been established. Synthesis of VA RNA I could not be detected in cells infected by viruses deleted of E1A coding sequences, such as *dl* 312 discussed previously (JONES and SHENK 1979). However, as temperature-sensitive mutations of viral replication proteins that prevent entry of the reproductive cycle into the late phase also reduced VA RNA I synthesis to undetectable levels in these experiments, it was concluded that VA RNA I synthesis is not specifically or directly dependent on E1A gene products (JONES and SHENK 1979).

# 3 E2E RNA Polymerase III Transcription Unit

## 3.1 Organization

The subgroup C adenovirus E2 gene, which was among the first viral transcription units to be identified, has been studied in considerable detail. This complex transcription unit, which is transcribed by RNAp II in response to viral E1A proteins during the early phase of infection (see BERK 1986; FLINT and SHENK 1989; SHENK and FLINT 1991, encodes the viral replication proteins (see SHARP 1984; CHALLBERG and KELLY 1989). Such early transcription is from a promoter, the E2 early (E2E) promoter, located near map coordinate 75 in the 1-strand of the viral genome. At least 90% of the very large E2 primary transcripts synthesized from this promoter are polyadenylated at a site near map coordinate 61.5 and spliced to yield the E2E mRNA encoding the 72-kDa single-stranded DNA-binding protein (see SHARP 1984; FLINT 1986). This mRNA comprises three exons (Fig. 2A), two short, noncoding exons and a third containing the 72-kDa protein-coding sequence. The remaining E2E primary transcripts are polyadenylated at a site near map coordinate 11.5 and alternatively spliced to generate two large mRNA species encoding the viral DNA polymerase and protein primer for viral DNA synthesis, the preterminal protein. With the onset of the late phase of infection, a second E2 promoter, the E2 late (E2L) promoter located near map coordinate 72.1, is activated and eventually accounts for at least 90% of all E2 mRNA made in infected cells (CHOW et al. 1979; GOLDENBERG et al. 1981; BERK 1986; FLINT 1986). As illustrated in Fig. 2A, E2 mRNA made from the E2L promoter are distinguished from those expressed from the E2E promoter only by their unique, 5'-terminal exon. As these are noncoding, activation and predominant utilization of the E2L promoter during the late phase of subgroup C adenovirus infection do not alter the coding capacity of the E2 transcription unit.

The elements that comprise the RNAp II E2E promoter are well characterized, as this promoter has recieved considerable attention as a model for activation of early promoters by E1A and other activating proteins. In both in vitro systems and infected cells, the E2E promoter directs initiation of RNAp II transcription from two sites, the major +1 and minor –26 sites shown in Fig. 2A (see BERK 1986; SHENK and FLINT 1991). Transcription from each of these sites depends on a TATA element-like sequence located an appropriate distance upstream, and efficient transcription from both requires the upstream binding sites for the cellular factors ATF and E2F (Fig. 2). The TATA-like element controlling initiation from the +1 site exhibits a relatively poor match to canonical TATA sequences, such as the TATAAAA sequence of the Ad2 major late promoter (Fig. 2B). Nevertheless, this E2E sequence functions effectively as a TATA element in assays for both specific binding of human TBP and TATA element-dependent transcription (H. CHEN and S.J. FLINT, unpublished observations). Thus, it will be designated the E2E TATA element. In infected cells, this sequence, as

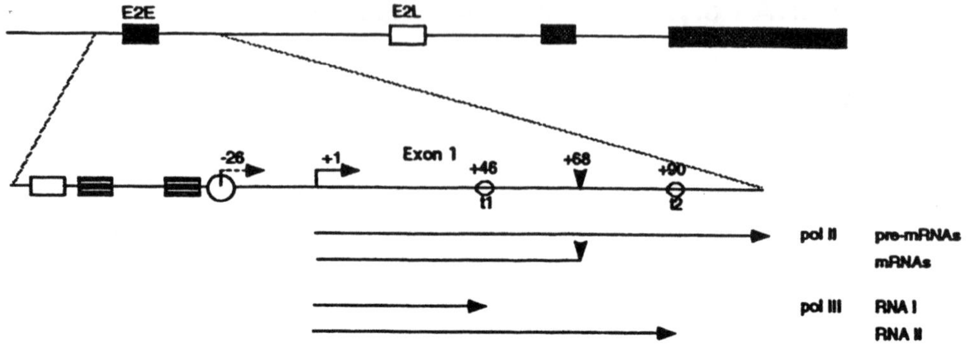


A

ATF E2F E2F TATA

GAGATGACGTAGTTTTCGCGCTTAAATTTGAGAAAGGGCGCGAAACTAGTCCTTAAGAGTCAGCG

-60 -40 -20

t1

CGCAGTATTTGCTGAAGAGAGCCTCCGCGTCTTCCAGCGTGCGCCGAAGCTGATCTTCGCTTTT

+1 +20 +40

t2

GTGATACAGGCAGCTGCGGGTGAGGGAGGGCAGAGACCTGTTTTTTATTTTCAGCTCTTGT

+60 +80 +100

B

**Fig. 2 A,B.** The subgroup C adenovirus E2 transcription units. **A** The organization of the Ad2 E2E RNAp II transcription is illustrated at the *top.* The first exons of RNA transcribed from the E2E and E2L promoters are shown by the *black* and *white rectangles*, respectively. All E2 mRNA include exons 2 and 3, shown in *gray.* The genomic positions of the 5' boundaries of each exon are shown below the line representing the viral genome. Features of the 5' end of the E2E transcription unit are shown below. The viral genome is represented by the *solid horizontal line*, and the major (+1) and minor (–26) sites of initiation by RNAp II by the *solid* and *dashed arrows*, respectively, drawn in the direction of transcription. The *vertical arrowhead* at position +68 represents the 5' splice site at the boundary between exon 1 and intron 1 of E2E pre-mRNA. The ATF- and E2F-binding sites and the TATA element controlling initiation from the +1 site are shown by the *open box, stripped boxes*, and *open circle* respectively. Sites at which RNAp III terminates E2E transcription in vitro are indicated by *open octagons.* The pre-mRNA and spliced exon 1 products of RNAp II transcription and the small RNA species transcribed by RNAp III are drawn below in the direction of transcription. **B** The sequence of the nontranscribed strand of the 5' terminal segment of the Ad2 E2E gene is shown, with the major initiation site represented by the *arrow* drawn in the direction of transcription. The upstream elements of the RNAp II E2E promoter are *highlighted*, and the termination sites for RNAp III transcription shown in *bold face*

well as the ATF and E2F sites shown in Fig. 2, are essential for both efficient basal activity of the E2E promoter and activation of this promoter in response to expression of E1A proteins (Manohar et al. 1990).

The organization of the E2E promoter summarized in Fig. 2 was originally elucidated using in vitro transcription systems, such as HeLa whole cell extracts, in which it directs accurate initiation by RNAp II. However, when conventional nuclear extracts were used as the source of transcriptional components, α-amanitine-resistant transcription from plasmids containing the E2E promoter was observed (Pruzan et al. 1992). Such transcription yields discrete RNA species of some 46 and 90 n, which have been termed E2E RNA I and RNA II, respectively

(PRUZAN et al. 1992). In vitro transcription from the E2E promoter in the presence of a concentration of α-amanitine that inhibits RNAp II terminates at the T-rich sites designated t1 and t2 in Fig. 2, whereas the RNAp II transcriptional machinery reads through these sequences (PRUZAN et al. 1992). These T-rich sequences closely resemble sites at which RNAp III terminates transcription of both cellular and viral genes (BOGENHAGEN and BROWN 1981; see GEIDUSCHEK and TOCCHINI-VALENTINI 1988). These properties, therefore, established that RNAp III present in conventional nuclear extracts can synthesize discrete, short RNA species from the 5'-terminal sequence of the E2E RNAp III transcription unit (PRUZAN et al. 1992).

Inefficient transcription by RNAp II from a vertebrate U6 snRNA promoter transcribed by RNAp III in vitro can be observed under certain in vitro conditions (MATTAJ et al. 1988). Conversely, artificial conditions that drive in vitro transcription from typical RNAp II promoters by RNAp III have been described (MITCHELL et al. 1992). These observations indicate that certain in vitro assay conditions can force inappropriate promoter recognition by the "wrong" eukaryotic RNA polymerase. Transcription from the E2E promoter by RNAp III in conventional nuclear extracts and by reconstituted RNAp III systems does not, however, represent simply another example of such physiologically irrelevant transcription: unlike any eukaryotic promoter analyzed previously, the E2E transcriptional control region directs transcription by both RNAp II and RNAp III in vivo (HUANG et al. 1994). The latter enzyme transcribes small E2E RNA species identical to those made in in vitro reactions in nuclei isolated from Ad2-infected cells, and small E2E RNA species with the properties of RNAp III transcripts made in in vitro reactions are present in the cytoplasm of infected cells (HUANG et al. 1994). The 3' termini of small, infected cell E2E RNA species, for example, map to the t1 and t2 termination sites initially defined in vitro (HUANG et al. 1994). Thus, the E2E transcriptional control region represents the first documented case of one that can specify transcription by more than a single RNA polymerase in vivo.

It has not yet been experimentally determined whether RNAp III also transcribes E2E genes of other adenoviruses. Nevertheless, the similar organization of the genomes of human adenoviruses and the conservation of important features of the E2E RNAp III transcription unit of the subgroup C serotypes Ad2 and Ad5 such as the t2 termination site, among the human adenoviruses for which the E2E sequence is available (a group that includes serotypes 40 and 41; S.J. FLINT, unpublished observations), strongly suggest that such superposition of RNAp II and RNAp III transcription units is unlikely to be a subgroup C-specific property.

## 3.2 E2E RNA Polymerase III Promoter

The E2E RNA species made by RNAp III in vitro or in infected cells are encoded by a viral DNA sequence that forms the 5' end of the RNAp II transcription unit (Fig. 2). Indeed both RNA polymerases initiate transcription in the vicinity of position +1 (PRUZAN et al. 1992; HUANG et al. 1994). Such superposition of the RNAp II and RNAp III E2E transcription units suggested that they might share promoter elements. Analysis of the activities of E2E promoters carrying precise substitutions of the upstream elements of the RNAp II promoter (Fig. 2) has established that efficient transcription by both enzymes in in vitro systems requires the TATA element described in the previous section (PRUZAN et al. 1992). Furthermore, mutation of the ATF-binding site (see Fig. 2) reduced the efficiency of transcription by both RNA polymerases by a factor of 2 to 3 (PRUZAN et al. 1992). Substitution of the E2F sites also induced modest changes in the efficiency of E2E transcription by RNAp III (PRUZAN et al. 1992). These experiments , therefore, established that upstream elements of the RNAp II promoter, in particular the TATA element, also function in the E2E RNAp III promoter.

A number of cellular and viral promoters whose transcription by RNAp III absolutely requires an upstream TATA element have been identified (see MURPHY et al. 1989; DAHLBERG and LUND 1991; HERNANDEZ 1992). Furthermore, it is well established that transcription of vertebrate U6 snRNA genes by RNA polymerase III rather than by RNAp II, which is responsible for transcription of the closely related U1 and U2 RNA genes, is specified by their TATA elements (see DAHLBERG and LUND 1991; HERNANDEZ 1992, 1993). Thus, the dependence of E2E transcription by RNAp III on an upstream TATA element is not a surprising property. Some TATA element-containing RNAp III promoters, notably vertebrate U6 RNA promoters, do not include internal elements analogous to the A and B box sequences of type 2 RNAp III promoters described in Sect. 2.2, whereas others require internal elements for efficient transcription (see MURPHY et al. 1989; HERNANDEZ 1992). In the case of the E2E promoter, one element located between positions +62 and +90 modestly stimulates transcription by RNAp III, whereas a second internal element, within the segment +30 to +60, is absolutely essential (PRUZAN et al. 1992; D. ELLSWORTH and S.J. FLINT, unpublished observations). The latter, by contrast to the upstream elements discussed previously, can be defined as an RNAp III-specific element, for its deletion altered neither the efficiency nor accuracy of E2E transcription by RNAp II. The E2E RNAp III promoter therefore appears to comprise both upstream and internal elements, a property in which it resembles the cellular 7SL RNA and Epstein-Barr virus (EBV) EBER RNA promoters (see GEIDUSCHEK and TOCCHINI-VALENTINI 1988; MURPHY et al. 1989; HERNANDEZ 1992).

Our current understanding of the organization of the promoter that directs E2E transcription by RNAp III is based solely on the results of in vitro assays. In general, such assays have accurately identified elements of both RNAp II and RNAp III promoters. In the specific case of the E2E promoter, in vitro systems reproduce both dependence of RNAp II transcription on two of the three types of upstream element that operate in infected cells and specific initiation and termination of RNAp III transcription (PRUZAN et al. 1992; HUANG et al. 1994). Nevertheless, such in vitro systems do not necessarily reproduce the exact quantitative contributions made by individual elements in vivo. The E2F sites are, for example, important for efficient E2E transcription by RNAp II in infected cells, but do not contribute a great deal to activity of this promoter in standard in vitro systems. A complete description of the E2E RNAp III promoter will, therefore, require analysis of the activities of mutated promoters in adenovirus-infected cells.

Although the full set of factors necessary for initiation of E2E transcription by RNAp III has not yet been defined, the inhibition of E2E RNAp III transcription observed when nuclear extracts were depleted of TBP-containing complexes using monoclonal antibodies raised against human TBP (CHATTERJEE et al. 1993) indicated that this set includes a TBP-containing factor (PRUZAN et al. 1992). Transcription by RNAp III could be fully restored by a TBP-containing complex immunoaffinity purified from the TFIIIB-containing phosphocellulose fraction described in Sect. 2.2, but not by a TFIID-containing fraction (PRUZAN et al. 1992). Indeed, the latter further inhibited RNAp III transcription when added to immunodepleted extracts. These observations established that a TBP-containing complex, biochemically and functionally distinguishable from the RNAp II factor TFIID, mediates transcription from the E2E transcriptional control region by RNA polymerase III (PRUZAN et al. 1992). Subsequently, a purified TBP-containing complex of some 220 kDa that is necessary for RNAp III transcription from the TATA element-containing E2E promoter has been shown to fulfill the requirement for TFIIIB when the VA RNA I promoter, which possesses no such element, is transcribed in reconstituted reactions containing partially purified RNAp III and TFIIIC (R. PRUZAN, A. UISHEVA and S.J. FLINT, unpublished observations). Thus, this complex can be functionally defined as TFIIIB or one essential component of TFIIIB activity.

Transcription from the E2E promoter by RNAP III depends on an internal control region, as discussed previously, and an activity present in partially purified, human TFIIIC (R. PRUZAN, unpublished observations). These properties suggest that initiation of RNAp III transcription from this promoter also requires TFIIIC. Any TFIIIC-binding site in the E2E RNAp III promoter must, however, be of relatively low affinity, as no sequences closely matching the A and B box consensus sequences of type 2 promoters are present in the E2E RNAp III transcription unit. We believe that it is the combination of such a nonconsensus TFIIIC-binding site with the unusual TATA element of the E2E transcriptional control region that is likely to permit recognition by both RNAp II and RNAp III. The promoter of the EBV EBER-2 gene comprises both consensus A and B box sequences and an upstream region (ROSA et al. 1981; HOWE and SHU 1989). The latter contains binding sites for ATF and SP1, closely preceding a TATA-like sequence, TATAGAG, that is very similar to the noncanonical TATA element of the E2E promoters, TTAAGAG. The EBER-2 promoter is, however, exclusively transcribed by RNAp III, unless its TATA-like sequence is replaced with the canonical TATA element of the Ad2 major late promoter (HOWE and SHU 1993). Thus, a noncanonical TATA sequence in conjunction with consensus TFIIIC-binding sites results in exclusive recognition by RNAp III, whereas typical A and B box sequences combined with a canonical TATA element permit transcription by both RNAp II and RNAp III, presumably because RNAp II-specific (e.g., TFIID) and

RNAp III-specific (e.g., TFIIIC) factors are on more or less equal footing in the competition for access to their binding sites in the promoter. The combination of a noncanonical TATA sequence with noncansensus internal elements, which we believe are present in the E2E RNAp III promoter, would therefore also be expected to permit transcription by both RNA polymerases. The hypothesis that it is the combination of RNAp II-specific and RNAp III-specific elements of similar strengths (both relatively weak in the case of the E2E genes) in a single transcriptional control region that is essential for recognition by the two RNA polymerases is consistent with our observation that substitution of the E2E TATA sequence with the canonical major late TATA element impairs transcription by RNAp III (D. Riley, R. Pruzan and S.J. Flint, unpublished observations).

## 3.3 Expression and Function of E2E RNA Polymerase III Transcripts in Adenovirus-Infected Cells

The presence of a third RNAp III transcription unit in the adenoviral genome was unexpected, for no hint of the small E2E RNA species described in the previous section was discerned in nearly 30 years of investigation of small, viral RNA species. However, the E2E RNA species made by RNAp III attain maximal concentrations orders of magnitude lower than those of VA RNA I or VA RNA II (see next paragraph) and are thus difficult to detect. The discovery of this transcription unit raises the issue of the part played by these previously unrecognized RNAp III transcripts during the adenovirus reproductive cycle. Although this question cannot be answered definitively, the properties exhibited by E2E RNAp III transcripts in infected cells and the unusual organization of the promoters that permit transcription by RNAp II or by RNAp III suggest that E2E transcription by RNAp III may serve to regulate production of E2E mRNA.

The results of both run-off transcription in nuclei isolated from Ad5-infected HeLa cells and quantitative analysis of steady state populations of infected-cell, cytoplasmic RNA indicate that E2E transcription by RNAp III begins during the early phase of infection (Huang et al. 1994; W. Huang and S.J. Flint, unpublished observations). Like VA-RNA (see Sect. 2.3), these RNAp III transcripts attain their maximal concentration following the onset of the late phase of infection. Throughout the period examined, E2E RNA RNAp III transcripts, also like the VA RNA, are concentrated in cytoplasmic fractions of infected cells. However, the small E2E RNA species accumulate to only very low concentrations in the cytoplasm, reaching a maximum of five to ten copies per cell 12 h after infection in the case of E2E RNA II (Huang et al. 1994). This value is not only some $10^7$-fold lower than that reported for VA RNA I (Söderlund et al. 1976), but also far below the concentrations of even the less abundant, small, nuclear, or cytoplasmic cellular RNA species, on the order of $10^4$ copies per cell (see Zieve and Sauterer 1990). Thus, these E2E RNAp III transcripts are present in the cytoplasm of

Ad5-infected cells at a much lower concentration than the E2E mRNA transcribed by RNA polymerase II. The results of preliminary experiments suggest, however, that these small E2E RNA species are less stable than E2E mRNA.

The small E2E RNA species made by RNAp III in productively infected cells might, like VA RNA I, participate in some reaction important for efficient virus reproduction. In this case, the superposition of the RNAp III promoter on that directing E2E transcription by RNAp II would simply appear to represent a device that had evolved to maximize the efficiency with which the genetic information of the virus was utilized. On the other hand, the superposition of the RNAp III and RNAp II transcription units, whose promoters share upstream elements, and the properties of E2E RNAp III transcripts and their synthesis in infected cells suggest that RNAp III transcription may serve to limit, by competition, the activity of the RNAp II E2E promoter. Such competition between RNAp II and RNAp III transcription from the E2E promoter may also account, at least in part, for the activation of the E2L promoter at the beginning of the late phase of infection (see Sect. 3.1.). The hypothesis that E2E transcription by RNAp III plays a regulatory role raises the question of why it might be necessary to restrain transcription from the E2E promoter by RNAp II during the early phase of infection. It is possible that production of too high a concentration of the 72-kDa protein, whose overexpression in the absence of other viral proteins is lethal to HeLa cells (KLESSIG et al. 1984), early in the infectious cycle is deleterious. Accumulation of the proteins encoded by E2E mRNA to a certain threshold concentration must trigger initiation of viral DNA synthesis and entry into the late phase of the cycle. Thus, it is also possible that competition between RNAp III and RNAp II for access to the E2E promoter represents a device to extend the length of the early phase, so that, for example, early gene products that function during the late phase (such as the E1B 55-kDa and E4 34-kDa proteins) are present at optimal concentrations when required.

It will be considerable interest to learn whether the E2E RNAp III transcription unit described in this section does indeed represent an unusual regulatory tool, especially as the superposition of RNAp II and RNAp III transcription units exemplified by the subgroup C adenovirus E2E region seems unlikely to be a virus-specific phenomenon: transcriptional control regions of the human *c-myc* gene appear likely cellular candidates for this arrangement, for they direct synthesis of small RNA species by RNAp III in vitro, in *Xenopus* oocytes, and in isolated nuclei (CHUNG et al. 1987; BENTLEY et al. 1989). Although analogues of these RNAp III transcripts have not been detected in intact cells (see SPENCER and GROUDINE 1991), no specific search has been made for putative RNAp III transcripts that might, like the small E2E RNA, be present at very low concentrations. Be that as it may, the answers to questions about the function and mechanism of E2E transcription by RNAp III raised in this section seem certain to provide new insights into both the adenovirus reproductive cycle and the mechanisms that govern promoter recognition by the eukaryotic RNA polymerases.

## References


Akusjärvi G, Mathews MB, Andersson P, Vennströmm B, Pettersson U (1980) Structure of genes for virus-associated RNAI and RNAII of adenovirus type 2. Proc Natl Acad Sci USA 77: 2424–2428

Bartholomew B, Kassavetis GA, Braun BR, Geiduschek EP (1990) The subunit structure of Saccharomyces cerevisiae transcription factor IIIC probed with a novel photo-crosslinking reagent EMBO J 9: 2197–2205

Bartholomew B, Kassavetis GA, Geiduschek EP (1991) The components of Saccharomyces cerevisiae transcription factor IIIB are stereospecifically located upstream of a tRNA gene and interact with the second-largest subunit of TFIIIC. Mol Cell Biol 11: 5181–5189

Bentley DL, Brown WL, Groudine M (1989) Accurate TATA-box dependent polymerase III transcription from promoters of the c-myc gene in injected Xenopus oocytes Genes Dev 3: 1179–1189

Berger SL, Folk WR (1985) Differential activation of RNA polymerase III-transcribed genes by the polyomavirus enhancer and the adenovirus E1A gene products. Nucleic Acids Res 13: 1413–1428

Berk AJ (1986) Adenovirus promoters and E1A transactivation. Annu Rev Genet 20: 45–79

Bhat RA, Thimmappaya B (1983) Two small RNAs encoded by Epstein-Barr virus can functionally substitute for the virus-associated RNAs in the lytic growth of adenovirus 5. Proc Natl Acad Sci USA 80: 4789–4793

Bhat RA, Thimmappaya B (1984) Adenovirus mutants with DNA sequence perturbations in the intragenic promoter of VA I RNA gene allow the enhanced transcription of VAII RNA gene in HeLa cells Nucleic Acids Res 12: 7377–7388

Bhat RA, Metz B, Thimmappaya B (1983) Organization of the noncontiguous promoter components of adenovirus VA I RNA gene is strikingly similar to that of eukaryotic tRNA genes. Mol Cell Biol 3:1996–2005

Bhat RA, Domer PH, Thimmappaya B (1985) Structural requirements of adenovirus VA I RNA for its translation enhancement function. Mol Cell Biol 5:187–196

Bogenhagen DF, Brown DD (1981) Nucleotide sequences in Xenopus 5S DNA required for transcription termination. Cell 24: 261–270

Bogenhagen DF, Sakonju S, Brown DD (1980) A control region in the center of the 5S RNA gene directs specific initiation of transcription II. The 3' border of the region. Cell 19: 27–35

Cannon RE, Wu G-J, Railey JF (1986) Functions of and interactions between the A and B blocks in the adenovirus type 2-specific VA RNA I gene. Proc Natl Acad Sci. USA 83: 1285–1289

Carey M, Gerrard SP, Cozzarelli NR (1986) Analysis of RNA polymerase III transcription complexes by gel filtration. J Biol Chem 261: 4309–4317

Celma ML, Pan J, Weissman SM (1977a) Studies of low molecular weight RNA from cells infected with adenovirus 2. II. Heterogeneity at the 5' end of VA-RNA I. J Biol Chem 252: 9043–9046

Celma ML, Pan J, Weissman SM (1977b) Studies of low molecular RNA from cells infected with adenovirus 2. I. The sequences at the 3' end of VA-RNA I J Biol Chem 252: 9032–9042

Challberg MD, Kelly TJ (1989) Animal virus DNA replication. Annu Rev Biochem 58: 671–717

Chatterjee PK, Pruzan RP, Flint SJ (1993) Purification of an active TATA-binding protein-containing factor using a monoclonal antibody that recognizes the human TATA-binding protein. Protein Expr Purif 4: 445–455

Chiang C-M, Ge H, Wang Z, Hoffman A, Roeder RG (1993) Unique TATA-binding protein-containing complexes and cofactors involved in transcription by RNA polymerases II and III. EMBO J 12: 2749–2762

Chow LT, Broker TR, Lewis J (1979) Complex splicing patterns of RNAs from the early regions of adenovirus-2 J Mol Biol 134: 265–303

Chung J, Sussman OJ, Zeller R, Leder P (1987) The c-myc gene encodes superimposed RNA polymerase II and III promoters. Cell 51: 1001–1008

Comai L, Tanese N, Tjian R (1992) The TATA-binding protein and associated factors are integral components of the RNA polymerase I transcription factor, SL1. Cell. 68: 965–976

Cormack BP, Struhl K (1992) The TATA-bindig protein is required for transcription by all three nuclear RNA polymerases in yeast cells. Cell 69: 697–702

Dahlberg J, Lund E (1991) How does III X II make U6? Science 254: 1452–1453

Datta S, Soong C-J, Wang DM, Harter ML (1991). A purified adenovirus 289-amino-acid E1A protein activates RNA polymerase III transcription in vitro and alters transcription factor TFIIIC. J Virol 65: 5297–5304

Dean N, Berk AJ (1988) Ordering promoter binding of class III transcription factors TFIIIC1 and TFIIIC2. Mol Cell Biol 8: 3017–3025
Dreyfuss G, Philipson L, Mattaj IW (1988) Ribonucleoprotein particles in cellular processes. J Cell Biol 106: 1419–1425
Engelke DR, Ng S-Y, Shastry BS, Roeder RG (1980) Specific interaction of a purified transcription factor with an internal control region of 5S RNA genes. Cell 19: 717–728
Engler JA, Hoppe MS, van Bree MP (1983) The nucleotide sequence of the genes encoded in early region 2b of human adenovirus type 7. Gene 21: 145–159
Flint SJ (1986) Regulation of adenovirus mRNA formation. Adv Virus Res 31: 169–228
Flint SJ, Shenk T (1989) Adenovirus E1A protein: paradigm viral transactivator. Annu Rev Genet 23: 141–161
Föhring B, Geis A, Kronly M, Raska KJ (1979) Adenovirus type 12 VA RNA. Virology 95: 295–302
Fowlkes DM, Shenk T (1980) Transcriptional control regions of the adenovirus VAI RNA gene. Cell 22: 405–413
Fuhrman SA, Engelke DR, Geiduschek EP (1984) HeLa cell RNA polymerase III transcription factors. Functional characterization of a fraction identified by its activity in a second template rescue assay. J Biol Chem 259: 1934–1943
Gabrielsen OS, Sentenac A (1991) RNA polymerase III(C) and its transcription factors. Trends Biochem Sci 16: 412–416
Gaynor RS, Feldman LT, Berk AJ (1985) Transcription of class III genes activated by viral immediate early proteins. Science 230: 447–450
Geiduschek EP, Tocchini-Valentini GP (1988) Transcription by RNA polymerase III. Annu Rev Biochem 57: 873–914
Goldenberg CJ, Rosenthal R, Bhaduri S, Raskas HJ (1981) Coordinate regulation of two cytoplasmic RNA species transcribed from early region 2 of the adenovirus 2 genome. J Virol 38: 932–939
Greenblatt J (1991) Roles of TFIID in transcriptional initiation by RNA polymerase II. Cell 66: 1067–1070.
Guilfoyle R, Weinmann R (1981) Control region for adenovirus VA RNA transcription. Proc Natl Acad Sci USA 78: 3378–3382
Guthrie C, Patterson B (1988) Spliceosomal RNAs. Annu Rev Genet. 22: 387–419
Hernandez N (1992) Transcription of vertebrate snRNA genes and related genes. In: McKnight SL, Tjian R (eds) Transcriptional regulation. Cold Spring Harbor Laboratory, New York, pp 281–313
Hernandez N (1993) TBP, a universal eukaryotic transcription factor? Genes Dev 7: 1291–1308
Hoeffler WK, Roeder RG (1985) Enhancement of RNA polymerase III transcription by the E1A gene product of adenovirus. Cell 41: 955–963
Hoeffler WK, Kovelman R, Roeder RG (1988) Activation of transcription factor TFIIIC by the adenovirus E1A protein. Cell 53: 907–920
Howe JG, Shu M-D (1989) Epstein-Barr virus small RNA (EBER) genes: unique transcription units that combine RNA polymerase II and III promoter elements. Cell 57: 825–834
Howe JG, Shu M-D (1993) Upstream basal promoter element important for exclusive RNA polymerase III transcription of the EBER2 gene. Mol Cell Biol 13: 2655–2665
Huang JT, Schneider RJ (1991) Adenovirus inhibition of cellular protein synthesis involves inactivation of cap-binding protein. Cell 65: 271–280
Huang W, Pruzan R, Flint SJ (1994) In vivo transcription from the adenovirus E2 early promoter by RNA polymerase III. Proc Natl Acad Sci USA 91: (in press)
Jaehning JA, Woods PS, Roeder RG (1977) Purification, properties and subunit structure of deoxyribonucleic acid-dependent ribonucleic acid polymerase III from uninfected and adenovirus 2-infected KB cells. J Biol Chem 252: 8762–8771
Jones N, Shenk T (1979) An adenovirus type 5 early gene function regulates expression of other early viral genes. Proc Natl Acad USA 76: 3665–3669
Kassavetis GA, Riggs DL, Negri R, Nguyen LH, Geiduschek EP (1989) Transcription factor IIIB generates extended DNA interactions in RNA polymerase III transcription complexes on tRNA genes. Mol Cell Biol 9: 2551–2556
Kassavetis GA, Joazerio CPA, Pisano M, Geiduschek EP, Colbert T, Hahn S, Bianco AJ (1992) The role of the TATA-binding protein in the assembly and function of the multisubunit yeast RNA polymerase III transcription factor TFIIIB. Cell 71: 1055–1064
Klessig DF, Brough DE, Cleghorn VG (1984) Introduction, stable integration and controlled expression of a chimeric adenovirus gene whose product is toxic to the recipient human cells. Mol Cell Biol 4: 1354–1362

Kovelman R, Roeder RG (1992) Purification and characterization of two forms of human transcription factor IIIC. J Biol Chem 267: 24, 446–24, 456
Larsson S, Svensson C, Akusjärvi G (1986a) Characterization of low molecular weight virus-associated (VA) RNA encoded by simian adenovirus type 7, which can functionally substitute for adenovirus type 5 VA-RNA I - J Virol 60: 635–644
Larsson S, Bellett A, Akusjärvi G (1986b) VA RNAs from avian and human adenoviruses: dramatic differences in length, sequence and gene location. J Virol 58: 600–609
Lassar AB, Martin PL, Roeder RG (1983) Transcription of class III genes: formation of preinitiation complexes. Science 222: 740–748
Lobo SM, Tanaka M, Sullivan ML, Hernandez N (1992) A TBP complex essential for transcription from TATA-less but not TATA-containing RNA polymerase III promoters is part of the TFIIIB fraction. Cell 71: 211–220
Manohar CF, Kratochvil J, Thimmappaya B (1990) The adenovirus EII early promoter has multiple E1A-sensitive elements, two of which function cooperatively in basal and virus-induced transcription. J Virol 64: 2457–2466
Mathews MB (1975) Genes for VA RNA in adenovirus 2. Cell 6: 223–229
Mathews MB (1995) Structure, function, and evolution of adenovirus virus-associated RNA. Springer, Berlin Heidelberg New York (Current Topics in Microbiology and Immunology, vol 199/II)
Mathews MB, Pettersson U (1978) The low molecular weight RNAs of adenovirus 2 infected cells. J Mol Biol 119: 293–328
Mattaj IW, Dathan NA, Parry HD, Carbon P, Krol A (1988) Changing the RNA polymerase specificity of U snRNA gene promoters. Cell 58: 55–67
Mitchell MT, Hobson GM, Benfield PA (1992) TATA box-mediated polymerase III transcription in vitro. J Biol Chem 267: 1995–2005
Murphy S, Moorefield B, Pieler T (1989) Common mechanisms of promoter recognition by RNA polymerases II and III. Tends Genet 5: 122–126
Ohe K, Weissman SM (1970) Nucleotide sequence of an RNA from cells infected with adenovirus 2. Science 167: 879–881
Pan J, Celma ML, Weissman SM (1977) Studies of low molecular weight RNA from cells infected with adenovirus 2. III. The sequence of the promoter for VA-RNA I. J Biol Chem 252: 9047–9054
Patel G, Jones NC (1990) Activation in vitro of RNA polymerase II and III directed transcription by baculovirus produced E1A protein. Nucleic Acids Res 18: 2909–2915
Pettersson U, Philipson L (1975) Location of sequences on the adenovirus genome coding for the 5.5S RNA. Cell 6: 1–4
Price R, Penman S (1972) Transcription of the adenovirus genome by an alpha-amanitin-sensitive RNA polymerase in HeLa cells. J Virol 9: 621–626
Pruzan R, Chatterjee PK, Flint SJ (1992) Specific transcription from the adenovirus E2E promoter by RNA polymerase III requires a subpopulation of TFIID. Nucleic Acids Res 20: 5705–5712
Railey JF, Wu GJ (1988) Organization of multiple regulatory elements in the control region of the adenovirus type 2-specific VA RNA I gene: fine mapping with linker-scanning mutants. Mol Cell Biol 8: 1147–1159
Reich PR, Forget BG, Weissman SM, Rose JA (1966) RNA of low molecular weight in KB cells infected with adenovirus type 2. J Mol Biol 17: 428–439
Rich A, RajBhandary UL (1976) Transfer RNA; molecular structure, sequence and properties. Annu Rev Biochem 45: 805–860
Rigby PWJ (1993) Three is one and one in three: it all depends on TBP. Cell 72: 7–10
Roeder RG (1976) Eukaryotic nuclear RNA polymerases In: Losick R, Chamberlin M (eds) RNA polymerase. Cold Spring Harbor Laboratory, New York, pp 285–329
Rohan RM, Ketner G (1987) A comprehensive collection of point mutations in the internal promoter of the adenoviral VA I gene. J Biol Chem 262: 8500–8507
Rosa MD, Gottlieb E, Lerner MR, Steitz JA (1981) Striking similarities are exhibited by two small Epstein-Barr virus-encoded ribonucleic acids and adenovirus-associated ribonucleic acids VA I and VA II. Mol Cell Biol 1: 785–796
Sakonju S, Bogenhagen DF, Brown DD (1980) A control region in the center of the 5S RNA gene directs specific initiation of transcription 1. The 5' border of the region. Cell 19: 13–25
Schneider HR, Waldschmidt R, Seifart KH (1989) Purification of human transcription factor IIIC and its binding to the gene for ribosomal 5S RNA. Nucleic Acids Res 17: 5003–5017
Schneider HR, Waldschmidt R, Jahn D, Seifart KH (1990) Human transcription factor IIIC contains a polypeptide of 55kDa speciafically binding to Pol III genes. Nucleic Acids Res 18: 4743–4750

Schultz MC, Reeder RH, Hahn S (1992) Variants of the TATA-binding protein can distinguish subsets of RNA polymerase I, II and III promoters. Cell 69: 697–702

Segall J, Matsui T, Roeder RG (1980) Multiple factors are required for the accurate transcription of purified genes by RNA polymerase III. J Biol Chem 255: 11986–11991

Sharp PA (1984) Adenovirus transcription. In: Ginsberg HS (ed) The adenovirus. Plenum, New York, pp 173–204

Sharp PA (1992) TATA-binding protein is a classless factor. Cell 68: 819–821

Shastry BS, Ng S-Y, Roeder RG (1982) Multiple factors involved in the transcription of class III genes in Xenopus laevis. J Biol Chem 257: 12,979–12,986

Shenk TE, Flint SJ (1991) Transcriptional and transforming activites of the adenoviruses E1A proteins. Adv Cancer Res 57: 47–83

Shu L, Hong JS, Wei Y-F, Engler J (1986) Nucleotide sequence of the genes encoded in early region 2b of human adenovirus type 12. Gene 46: 187–195

Simmen KA, Bernues J, Lewis JD, D, Mattaj IW (1992) Confractionation of the TATA-binding protein with the RNA polymerase III transcription factor TFIIIB. Nucleic Acids Res 20: 5889–5898

Snouwaert J, Bunick D, Hutchison C, Fowlkes DM (1987) Large numbers of random point and cluster mutations within the VAI gene allow characterization of sequences required for efficient transcription. Nucleic Acids Res 15: 8293–8303

Söderlund H, Pettersson U, Vennströmm B, Philipson L, Mathews MB (1976) A new species of virus-coded, low molecular weight RNA from cells infected with adenovirus type 2. Cell 7: 585–593

Sollerbrant K, Akusjärvi G, Svensson C (1993) Repression of RNA polymerase III transcription by adenovirus E1A. J Virol 67: 4195–4204

Spencer C, Groudine M (1991) Control of c-myc regulation in normal and neoplastic cells. Adv Cancer Res 56: 1–48

Steitz JA, Black DL, Gerke V, Parker KA, Kramer A, Frendeway D, Keller W (1988) Functions of the abundant U snRNPs. In: Birnstiel ML (ed) Structure and function of major and minor snRNPs. Springer, Berlin Heidelberg New York, pp 115–154

Taggart AKP, Fisher TS, Pugh BF (1992) The TATA-binding protein and associated factors are components of pol III transcription factor TFIIIB. Cell 71: 1015–1028

Thimmappaya B, Jones N, Shenk T (1979) A mutation which alters initiation of transcription by RNA polymerase III on the Ad5 chromosome. Cell 16: 946–954

Timmers HTM, Sharp PA (1991) The mammalian TFIID protein is present in two functionally distinct complexes. Genes Dev 5: 1946–1956

Timmers HTM, Meyers RE, Sharp PA (1992) Composition of transcription factor B-TFIID. Proc Natl Acad Sci USA 89: 8140–8144

van Dyke MW, Roeder RG (1987) Multiple proteins bind to VA RNA genes of adenovirus type 2. Mol Cell Biol 7: 1021–1031

Varrichio F, Raska K (1976) Synthesis of VA RNA in productive and abortive infections with adenovirus type 12. Virology 71: 450–458

Vennstrom B, Pettersson U, Philipson L (1978a) Initiation of transcription in nuclei isolated from adenovirus infected cells. Nucleic Acids Res 5: 205–219

Vennström B, Pettersson U, Philipson L (1978b) Two sites for initiation of adenovirus 5.5S RNA. Nucleic Acids Res 5: 195–204

Weinmann R, Raskas HJ, Roeder RG (1974) Role of DNA-dependent RNA polymerases II and III in transcription of the adenovirus genome late in productive infection. Proc Natl Acad Sci USA 71: 3426–3430

Weinmann R, Brendler HJ, Raskas HJ, Roeder RG (1976) Low molecular weight viral RNAs transcribed by RNA polymerase III during adenovirus 2 infection. Cell 7: 557–566

White RJ, Jackson SP (1992) Mechanism of TATA-binding protein recruitment to TATA-less class III promoter. Cell 71: 1041–1053

White RJ, Jackson SP, Rigby PWJ (1992) A role for the TATA-box-binding protein component of the transcription factor TFIID as a general RNA polymerase III transcription factor. Proc Natl Acad Sci USA 89: 1949–1953

Wolin SL, Walter P (1991) Small ribonucleoproteins. Current Opin Struct Biol 1: 251–257

Wu G-J (1978) Adenovirus DNA-directed transcription of 5.5S RNA in vitro. Proc Natl Acad Sci USA 75: 2175–2179

Wu G-J, Cannon RE (1986) Termination sequences in the control region of the Ad2-specific VA RNA 2 gene. J Biol Chem 261: 12, 633–12, 642

Wu G-J, Railey JF, Cannon RE (1987) Defining the functional domains in the control region of the adenovirus type 2 specific VA RNAI gene. J Mol Biol 194: 423–442
Yoshinaga S, Dean N, Han M, Berk AJ (1986) Adenovirus stimulation of transcription by RNA polymerase III. evidence for an E1A-dependent increase in transcription factor IIIC concentration. Embo J 5: 343–354
Yoshinaga SK, Boulanger PA, Berk AJ (1987) Resolution of human transcription factor TFIIIC into two functional components. Proc Natl Acad Sci USA 84: 3585–3589
Yoshinaga SK, L'Etoile ND, Berk AJ (1989) Purification and characterization of transcription factor III C2. J Biol Chem. 264: 10726–10731
Young HF (1991) Ribosomal RNA and translation. Annu Rev Biochem 60: 191–227
Zieve GW, Sauterer RA (1990) Cell biology of the snRNP particles. Crit Rev Biochem Mol Biol 25: 1–46

# Adenovirus Endopeptidase and Its Role in Virus Infection

J.M. Weber



## 1 Introduction

Although adenoviruses were first described in 1953, significant progress on the study of viral proteins was only realized after the application of the newly discovered protein separation technique by sodium dodecyl sulfate polyacrylamide gel electrophoresis (SDS-PAGE). Metabolic labeling experiments using SDS-PAGE demonstrated the processing of several slightly larger precursor proteins into smaller and stable products (Anderson et al. 1973). Processing of RNA virus polyproteins was well known by this time, but the processing of individual viral proteins had only been described in bacteriophages (Murialdo and Siminovitch 1972). Studies with adenovirus temperature-sensitive mutants showed that processing was coordinately regulated and was linked to some late event in virus assembly, because all mutants were uniformly defective for processing. As most mutants fail to assemble virus particles, assembly appears to be a minimal requirement for precursor processing. That it is not a sufficient requirement is shown by some fiber mutants which assemble fiberless capsids in the absence of precursor processing (Falgout and Ketner 1988). The *ts*1 mutant was to become the paradigm of this late event. At the nonpermissive temperature, *ts*1 assembles virus particles which contain the genome and six unprocessed precursor proteins, namely pVI, pVII, pVIII, pIIIa, pμ, and preterminal

Sherbrooke University, Faculty of Medicine, Department of Microbiology, Sherbrooke, Québec J1H 5N4, Canada

protein, pTP (WEBER 1976, 1990). Significantly, these particles are not infectious because they fail to uncoat (MIRZA and WEBER 1980; MILES et al. 1980).

The *ts*1 mutation was subsequently mapped and sequenced within the distal-most L3 open reading frame as a T to C transition mutation causing the replacement of a Pro by a Leu residue at position 137 (YEH-KAI et al. 1983). Studies with *ts*1 revertants confirmed the identity of the open reading frame and the *ts*1-proteinase gene (WEBER and HOUDE 1987). Formal proof of the adenoviral proteinase came with the demonstration of in vitro proteinase activity of the purified recombinant enzyme expressed in *Escherichia coli* (ANDERSON 1990; HOUDE and WEBER 1990).

## 2 Proteinase Sequences

The sequence of 12 proteinase genes has been determined and submitted to the European Molecular Biology Laboratory (EMBL)/GenBank thus far. They were all located in the rightward transcribed strand in the L3 region between the hexon and the DNA-binding protein gene approximately in the middle of the genome. The protein sequences are translations from DNA sequences, direct verification having been obtained partially for H2 (adenovirus type 2, Ad2) only (ANDERSON 1990). A multiple sequence alignment is displayed in Fig. 1. The sequences vary in length between 201 and 214 residues including the initiating methionine. Homology as percentage similarity varies between 93.6% for the closest pair (H40 and H41) to 54.5% for the most distant pair of enzymes (H4 and Bav7). Because the amino acid sequence of H5 differs from that of H2 by only one residue, H5 is not treated as a distinct enzyme. Only 17.6% of the residues are identical across the 12 sequences in this particular alignment. The catalytic site of cysteine proteinases is generally composed of a cysteine, a histidine, and an aspartic acid residue. Among the adenovirus proteinases, H54 is the only conserved His residue. The two conserved Cys residues, C104 and C122, appear to be separated by a variable region which might be a loop structure. Neither of these cysteines is embedded in the GXCGG motif characteristic of other cysteine proteinases. Similarly, the only conserved serine residue, S182, is also not embedded in the signature serine proteinase motif of GDSGG. Candidates for a possible third residue in the active site are D26 and D71.

Homology searches of the DNA databanks (with BLAST, TFASTA) with the H2 sequence have not identified any significant relationship to known sequences (ALTSCHUL et al. 1990). Once the three-dimensional structure of this enzyme becomes available it will be possible to determine whether it is indeed a unique structure as it appears at this time. A search with the PROSITE library of protein motifs failed to identify a single conserved motif in the alignment. This might suggest that the protein does not undergo any post-translational modifications.

```
                             26                   44        54                71
AAV-1   MS-----GTTETQLRDLLSSMHLRHRFLGVFDKSFPGFLDPHVPASAIVNTGSRASGGMHWIGFAFDPAAGRCYMFDPFGWSDQKLWELYRVKYNAFMRR
BAV-3   M------GSREEELRFILHDLGVGPYFLGTFDKHFPGFISKDRMSCAIVNTAGRETGGVHWLAMAWHPASQTFYMFDPFGFSDQKLKQIYNFEYQGLLKR
BAV-7   MS-----GLSEKEVFLLLSSLQCTHGFLGTFDCRFPGFINKVKVQTAIINTGPREQGGIHWIALAWDPKSYQMFIFDPLGWKNDQLMKYYKFSYSNLIKR
CAN-1   MAEG---GSSEEELRAIVRDLAVTPFFLGTFDKRFPGFISSQRITCAVVNTAGRETGGVHWLAMAWNPRSKTFYMFDPFGFSDSKLKQVYSFEYEGLLRR
H12     M------GSSEQELTAIVRDLGCGPYFLGTFDKRFPGFVSRDRLSCAIVNTAGRETGGVHWLAFGWNPKSHTCYLFDPFGFSDQRLKQIYQFEYESLLRR
H2      M------GSSEQELKAIVKDLGCGPYFLGTYDKRFPGFVSPHKLACAIVNTAGRETGGVHWMAFAWNPRSKTCYLFEPFGFSDQRLKQVYQFEYESLLRR
H3      MTCGSGNGSSEQELKAIVRDLGCGPYFLGTFDKRFPGFMAPDKLACAIVNTAGRETGGEHWLAFGWNPRYNTCYLFDPFGFSDERLKQIYQFEYEGLLRR
H4      MAA----GSGEQELRAIIRDLGCGPYFLGTFDKRFPGFMAPHKVACAIVNTAGRETGGEHWLAFAWNPRSNTCYLFDPFGFSDQRLKQIYQFEYEGLLRR
H40     M------GSSEQELVAIVRELGCGPYFLGTFDKRFPGFMAPHKLACAIVNTAGRETGGVHWLALAWNPKNRTCYLFDPFGFSDERLKQIYQFEYEGLLKR
H41     M------GSSEQELVAIARDLGCGSYFLGTFDKRFPGFMAPNKLACAIVNTAGRETGGVHWLALAWNPKSHTCYLFDPFGFSDERLKQIYQFEYEGLLKR
MAV-1   M------GSSETELRQLVADLGIGS-FLGIFDKHFPGFISVNKPACAIVNTASRETGGVHWLAMAWYPTSSTFYLFDPFGFSDRKLQQVYKFEYERLLKR
        *      * * ..  .  .    *** .* ****.         *..**. * ** **.. .  *      ..*.*.*  . .*   *   *  ...*

              104                   122       ↓ts1                                     170       182
AAV-1   TGLRQ-PDRCFTLVRSTEAVQ-----CPCSAACGLFSALFIVSFDRYRSKPMDGNPVIDTVVGVKHENMNSPPYRDILHRNQERTYYWWTKNSAYFRAHQ
BAV-3   SALTSTADRCLTLIQSTQSVQGPN-----SAACGLFCCMFLHAFVRWPLRAMDNNPTMNLIHGVPNNMLESPSSQNVFLRNQQNLYRFLRRHSPHFVKHA
BAV-7   SAL-SSPDKCVKVIKNSQSVQ-----CTCAGSCGLFCVFFLYCFYKYKSNAFKNCLFQSLYGSIPS---LTPPNPTNLHKNQDFLYKFFKEKSLYFRQNE
CAN-1   SAIASTPDRCVTLAKSNETIQGPN-----SAACGLFCCMFLHAFVNWPDNPFNHNPTMGPLKSVPNYKLYDPTVQHVLWENQEKLYKFLEKNSAYFRAHA
H12     SALAATKDRCVTLEKSTQTVQGP-----FSAACGLFCCMFLHAFTHWPDHPMDKNPTMDLLTGVPNCMLQSPQVVGTLQRNQNELYKFLNSLSPYFRHNR
H2      SAIASSPDRCITLEKSTQSVQGP-----NSAACGLFCCMFLHAFANWPQTPMDHNPTMNLITGVPNSMLNSPQVQPTLRRNQEQLYSFLERHSPYFRSHS
H3      SALA-TKDRCITLEKSTQSVQGP-----RSAACGLFCCMFLHAFVHWPDRPMDGNPTMKLVTGVSNSMLQSPQVQPTLRRNQEVLYRFLNTHSSYFRSHR
H4      SALA-TKDRCVTW-KSHQTCRVRVGRCGFSAACSTAC--------AWPT-PMDKNPTMNLLTGVPNGMLQSPQVEPTLRRNQEALYRFLNSHSAYFRSHR
H40     SALASTPDHCITLIKSTQTVQGP-----FSAACGLFCCMFLHAFVNWPTSPMERNPTMDLLTGVPNSMLQSPQVVPTLRHNQERLYRFLAQRSPYFQRHC
H41     SALASTPDHCITLVKSTQTVQGP-----FSAACGLFCCMFLHAFIHWPSNPMEQNPTMDLLTGVPNSMLQSPQVEPTLRRNQERLYRFLTQHSPYFRRHR
MAV-1   SAVSSSSSKCVTLVKSHQTVQGPH-----SAACGLFCVLFLAAFGKYPQNPMNNNPIMGPIEGVPNDQMFNPCYTKTLYRNQQWVYSYLNKNSLYFRLHV
        ...      *    .. .. .         ...*.            .              ..            *     .  **.  *      *  * .

AAV-1   EELRRETALNALPENHV--------   206 a.a.
BAV-3   AQIEADTAFDKMLTN----------   204 a.a.
BAV-7   EYIVSNTKIGLIKSHI---------   202 a.a.
CAN-1   AAIKTRTAFNK-LKQ----------   206 a.a.
H12     ERIEKATSFTKMQNGLK--------   206 a.a.
H2      AQIRSATSFCHLKNM----------   204 a.a.
H3      ARIERATAFDRMDMQ----------   209 a.a.
H4      ARIEKATAFDRMNQDM---------   201 a.a.
H40     ERIKKATAFDQMKNNM---------   205 a.a.
H41     ERIEKATAFDQMKNAQVLFHNKIFY   214 a.a.
MAV-1   ELIKKNTAFDKLLVRK---------   204 a.a.
          .  * .
```

**Fig. 1.** Multiple sequence alignment of the human (H), bovine (*BAV*), canine (*CAN*), murine (*MAV*), and avian (*AAV*) endoproteinases. The alignment was constructed by the CLUSTAL V program (HIGGINS and SHARP 1989), using fixed and varying gap penalties of 10. The numbering is according to the H2 sequence. As H5 only differs from H2 in having an H at position 63, it is not included. The catalytic site residues are H54 and C104. The *arrow* locates the *ts*1 mutation (P to L) in H2. *Asterisks* indicate a match across all sequences; *dots* indicate conservative substitutions

## 3 Isolation, Purification, and Enzyme Assays

Isolation, purification, and enzyme assays have been reviewed recently in WEBER and TIHANYI (1994) and will therefore not be discussed here.

## 4 Properties of the Enzyme

Approximately 20 proteinase molecules are packaged in virions, and enzyme activity can be recovered from disrupted particles (TREMBLAY et al. 1983; ANDERSON 1990). *Ts*1 does not package its enzyme at the nonpermissive temperature (ANDERSON 1990; J.M. WEBER, unpublished observations). Consistent with its basic nature (isoelectric point, pI,10.59), the protein copurifies with viral cores and indeed binds to DNA directly, though nonspecifically (EVERITT and INGELMAN 1984; ANDERSON 1990; WEBSTER et al. 1993; J.M. WEBER, unpublished observations). MANGEL et al. (1993) have also claimed tenfold enhanced enzyme activity in the presence of DNA or other negatively charged polymers. It would be interesting to determine whether this DNA-binding property has any role in packaging the enzyme with the genome.

Early work on crude preparations of the proteinase suggested that it was a chymotrypsin-like neutral, nonmetalo, serine proteinase (BHATTI and WEBER 1979a, b). More recently, WEBSTER et al. (1989) have developed a peptide-based enzyme assay and, using a series of inhibitors, have concluded that the enzyme is most likely a cysteine proteinase. This conclusion was confirmed using purified recombinant enzyme and the H2-pVII protein as substrate (TIHANYI et al. 1993). An unusual property of the enzyme is its apparent requirement for activation by its substrate. Two groups have simultaneously shown that the 11-mer peptide, GVQSLKRRRCF, isolated from virions was required to render the purified enzyme active (MANGEL et al. 1993; WEBSTER et al. 1993). The peptide represents a cleavage fragment from the C terminus of pVI. A similar peptide from the pVI of H12, GVKSLKRRRCY, can also activate the H2 enzyme (FREIMUTH and ANDERSON 1993). The peptide is highly conserved in other serotypes as well: GVKSLK-RRRCY in H40 and H41; GLQPIKRRRCF in Mav1 (WEBER et al. 1994). The activation mechanism may be a general feature common to all adenoviruses. It is less certain whether activation is an absolute requirement for enzyme activity. The requirement appears to be absolute when the substrate is a peptide, but not when the substrate is a protein: the purified enzyme alone was sufficient to cleave actin, ovalbumin, and fibrin (TIHANYI et al. 1993). The complexity of this question is also underlined by the observation that the cleaved form of L1-52K is associated with empty capsids that contain pVI and other uncleaved precursors, suggesting that this protein can also be cleaved in the absence of pVI-mediated activation (HASSON et al. 1992).

In search of an identification of the active site residues, RANCOURT et al. (1994) studied the effect of mutating conserved residues on enzyme activity. These studies confirm H54 and C104 as catalytic site residues. Mutation of C122 and C126 had little effect on the enzyme. Interestingly, the mutant C104S restored activity, but the enzyme lost sensitivity to cysteine proteinase inhibitors and became sensitive to serine proteinase inhibitors. This interchangeability of Cys and Ser at the active site has been observed in other proteinases as well (LAWSON and SEMLER 1991).

The H2 enzyme contains eight cysteines, but apparently none of them participate in disulfide bonds. Gel filtration gives a molecular weight of approximately 25 000, suggesting that the enzyme is a monomer like most other proteolytic enzymes (TIHANYI et al. 1993).

## 5 Substrate Specificity

Six and possibly seven viral proteins are cleaved by the proteinase in the course of virion assembly: pVI, pVIII, and pIIIa are located in the shell of the capsid; pVII and pμ (also called 11K or L1-79R) are DNA-binding core proteins; and pTP is bound covalently to the 5' ends of the viral DNA. There is strong evidence that the nonstructural L1-52K protein, which is thought to have a scaffolding function during virus assembly, is also cleaved at a single site (HASSON et al. 1992). N-terminal sequencing of the cleaved proteins has been carried out only in H2 and H12. Using a series of octapeptides and in vitro cleavage reactions with enzyme derived from virions, WEBSTER et al. (1989) have proposed the following consensus cleavage sites: (M,L)XGG-X or (M,L)XGX-G. These were subsequently updated to (M, L, I)XGG-X and (M, L, I)XGX-G (ANDERSON 1990). The GenBank/EMBL nucleotide databanks were screened for adenovirus proteinase substrate proteins and putative cleavage sites in homologous regions compiled (Table 1). In the case of pTP, up to four consensus sites are present, but it is not known how many are actually cleaved by the enzyme. In H2, the first site appears to be cleaved rapidly, while the second site may be rate limiting (TREMBLAY et al. 1983; WEBSTER et al. 1993). To better display the range of amino acids in the four positions on either side of the cleavage site, the data in Table 1 was summarized in Table 2. Several interesting features emerge: (a) Cys is the only amino acid absent; (b) the P1' position does not appear to tolerate acid, acid amide, or basic residues; (c) residues V, T, G, F, A, S, R, and L appear to be most frequently represented. To completely define the range of amino acids acceptable around the cleavage site and the effect on the efficiency of cleavage, an approach involving random peptide libraries will have to be used (MATTHEWS and WELLS 1993).

As alluded to above, the recombinant enzyme may also cleave nonviral proteins which contain a consensus cleavage sequence. Limited cleavage could

**Table 1.** Adenovirus substrate proteins and putative cleavage sites

| Virus | Protein[a] | Cleavage site | | P1 | Peptide cleaved[b] | |
|---|---|---|---|---|---|---|
| | | | | | in vitro | in vivo |
| H2[c] | pVII (197) | MFGG | AKKR | 24 | Yes | Yes |
| | pVIII (227) | IGGA | GRSS | 157 | – | Yes |
| | pVIII (227) | LAGG | FRHR | 111 | Yes | – |
| | pVI (250) | MSGG | AFSW | 33 | Yes | Yes |
| | pVI (250) | IVGL | GVQS | 239 | – | Yes |
| | 11-kDa (79) | MAGH | GLTG | 26 | – | – |
| | | LTGG | MRRA | 31 | – | Yes |
| | | MRGG | ILPL | 50 | – | Yes |
| | pTP (668) | MRGF | GVTR | 172 | Yes | – |
| | | MGGR | GRHL | 180 | Yes | – |
| | | LGGG | VPTQ | 314 | No | – |
| | | MTGG | VFQL | 346 | No | – |
| | pIIIa (585) | LGGS | GNPF | 570 | – | – |
| | L1 52-kDa (415) | LAGT | GSGD | 351 | – | – |
| H12 | pVI (265)[d] | LNGG | AFNW | 33 | – | Yes |
| | pVI (265)[d] | IVGL | GVKS | 254 | – | Yes |
| | 11-kDa (72)[d] | LTGN | GRFR | 27 | – | Yes |
| | 11-kDa (72)[d] | MKGG | VLPF | 42 | – | Yes |
| | pVII (188)[d] | MYGG | AKTR | 23 | – | Yes |
| | pIIIa (582)[d] | LGGS | GNPF | 565 | – | Yes |
| | pTP (606)[e] | LQGY | GSTH | 140 | – | – |
| | | LRGG | VFEL | 289 | – | – |
| | pVIII (233)[d] | IRGK | GIQL | 138 | – | Yes |
| | | LGGS | GRSS | 163 | – | Yes |
| | L1 52-kDa (373)[e] | LTGA | GTED | 328 | – | – |
| H40[f] | pVI (267) | LNGG | AFSW | 33 | – | – |
| | pVI (267) | IVGL | GVKS | 256 | – | – |
| | pVIII (233) | LAGG | ARHV | 111 | – | – |
| | pVII (185) | MYGG | AKRR | 23 | – | – |
| | pTP (643) | LRGF | GSTR | 172 | – | – |
| | pTP (643) | LSGS | GMQG | 318 | – | – |
| | pTP(643) | MQGG | VFEL | 323 | – | – |
| | L1 52-kDa (380) | LTGT | GTDA | 336 | – | – |
| | 11-kDa (70) | MAGS | GRRR | 27 | – | – |
| | 11-kDa (70) | IKGG | FLPA | 40 | – | – |
| | IIIa (576) | LGGT | GAAS | 556 | – | – |
| H41 | pVIII (233)[g] | LAGG | SRHV | 111 | – | – |
| | pVI[h] | IVGL | GVKS | – | – | – |
| CAN1 | pVII (132)[i] | LFGG | AKQK | 23 | – | – |
| MAVI | pVI[j] | IMGL | GLQP | – | – | – |

[a] The number of amino acids in the precursor protein is given in brackets.
[b] A dash indicates that the cleavage site has not been confirmed by direct means, such as sequencing or in vitro cleavage of the cognate peptide.
[c] The cleavage sites in H5 proteins pVI, pVII, pVIII, and L1 52-kDa protein are identical to those in H2 (Chroboczek et al. 1992).
[d] Freimuth and Anderson 1993.
[e] W. Doerfler, GenBank file AT12CGA.
[f] V. Mautner, GenBank file Adrgenome.
[g] Pieniazek et al. 1989.
[h] Toogood et al. 1989.
[i] Cai and Weber 1993.
[j] Weber et al. 1993.

**Table 2.** Amino acids at the proteinase cleavage sites[a]

| N ....... P4 | P3 | P2 | P1 | P1′ | P2′ | P3′ | P4′ | ..... C |
|---|---|---|---|---|---|---|---|---|
| | **A** | **G** | **G** | **A** | **T** | **S** | **A** | Small |
| | **G** | | **S** | **G** | S | **T** | **S** | hydrophilic |
| | **S** | | T | S | P | **P** | P | |
| | **T** | | **A** | | A | G | G | |
| | | | | | | A | A | |
| | | | | | | D | | |
| | **N** | | **N** | | **N** | **N** | D | Acid, |
| | Q | | | | | **Q** | Q | acid amide, |
| | | | | | | E | | hydrophilic |
| | **R** | | **R** | | **R** | **R** | **R** | |
| | **K** | | **H** | | **K** | **K** | **K** | Basic |
| | | | **K** | | | **H** | H | |
| | | | | | M | | | |
| **M** | **V** | | **L** | **M** | **V** | | **L** | |
| **I** | M | | | **I** | **L** | | V | Small hydrophobic |
| **L** | | | | **V** | **I** | | | |
| | **F** | | **F** | **F** | **F** | **F** | **F** | Aromatic |
| | **Y** | | Y | | | | **W** | |

[a] Data from Table 1. Residues confirmed by amino acid sequencing appear in bold.

be demonstrated of denatured ovalbumin and actin as well as cleavage of fibrinogen in a clot-solubilizing assay (TIHANYI et al. 1993). Denaturation was probably required to expose otherwise inaccessible cleavage sites. For this reason actin and ovalbumin would probably not be affected by the enzyme in cells. A recent report demonstrates the cleavage of cytokeratin K18 and provides suggestive evidence for the cleavage of cytokeratin K7 (CHEN et al. 1993). These cleavages take place late in infection with H2 and result in the disassembly of the cytokeratin network. Cells infected with *ts*1 at 39°C do not experience these cleavages. Breakdown of the cytoskeletal network in the late phase of infection may have a function in accelerating cell lysis and thereby promoting viral spread.

# 6 Conclusion

While in the early 1980s adenovirus was the only eukaryotic DNA virus known to code for a proteinase, today several other viruses are known to possess proteinases. The importance of these enzymes to DNA virus replication appears to be becoming as important as it was demonstrated to be for RNA viruses. Based on the phenotype of the *ts*1 mutant and the nature of the substrate proteins identified to date, the function of the enzyme in virus assembly has clearly been established. The substrate-mediated enzyme activation mechanism appears to ensure that cleavage does not occur prematurely and that it is tightly coupled to assembly events. A working model for assembly might be built on three independent observations: (a) young virions tend to copurify with the nuclear matrix (KHITTOO et al. 1986); (b) viral DNA replicates while linked to the nuclear matrix via the pTP (FREDMAN and ENGLER 1993); (c) the proteinase cleaves cytokeratin K18, a constituent of the matrix, late in infection (CHEN et al. 1993). According to this model virions would assemble around the nuclear matrix-linked viral DNA, and the proteinase cleaves the first site on pTP, thereby releasing the tethered young virions, and subsequently also cleaves cytokeratin K18 (and K7?), thereby further facilitating release and also promoting cell lysis.

Because of the absence of identifiable protein motifs, functions other than proteolytic ones appear at present unlikely. Major unanswered questions include the relationship to other proteinases (perhaps via the determination of the crystal structure), the possible third residue of the active site, the substrate-binding pockets, the precise mechanism of activation, and the mechanism of encapsidation.

The unique nature of the enzyme and the likelihood that all adenovirus proteinases act similarly offers an ideal target for the development of specific chemotherapeutic agents active against all adenovirus.

*Acknowledgments.* I would like to thank Carole Picard and Andree Houle for help with the preparation of the manuscript and Guy Guillemette for stimulating discussions. Financial support was provided by grant MT4164 from the Medical Research Council of Canada.

## References

Altschul SF, Gish W, Miller W, Myers EW, Lipman DJ (1990) Basic local alignment search tool. J Mol Biol 215: 403–410

Anderson CW (1990) The proteinase polypeptide of adenovirus serotype 2 virions. Virology 177: 259–272

Anderson CW, Baum PR, Gesteland RF (1973) Processing of adenovirus 2-induced proteins. J Virol 12: 241–252

Bhatti AR, Weber J (1979a) Protease of adenovirus type 2: partial characterization. Virology 96: 478–485

Bhatti AR, Weber JM (1979b) Protease of adenovirus type 2: subcellular localization. J Biol Chem 254: 12265–12268

Cai F, Weber JM (1993) Organization of the avian adenovirus genome and the structure of its endopeptidase. Virology 196: 358–362

Chen PH, Ornelles DA, Shenk T (1993) The adenovirus L3 23-kilodalton proteinase cleaves the amino-terminal head domain from cytokeratin 18 and disrupts the cytokeratin network of HeLa cells. J Virol 67: 3507–3514

Chroboczek J, Bieber F, Jacrot B (1992) The sequence of the genome of adenovirus type 5 and its comparison with the genome of adenovirus type 2. Virology 186: 280–285

Everitt E, Ingelman M (1984) Core and chromatin association of the adenovirus type 2 specified endoproteinase. Microbios Lett 25: 75–82

Falgout B, Ketner G (1988) Characterization of adenovirus particles made by deletion mutants lacking the fiber gene. J Virol 62: 622–625

Fredman JN, Engler JA (1993) Adenovirus precursor to terminal protein interacts with the nuclear matrix in vivo and in vitro. J Virol 67: 3384–3395

Freimuth P, Anderson CW (1993) Human adenovirus serotype 12 virion precursors pMu and pVI are cleaved at amino-terminal and carboxy-terminal sites that conform to the adenovirus 2 endoproteinase cleavage consensus sequence. Virology 193: 348–355

Hasson TB, Ornelles DA, Shenk T (1992) Adenovirus L1 52- and 55-kilodalton proteins are present within assembling virions and colocalize with nuclear structures distinct from replication centers. J Virol 66: 6133–6142
Higgins DG, Sharp PM (1989) Fast and sensitive multiple sequence alignments on a microcomputer. CABIOS 5: 151–153
Houde A, Weber JM (1990) Adenovirus proteinases. Comparison of aminoacid sequences and expression of the cloned cDNA in Escherichia coli. Gene 88: 269–273
Khittoo G, Delorme L, Dery CV, Tremblay ML, Weber JM, Bibor-Hardy V, Simard R (1986) Role of the nuclear matrix in adenovirus maturation. Virus Res 5: 391–403
Lawson MA, Semler BL (1991) Poliovirus thiol proteinase 3C can utilize a serine nucleophile within the putative catalytic triad. Proc Natl Acad Sci USA 88: 9919–9923
Mangel WF, McGrath WJ, Toledo DL, Anderson CW (1993) Viral DNA and a viral peptide can act as cofactors of adenovirus virion proteinase activity. Nature 361: 274–275
Matthews DJ, Wells JA (1993) Substrate phage: selection of protease substrates by monovalent phage display. Science 260: 1113–1117
Miles BD, Luftig RB, Weatherbee JA, Weihing RR, Weber J (1980) Quantitation of the interaction between adenovirus types 2 and 5 and microtubules inside infected cells. Virology 105: 265–269
Mirza MAA, Weber J (1980) Infectivity and uncoating of adenovirus cores. Intervirology 13: 307–311
Murialdo H, Siminovitch L (1972) The morphogenesis of bacteriophage lambda. IV. Identification of gene products and control of the expression of the morphogenic information. Virology 48: 785–823
Pieniazek N, Velarde J, Pieniazek D, Luftig RB (1989) Nucleotide sequence of human enteric adenovirus type 41 hexon-associated protein VIII precursor (pVIII) including the early region E3 promoter. Nucleic Acids Res 17: 5398
Rancourt C, Tihanyi K, Bourbonnière M, Weber JM (1994) Identification of active site residues of the adenovirus endopeptidase. Proc Natl Acad Sci USA 91: 844–847
Tihanyi K, Bourbonniere M, Houde A, Rancourt C, Weber JM (1993) Isolation and properties of the adenovirus type 2 proteinase. J Biol Chem 268: 1780–1785
Toogood CIA, Murali R, Burnett RM, Hay RT (1989) The adenovirus type 40 hexon: sequence predicted structure and relationship to other adenovirus hexons. J Gen Virol 70: 3203–3214
Tremblay ML, Déry CV, Talbot BG, Weber J (1983) In vitro cleavage specificity of the adenovirus type 2 proteinase. Biochim Biophys Acta 743: 239–245
Weber J (1976) Genetic analysis of adenovirus type 2. III. Temperature sensitivity of processing of viral proteins. J virol 17: 462–471
Weber J (1983) Genetic identification of an endoproteinase encoded by the adenovirus genome. J Mol Biol 167: 217–222
Weber JM (1990) The adenovirus proteinase. Semin Virol 1: 379–384
Weber JM, Houde A (1987) Spontaneous reversion of a C/T transition mutation in the adenovirus endoproteinase gene. Virology 156: 427–428
Weber JM, Tihanyi K (1994) Adenovirus endopeptidases. Proteolytic enzymes, pt D. Methods Enzymol 244: 595–604
Weber JM, Cai F, Murali R, Burnett RM (1994) Sequence and structural analysis of murine adenovirus type 1 hexon. J Gen Virol 75: 141–147
Webster A, Russell WC, Kemp GD (1989) Characterization of the adenovirus proteinase; substrate specificity. J Gen Virol 70: 3215–3223
Webster A, Hay RT, Kemp G (1993) The adenovirus protease is activated by a virus-coded disulphide-linked peptide. Cell 72: 97–104
Yeh-Kai L, Akusjärvi G, Aleström P, Pettersson U, Tremblay M, Weber J (1983) Genetic identification of an endo-proteinase encoded by the adenovirus genome. J Mol Biol 167: 217–222

# E3 Transcription Unit of Adenovirus

W.S.M. Wold, A.E. Tollefson, and T.W. Hermiston



## 1 Region E3 May Encode Products that Counteract Host Defense Mechanisms

Mammals have innate and adaptive defenses that protect them from virus infections. Viruses, in particular the large DNA viruses (adenoviruses, herpesviruses, poxviruses), have evolved mechanisms that counteract the host's antiviral defenses. In general, these viruses prevent killing of infected cells by cytotoxic T lymphocytes (CTL), block the inflammatory response, inhibit complement fixation, prevent shut-off of cellular protein synthesis in response to interferon, and block apoptosis (which may be a host defense against virus infection). Reviews have appeared recently on how some of these proteins function (Gooding 1992; G.L. Smith 1994). Here we will focus on the human adenoviruses, in particular the E3 transcription unit, which appears to be, at least in part, a cassette of genes that functions to counteract the host's antiviral defenses. Recent reviews on this topic are by Pääbo et al. (1989), Wold and Gooding (1989, 1991), Gooding and Wold (1990), Gooding (1992), and Wold (1993).

Department of Molecular Microbiology and Immunology, St. Louis University School of Medicine, 1402 South Grand Blvd., St. Louis, MO 63104, USA

None of the E3 proteins is required for adenovirus replication in cultured cells (KELLY and LEWIS 1973) or in the lungs of hamsters (MORIN et al. 1987) or cotton rats (GINSBERG et al. 1989). This is consistent with the proposed role of E3 in mediating host responses. The nonessential nature of the E3 proteins has permitted the construction of many viable virus E3 mutants (WOLD et al. 1986; BRADY et al. 1992). Also adenovirus vectors that lack E3 or have foreign genes inserted into E3 have been constructed and are being used for vaccine development and gene therapy (GRUNHAUS and HORWITZ 1992).

## 2 Features of the E3 Transcription Unit in Adenovirus Types 2 and 5 (Subgroup C)

### 2.1 E3 Promoter

The E3 transcription unit of adenovirus type 2 (Ad2) and Ad5 is located at genome position 76–86 and is transcribed off r-strand. E3 is a delayed early transcription unit whose transcription is induced by the 289R EIA protein (BERK et al. 1979; JONES and SHENK 1979). The E3 promoter is located within the coding sequence for virion protein VIII, an essential protein which is highly conserved among serotypes in different subgroups. The E3 promoter has a TATA box as well as upstream sites for the ATF, AP1, and NF1/CTF transcription factors (HURST and JONES 1987; GARCIA et al. 1987; BUCKBINDER et al. 1989; KORNUC et al. 1990; WILLIAMS et al. 1990). Mutagenesis studies indicate that the AP1 and ATF sites are important for basal and EIA-inducible activity. The E3 promoter also has two binding sites for NFκ B (WILLIAMS et al. 1990).

### 2.2 Mapping of the E3 mRNA and Proteins

During early stages of infection, the E3 promoter drives expression of about nine alternatively spliced mRNA that are polyadenylated at one of two sites, E3A or E3B (Fig.1). The structures and abundance of the E3 mRNA from Ad2 were determined by electron microscopy (CHOW et al. 1979a, b; KITCHINGMAN and WESTPHAL 1980; PERSSON et al. 1980b), the nuclease protection method (BERK and SHARP 1978), cDNA cloning followed by DNA sequencing (STÅLHANDSKE et al. 1983), and RNA blotting (BHAT et al. 1986). The Ad5 E3 mRNA were mapped by nuclease protection (CLADARAS et al. 1985). E3 proteins were identified using antisera against synthetic peptides corresponding to the predicted proteins, TrpE fusion proteins expressed in *Escherichia coli* or purified protein (gp19K, PERSSON et al. 1979; 14.7K, PERSSON et al. 1978), all in conjunction with cell-free translation of hybridization-purified mRNA and the employment of appropriate virus E3 deletion mutants. The following proteins have been identified: 12.5K (HAWKINS and

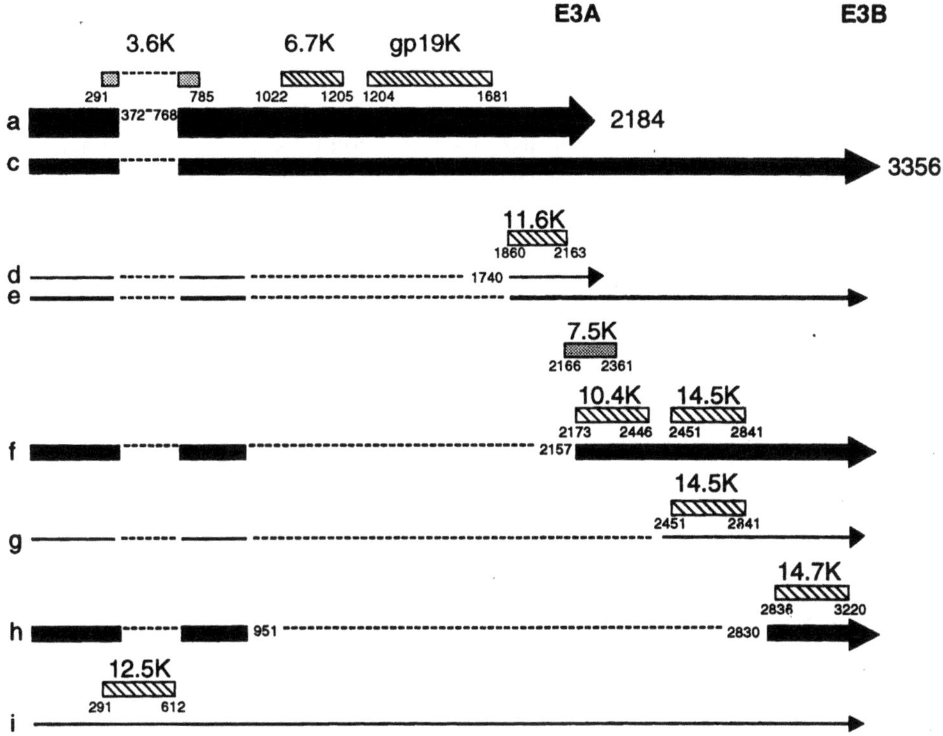


**Fig. 1.** E3 transcription unit of adenovirus type 2 (Ad2). The *split arrows* depict the spliced structures of the mRNA; exons are *solid* and the *thickness* implies the relative abundance. E3A and E3B are polyadenylation sites. The 3' end of mRNA *i* may be at the E3A or E3B site. The DNA sequence was determined by HÉRISSÉ et al. (1980) and HÉRISSÉ and GALIBERT (1981). Nucleotide 1 is the transcription initiation site (BAKER and ZIFF 1981); this corresponds to nucleotide 27 609 in the sequence of Ad2 (ROBERTS et al. 1986). The organization of E3 in Ad5 is similar (CLADARAS and WOLD 1985; CLADARAS et al. 1985). *gp*, glycoprotein

WOLD 1992), 6.7K (WILSON-RAWLS et al. 1990), gp19K (PERSSON et al. 1980a, b; ROSS and LEVINE 1979), 11.6K (WOLD et al. 1984; TOLLEFSON et al. 1992), 10.4K (TOLLEFSON et al. 1990b), 14.5K (TOLLEFSON et al. 1990a), and 14.7K (PERSSON et al. 1978, 1980a,b; TOLLEFSON and WOLD 1988; WANG et al. 1988).

Splicing determines which E3 protein is made (Fig.1). The following discussion will be limited to Ad2, but the points made are also true for Ad5. $ATG_{291}$ and $ATG_{1022}$ are the only ATGs in any reading frame upstream of $ATG_{1204}$, the initiation codon for gp19K (HÉRISSÉ et al. 1980; CLADARAS and WOLD 1985). The 12.5K protein, which is predicted to initiate at $ATG_{291}$, is translated from the unspliced minor mRNA *i* (HAWKINS and WOLD 1992). All the other E3 mRNA have the 372–768 splice. A 3.6K protein (33 residues) which initiates at $ATG_{291}$ and terminates at $TAA_{785}$ should be synthesized from these mRNA; attempts to identify this protein have been unsuccessful (unpublished results). $ATG_{291}$ has the sequence context AC*TGAATGA*, which is predicted by the KOZAK (1987) rules to be inefficient. Consistent with this, a virus mutant that has $ATG_{291}$ converted to CC*ACC*A*TG*A (an optimal Kozak sequence) synthesizes increased 12.5K (L. HAWKINS and W. WOLD, unpublished results). Also, synthesis of the other E3 proteins is reduced. These results are in accord with the idea that, in common with most eukaryotic mRNA, the E3 mRNA are translated by a scanning mechanism that initiates at the mRNA 5'-terminal cap. Thus, in all E3 mRNA except mRNA *i*, $ATG_{291}$ is probably bypassed nearly all the time by the translation machinery.

The 6.7K and gp19K proteins are coded by mRNA *a* and *c*, i.e., mutants that overexpress or underexpress mRNA *a* and *c* correspondingly overexpress or underexpress these proteins (WILSON-RAWLS et al. 1990). Also, gp19K has been translated from sucrose gradient fractions that contain mRNA *a* and *c* (PERSSON et al. 1980b). $ATG_{1022}$, which is predicted to initiate the 6.7K protein, has a putative inefficient initiation codon (AG*T*AT*ATG*A); this is consistent with the relatively low abundance of 6.7K considering that mRNA *a* and *c* account for more than 50% of total E3 mRNA. It is probable that $ATG_{1022}$ is usually bypassed by the scanning translation machinery. Translation of gp19K has been proven to initiate efficiently at $ATG_{1204}$ (CC*A*AG*ATG*A; KÄMPE et al. 1983), which is predicted to be an efficient initiation codon.

There is an open reading frame (ORF) in Ad2, $ATG_{1664}$ to $TGA_{1961}$, which could encode a 14K protein (HÉRISSÉ et al. 1980). This ORF is not conserved in Ad5 (CLADARAS and WOLD 1985). It is not known whether this protein is expressed.

Another 11.6K protein (Fig.1) is predicted to initiate at $ATG_{1860}$ in mRNA *d* and *e*. This 11.6K protein is made in very small quantities at early stages of infection, but in very large quantities, comparable to protein VIII, at late stages of infection (TOLLEFSON et al. 1992). At late stages, 11.6K is translated from new mRNA *d'* and *e'*, which are derived from the major late promoter (see Sect. 2.4).

The 10.4K and 14.5K proteins both appear to be coded by mRNA *f*, i.e., they are coordinately increased or decreased in mutants that overexpress or underexpress mRNA *f* (TOLLEFSON et al. 1990a,b). The 10.4K (KRAJCSI et al. 1992a) and 14.5K (KRAJCSI et al.1992b) proteins have been proven to initiate at $ATG_{2173}$ and $ATG_{2451}$, respectively. There are 2 bp between the stop codon for 10.4K and the start codon for 14.5K; most likely, translation of 10.4K is terminated and then translation of 14.5K is reinitiated from the same mRNA. The 14.5K protein may also be translated from mRNA *g* (Fig.1),a minor mRNA that has been observed by electron microscopy (CHOW et al. 1979a,b). We have never detected mRNA *g* using the nuclease protection assay or RNA blots.

There is an ORF for a 7.5K protein which would initiate at $ATG_{2166}$; this ATG is predicted to be inefficient, so it is presumably usually bypassed during translation scanning. A specific antiserum immunoprecipitated the 7.5K protein from cells infected with a mutant engineered to overexpress 7.5K, but the 7.5K protein could not be detected in cells infected with wild-type adenovirus (S. YEI and W. WOLD, unpublished results). Therefore, it is unclear whether 7.5K is a bona fide E3 protein.

The 14.7K protein, which is predicted to initiate at $ATG_{2836}$, is translated from mRNA *h* (PERSSON et al. 1980b; TOLLEFSON and WOLD 1988). The Ad2 14.7K protein migrates as three bands in sodium dodecyl sulfate polyacrylamide gel electrophoresis (SDS-PAGE), whereas the Ad5 14.7K protein migrates as two bands (TOLLEFSON and WOLD 1988); the reason for this is unknown.

## 2.3 Control of Constitutive Splicing of E3 mRNA

The E3 proteins must be required in different concentrations, depending on their function. For example, gp19K, which is required in stoichiometric amounts (Sect. 3.1), is encoded by mRNA *a* and *c*, which make up more than 50% of the E3 mRNA. Use of inefficient versus efficient initiation codons is one mechanism that affects the synthesis of the E3 proteins. Another factor is the abundance of the E3 mRNA, so mechanisms must exist that determine the abundance of the mRNA. In order to identify such mechanisms, we constructed a large number of virus mutants with deletions throughout E3, and we determined how the mutations affect the accumulation of the major mRNA. The parental virus used for these studies is rec700, an Ad5-Ad2-Ad5 recombinant. In rec700, mRNA *a* is roughly 40% of total E3 mRNA, mRNA *c* is 15%, mRNA *f* is 15%, and mRNA *h* is 25% (BHAT et al. 1986). In Ad2, mRNA *a* is somewhat higher and mRNA *c* is lower in abundance, and the reverse is true in Ad5; mRNA *h* is more abundant in Ad5 than Ad2.

We found that deletion of 5' or 3' splice sites had the predicted effect, i.e., the splice did not occur. None of the deletions affected the transport of mRNA into the cytoplasm, but some of the mutants had dramatic effects on the relative abundance of the different E3 mRNA. One interesting set of mutants identified a region, termed "region I" (DEUTSCHER et al. 1985; BHAT and WOLD 1985, 1986a, b, 1987; BHAT et al. 1986; BRADY et al. 1992), which is located within nt 1809–1937, an area of 129 nt (SCARIA and WOLD 1994). With mutants that delete region I, mRNA *f* and *h* account for nearly 50% each of the total E3 mRNA, and mRNA *a* and *c* are barely detectable. It is noteworthy that region I is not near a conventional splice or polyadenylation site (see Fig. 1). Region I lies within the 11.6K gene (Fig. 1); however, some mutants that do not synthesize functional 11.6K do not have a region I phenotype (A. SCARIA, A. TOLLEFSON and W. WOLD, unpublished results), which excludes the possibility that 11.6K is a *trans*-acting factor that controls the abundance of E3 mRNA.

We have hypothesized that region I functions in wild-type adenovirus to suppress splicing such that mRNA *a* and *c* (spliced once) accumulate to higher levels than do mRNA *f* and *h*, which are spliced twice and whose second intron contains the region I sequence (DEUTSCHER et al. 1985; BHAT and WOLD 1987; SCARIA and WOLD 1994). Such splice suppression would allow for gp19K to be synthesized in high amounts. We do not know how region I functions. Computer modeling has not revealed any obvious secondary structure. Most likely, region I binds a cell-coded *trans*-acting factor.

The second interesting set of mutants (region II mutants) have lesions either in the E3A ATTAAA (AUUAAA in RNA) polyadenylation signal at nt 2161 or they have deletions in the GT-rich component of the E3A signal located at approximately nt 2184–2220, downstream from the site where the 3'ends for mRNA *a* are formed (BRADY and WOLD 1987, 1988). Point mutations that alter the ATTAAA element, and small deletions within nt 2184–2220, result in nearly exclusive synthesis of mRNA*f* at the expense of the other E3 mRNA (BRADY and WOLD 1987, 1988). To explain these results, we formulated the "factor competition model," whose basic feature is that the 3' splice site for mRNA*f* (nt 2157) is only 4 bp upstream from the ATTAAA (nt 2161). We hypothesized that because of this proximity, in wild type, factors that bind to the ATTAAA sequence could not bind to the same pre-mRNA as factors that bind to the 2157 3' splice site. Thus, there would be a competition among splicing and cleavage/polyadenylation factors for binding to a given pre-mRNA. In wild type, the cleavage/polyadenylation factors must usually out-compete the splicing factors, because mRNA *a* is about three times more abundant than mRNA*f*. With deletion and ATTAAA point mutants, the polyadenylation factors should no longer bind; thus, the splicing factors should bind efficiently to the 2157 3' splice site, and mRNA*f* should be made very efficiently. Consistent with this model, insertion of 140 bp between the 2157 3' splice site and the ATTAAA, which should relieve the competition between the splicing and polyadenylation factors, gave the region II phenotype, i.e., mRNA*f* was virtually the only E3 mRNA made (BRADY and WOLD 1988). Thus, this specific arrangement of the 2157 3' splice site and the E3A polyadenylation signal apparently evolved in E3 of Ad2 to control the balance of E3 mRNA.

## 2.4 Regulation of E3 mRNA Synthesis

E3 is embedded within the major late transcription unit. The AATAAA polyadenylation signal for the L4 family of late mRNA is at nt 597 in E3, and the pre-mRNA for the L5 (fiber) mRNA must transcribe through E3. The location of E3 within the major late transcription unit allows for both transcriptional and post-transcriptional regulation of the E3 genes. Transcription occurs from the E3 promoter at early stages of infection, but it is reduced at late stages (BHAT and WOLD 1986a; BINGER and FLINT 1984; NEVINS et al. 1979; SHAW and ZIFF 1980; STEIN and ZIFF 1984). Nevertheless, all known E3 proteins continue to be synthesized at late stages, either from residual E3 mRNA or from new E3 mRNA that arise from the major late promoter at map position 16. These new mRNA, which were observed by electron microscopy (CHOW and BROKER 1978; CHOW et al. 1977, 1979a,b), contain the major late tripartite leader (TPL) spliced to the y-leader (the nt 768–951 exon in Fig. 1) in E3 and they terminate at the E3A or E3B polyadenylation sites. These TPL-E3 mRNA are reported to be less abundant than other late mRNA, suggesting that the E3 splicing and polyadenylation signals are not used as efficiently at late stages as the corresponding signals in the L1-L5 family of mRNA. However, the splices to form the y-leader are an exception, because the y-leader is present in 25%–50% of L5 mRNA (CHOW and BROKER 1978; UHLEN et al. 1982). Another exception is the 1740 3' splice site for mRNAs *d* and *e* (see Fig.1), which is abundantly spliced to the y-leader and the TPL (BHAT and WOLD 1986a; TOLLEFSON et al. 1992). The new mRNA formed, termed *d'* and *e'* are about as abundant as the L4 family of late mRNA. Thus, 11.6K (coded by these mRNA) is not only an "early" E3 protein, but also a "major late" protein, i.e., its mRNA represents the sixth family, in addition to L1-L5, of major late mRNA. Some of the other E3 mRNA are also spliced to the TPL at late stages, but at not nearly the abundance of mRNA *d'* and *e'*.

## 3 Properties and Functions of the E3 Proteins of Adenovirus Types 2 and 5

Five of the E3 proteins of Ad2 and Ad5 are integral membrane proteins; these are gp19K, 14.5K, 10.4K, 11.6K, and 6.7K. E3 is the only part of the adenovirus genome that encodes integral membrane proteins. Other adenovirus transcription units also have genes that have similar properties or functions, e.g., all the E2 proteins function in adenovirus DNA replication. Thus, we might expect that the E3 proteins may have related functions, namely to counteract host defenses and/or to affect signal transduction (Table 1).

**Table 1.** Functions of E3 proteins

| Host response | Protein | Intracellular location | Function/mechanism |
|---|---|---|---|
| CTL | E3 gp19K | Membrane of endoplasmic reticulum | Binds to MHC class I antigens, blocks their transport from the ER to the cell surface, and prevents killing of cells by CTL |
| TNF[a] | E3-14.7K | Nucleus and cytoplasm | Prevents TNF cytolysis, probably by blocking TNF activation of $cPLA_2$; probably inhibits TNF-induced inflammatory response by preventing synthesis of leukotrienes and prostaglandins by $cPLA_2$ |
| TNF | E3 10.4K/14.5K | Plasma membrane | Prevents TNF cytolysis and TNF activation of $cPLA_2$ but probably by a different mechanism than E3-14.7K and E1B-19K |
| EGF | E3 10.4K/14.5K | Plasma membrane | Downregulate the EGF receptor; mechanism and biological significance unknown |
| Cell death | E3 11.6K | Nuclear membrane | Promotes release of virus from lysed cells; mechanism unknown. |

CTL, cytotoxic T lymphocytes; MHC, major histocompatibility complex; ER endoplasmic reticulum; TNF, tumor necrosis factor; $cPLA_2$, cytosolic phospholipase $A_2$.

[a] The EIB 19K protein prevents TNF cytolysis and TNF activation of $cPLA_2$ in human but not mouse cells, but probably by a different mechanism than E3-14.7K and E3-10.4K/14.5K.

## 3.1 E3-Gp19K Inhibits Killing of Adenovirus-Infected Cells by Cytotoxic T Lymphocytes

Gp19K (also called E19 or E3/19) is perhaps the most abundant adenovirus early protein. It is a type I integral membrane protein (Fig.2). In Ad2 (KÄMPE et al. 1983) and Ad5 (WOLD et al. 1985), there is a cleaved $NH_2$-terminal signal sequence that directs gp19K into the membrane of the endoplasmic reticulum (ER). A deletion in this signal precludes insertion of gp19K into the ER (WILSON-RAWLS et al. 1994). Near the COOH terminus, there is a hydrophobic transmembrane domain followed by a 15-amino acid cytoplasmic domain. Proteolytic digestion of gp19K in membrane fractions reduces its molecular weight by about 500 in SDS-PAGE

```
                 -17
                  •          Signal Sequence        |                                   CBH
Subgroup C Ad2   M • • R Y M I L G L L A L A A V C S A A • • • • K K V E F K • • E P A • C N V T F K S E A
           Ad5   - • I - - I - - - - - - T - - S A H G T T • • • • Q - - D - - • • - - - - • - - - - - - A A - -
Subgroup B Ad3   - G A I L V V V A - - S - L G L G - - N L N P L D H D P C L D F D - E N - T L - - A P D T
           Ad7   - G A I L V L - A - - S - L G L G - - N L N P L D H D P C L D F D - E N - T L - - A P D T
           Ad35  - G P I L V L - V - - S - L E P G - - N Y • • • • • D P C L D F D - E N - T L - - A P D T
           Ad11  - G P I L V L - V - - S - L E P G - - N Y • • • • • D P C L D F D - E N - T L - - A P D T
                                                                                        CBH

                                           ER Luminal Variable Domain
                             25
                              •                                                                          CBH
Subgroup C Ad2   N E • C T T L I K C T T E H E K L I I R H K D K I G K Y A V Y A I W Q P G D T N D Y N V T
           Ad5   - - • - - - - - - - - - - - - - L - - - - - N - - - - - - - - - - - - - - - - - T E - - - -
Subgroup B Ad3   S R L - G V - - - - G W D C R S V E - T - N N - T W N N T L S T T - E - - V P Q W - T - S
           Ad7   S R L - G V - - - - G W D C R S V E - T - N N - T W N N T L S T T - E - - V P E W - T - S
           Ad35  S R I - G F - L S A D G K C R S V E - T - N N - T W N N T L S T T - E - - V P E W - T - S
           Ad11  S R I - G V - - - - G W - C R S V E - T - N N - T W N N T L S T T - E - - V P C W - T - S
                                                     CBH   CBH

                                                                 ER Luminal              ER Luminal
                                                              Conserved Domain         Spacer Domain
                                     75                                                100
                                      •                 |                           |   •
Subgroup C Ad2   V F Q G E N R K T F M Y K • F P F Y E M C D I T M Y M S K Q Y K L W P P Q K C L E N T G T
           Ad5   - - - - K S H - - - - - T • - - - - - - - - - - - - - - - - - - - - - - - N - V - - - - -
Subgroup B Ad3   - R G P D G S I R I S N N T - I - S - - - - L A - F - - R - - D - - - - S - • • - - I V A
           Ad7   - R G P D G S I R I S N N T - I - S - - - - L A - F - - R - - D - - - - S - • • - D I V A
           Ad35  - R G P D G S I R I S N N T - I - S - - - - L A - F - - - - - S - - - - S - • • D - I V -
           Ad11  - R G P D G S I R I S N N T - I - S A - - - L A - F - - - - - S - - - - C - • • D - I T -
                                       CBH

                                                                       Cytoplasmic Domain
                                                         125      |    Stop        |    ER
                |            Transmembrane Domain          •      |  Transfer      | Retention
Subgroup C Ad2   F C S T A • L L • • • • I T A L A L V C T L L Y L K Y K S R R S F I D E K K M P
           Ad5   - - C - - • M - • • • • - - V - - - - - - - - - - I - - - - - - - - - - E - - - - -
Subgroup B Ad3   - • - I - Y C - V T C I - - - I I C - - I H - L I V I R P - Q - N E E K E - - -
           Ad7   - • - I - Y C - V T C I - - - I I C - - I H - L I V I R P - Q - N K E K E - - -
           Ad35  - • - I - Y C - C A C L L - - - L C - - I H - L V T T R I K N A N N • K E - - -
           Ad11  - • - I - Y C - C A C L L - - - L C - - I H - L V T T R T K N A N N • K E - - -
```

**Fig. 2.** Sequence of the E3 gp19K protein. The sequence of gp19K from the serotypes shown is compared to that of adenovirus type 2 (Ad2); *dashes* indicate that the amino acid is identical. *Dots* indicate that the amino acid is not present. The N-linked glycosylation sites are *underlined.* The amino acids are *numbered* beginning with the first residue following the cleaved signal sequence. *ERL-VD,* endoplasmic reticulum luminal variable domain; *ERL-CD,* ER luminal conserved domain; *ERL-SD,* ER luminal spacer domain (HERMISTON et al. 1993a, 1993b). The amino acid sequence is predicted from the DNA sequence as determined for Ad2 (HÉRISSÉ et al. 1980), Ad5 (CLADARAS and WOLD 1985), Ad3 (SIGNÄS et al. 1986), Ad7 (HERMISTON et al. 1993a), and Ad11a (respiratory isolate; MEI and WADELL 1992)

(PERSSON et al. 1980a), indicating that the $NH_2$-terminal part of the molecule extends into the lumen on the ER.

Gp19K is localized exclusively in the membrane fraction of fractionated cells (PERSSON et al.1979, 1980a; TOLLEFSON et al. 1990b), and it is enriched in the ER fraction (PÄÄBO et al. 1987). Immunofluorescence indicates that gp19K is localized primarily in the ER (PÄÄBO et al. 1987; HERMISTON et al. 1993b). It has been detected on the cell surface by iodination (KVIST et al. 1978; PÄÄBO et al. 1983; A. TOLLEFSON and W.WOLD, unpublished results); this undoubtedly is a minor fraction of the protein and its significance is unclear.

The two potential sites for N-linked glycosylation in Ad2 gp19K are modified (WOLD et al. 1985) with exclusively high-mannose oligosaccharides, mainly $Man_8GlcNA_c$ (KORNFELD and WOLD 1981). The absence of complex oligosaccharides, which are added to proteins in the Golgi complex, is consistent with gp19K's localization in the ER. Gp19K from subgroup B serotypes has four potential N-glycosylation sites (Fig. 2), at least some of which are glycosylated with high-mannose oligosaccharides (KAPOOR et al. 1981; FLOMENBERG et al. 1987).

Gp19K is retained in the membrane of the ER by a specific signal located at its COOH terminus (PÄÄBO et al. 1987; COX et al. 1991). Studies in which the gp19K COOH terminus was fused to the COOH terminus of reporter molecules such as CD8 that normally exit the ER (NILSSON et al. 1989) indicate that the signal is at the extreme COOH terminus and has the motif KKXX or KXKXX (X is any amino acid) (JACKSON et al. 1990). Ad2 and Ad5, subgroup C, have the KKXX motif, and Ad7, 3, 11, and 35, subgroup B, have the KXKXX motif (Fig. 2). Many type I transmembrane ER proteins have one of these motifs (JACKSON et al. 1990). This signal mediates retrieval of proteins from the intermediate compartment back into the ER rather than direct retention of the protein in the ER (JACKSON et al. 1993). A peptide corresponding to the COOH terminus of gp19K has been shown to bind to microtubules (DAHLLÖFF et al. 1991), and perhaps this is important in the retrograde transport of gp19K from the intermediate compartment to the ER.

In 1978, gp19K expressed in Ad2-transformed rat cells (T2C4 cells) was shown to bind major histocompatibility complex (MHC) class I antigens (KVIST et al. 1978), molecules that function in cell-mediated immunity. Ad2 gp19K also binds to class I antigens of human (SIGNÄS et al. 1982; KÄMPE et al. 1983; SEVERINSSON et al. 1986) and mouse (SIGNÄS et al. 1982; BURGERT and KVIST 1987). Binding was detected by coimmunoprecipitation using antisera to gp19K, class I $\alpha$-chains, or $\beta_2$-microglobulin. The gp19K–class I complex is noncovalent and does not require oligosaccharides (BURGERT and KVIST 1987). Complex formation does not require other adenovirus proteins, because it is observed when gp19K is expressed in HeLa cells from an expression vector (PÄÄBO et al. 1983, 1986b; ANDERSSON et al. 1985). Also, the gp19K–class I complex forms when gp19K and class I $\alpha$ chain (HLA-B27) are cotranslated in vitro in a reticulocyte system or when they are coexpressed from a baculovirus vector in insect cells (GABATHULER et al. 1990).

The importance of gp19K binding to MHC class I antigens is illustrated in Fig. 3. The class I antigen is expressed in nearly all cell types and is a heterodimer

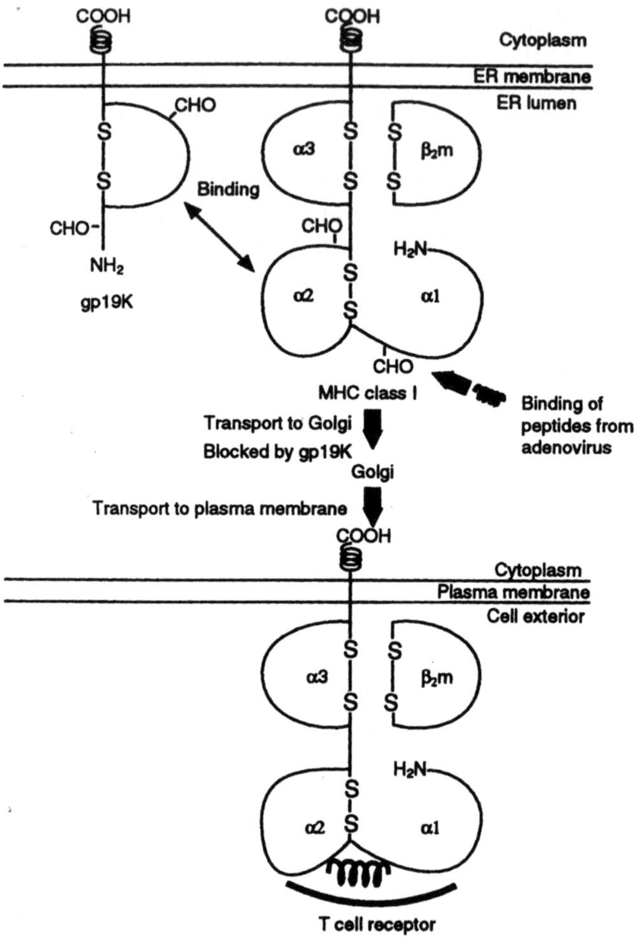


**Fig. 3.** Mechanism of action of *gp19K*. Class I antigens synthesized in the endoplasmic reticulum (*ER*) form complexes with adenovirus peptides, transport them through the Golgi complex to the cell surface, and present them to the receptor on cytotoxic T lymphocytes (CTL). The luminal domain of gp19K binds to the $\alpha_1$- and $\alpha_2$-domains of class I antigens (the peptide-binding domains) and retains the class I-peptide complex in the ER by virtue of an ER retention signal at the COOH terminus of gp19K. Consequently, adenovirus-infected cells are not killed by CTL. *MHC*, major histocompatibility complex

consisting of an integral membrane glycoprotein (the heavy or α-chain) and a noncovalently associated light chain ($\beta_2$-microglobulin; reviewed in RAMMENSEE et al. 1993). There are three domains in the α-chain, $\alpha_1$, $\alpha_2$, and $\alpha_3$, each of about 90 amino acids and coded by a separate exon. The $\alpha_1$- and $\alpha_2$-domains form a cleft that is occupied by an eight- to nine-amino acid peptide (e.g., from a viral protein). The class I–peptide complex is transported to the cell surface and displayed on the cell exterior. CTL bearing receptors capable of recognizing the peptide in the context of the class I antigen cause lysis of the cell displaying the class I–peptide complex. CTL are considered to be a major means of eliminating acute virus infections, including adenovirus infections (see ROUTES et al. 1993).

Gp19K has two key properties that allow it to affect the metabolism of class I antigens: it is retained in the ER, and it forms a complex with class I antigens. The consequence of these properties is that class I antigens are retained in the ER in the form of a complex with gp19K, rather than being transported to the cell surface (ANDERSSON et al. 1985; BURGERT and KVIST 1985). Retention of class I antigens in the ER has typically been demonstrated by two methods. First, class I antigens that are retained in the ER have exclusively high-mannose oligosaccharides that are sensitive to endoglycosidase H (endo H), whereas those that enter the Golgi apparatus acquire complex oligosaccharides that are resistant to endo H. The second method is quantitation of class I antigens on the cell surface by flow cytometry; however, flow cytometry is a less sensitive assay for gp19K function because it will detect class I antigens that were on the surface of cells prior to infection. If gp19K prevents cell surface expression of class I antigens, then it should protect cells from lysis by CTL. In accord with this prediction, gp19K-expressing cells are not killed by adenovirus-specific CTL that were raised in mice by immunization with human Ad5 (RAWLE et al. 1989, 1991). Also, adenovirus-infected cells (ANDERSSON et al. 1987) and human 293 cells stably transfected with the gp19K gene (BURGERT et al. 1987) are not killed by alloreactive CTL. Gp19K expressed in cells coinfected with adenovirus and transformed by SV40 reduces sensitivity to SV40-specific CTL (TANAKA and TEVETHIA 1988). Finally, when gp19K was expressed in cells from a vaccinia virus vector, it prevented lysis of the infected cells by CTL specific to vaccinia virus or to influenza virus (cells were coinfected with vaccinia vectors expressing influenza virus proteins) (COX et al. 1990, 1991).

Gp19K also prevents phosphorylation of class I antigens (LIPPÉ et al. 1991), which occurs within the cytoplasmic tail of class I antigens. This result suggests that phosphorylation of class I antigens occurs in a post-ER compartment. The role that phosphorylation plays in class I antigen function is unknown.

These observations have led to the hypothesis that gp19K may prevent killing of adenovirus-infected cells in humans, thereby prolonging acute infections, but in particular allowing the virus to establish persistent infections (ANDERSSON et al. 1985; BURGERT and KVIST 1985; PÄÄBO et al. 1989). This is an attractive hypothesis, but it has yet to be verified in an animal model and especially in humans. Perhaps the best supporting in vivo evidence is the observation by GINSBERG et al. (1989) that mutants in the gp19K gene display more pathology in the lungs of cotton rats than does wild-type Ad5. This pathology is characterized by enhanced infiltration of lymphocytes, macrophages, and monocytes into the infected area. However, there is no proof that this enhanced pathology is attributable to an inability of the gp19K mutant to block cell surface expression of class I antigens. Also, GRUNHAUS et al. (1994), using a vaccinia virus vector that expresses gp19K, did not observe an effect of gp19K on vaccinia virus replication or lethality in mice.

There is other suggestive evidence that an ability to downregulate class I antigen expression is important in the biology of human adenoviruses. In addition to the subgroup C serotypes, Ad2 and Ad5, DNA sequencing has identified the gp19K gene in several subgroup B serotypes, namely Ad3 (SIGNÄS et al. 1986),

Ad7 (Hermiston et al. 1993a), Ad11 (Mei and Wadell 1992), and Ad35 (Flomenberg et al. 1988), and also in the single subgroup E serotype, Ad4 (Hermiston et al. manuscript in preparation; (Fig. 2)). Actual binding of gp19K to human class I antigens (shown by coimmunoprecipitation of gp19K by the W6/32 monoclonal antibody which reacts with all human class I antigens) has been demonstrated for the subgroup B serotypes Ad3 (Pääbo et al. 1986a; Hermiston et al. 1993a), Ad7 (Hermiston et al. 1993a), Ad11, Ad34 (Pääbo et al. 1986a), and Ad35 (Flomenberg et al. 1987). Ad35 gp19K has also been shown to bind to monkey class I antigens (Flomenberg et al. 1992). Gp19K-class I antigen complex formation also occurs for Ad19 (subgroup D) and Ad4 (subgroup E) (Pääbo et al. 1986a). Since gp19K is conserved in these subgroups, it must be an important gene.

Interestingly, Ad12 and Ad31, two subgroup A serotypes, do not express gp19K or an analogous protein that can block transport of class I antigens to the cell surface (Pääbo et al. 1986a), and the absence of gp19K in the E3 region of Ad12 has been proved by DNA sequencing (Sprengel et al. 1994). However, the E1A region of Ad12 is known to inhibit transcription of MHC class I genes in rodent (Schrier et al. 1983; Friedman and Ricciardi 1988; Shemesh et al. 1991; Ozawa et al. 1993) and human (Vasavada et al. 1986) cells. Also, cells expressing E1A of Ad12 have decreased susceptibility to killing by CTL (Bernards et al. 1983; Yewdell et al. 1988). The E3 region of Ad40 (subgroup F) also lacks gp19K as determined by DNA sequencing (Davison et al. 1993). However, as is the case with Ad12, there is evidence that the E1A genes of Ad40 downregulate class I antigens in transformed rodent cells (Cousin et al. 1991). Thus, the subgroup A and F serotypes may inhibit class I antigen expression at the transcription level.

These observations suggest an interesting possibility regarding the tissue specificity of human adenoviruses. The subgroup F serotypes are significant causes of infantile diarrhea, and the subgroup A serotypes have most commonly been isolated from anal swabs (Brandt et al. 1969; Horwitz 1990). On the other hand, subgroup B, C, and E serotypes primarily cause respiratory infections (although they can also infect other tissues). Thus, downregulation of class I expression at the transcription level may facilitate gastrointestinal (GI) infections, whereas inhibition of class I antigen transport to the cell surface, via the functioning of gp19K, may facilitate respiratory infections.

Studies have been done to identify which sequences in gp19K are important in binding to class I antigens. Binding to class I antigen α-chain was observed with an Ad2 gp19K polypeptide consisting of residues 1–97 (beginning with the first residue following the cleaved signal), either immobilized on agarose (Pääbo et al. 1986b), translated in vitro, or expressed in insect cells from a baculovirus vector (Gabathuler et al. 1990). Thus, the luminal domain (residues 1–107) of gp19K is sufficient for binding to class I antigens. Anecdotal evidence suggests that the transmembrane and cytoplasmic domains of gp19K may increase the avidity of binding (Gabathuler et al. 1990; Pääbo et al. 1986b, 1987). In a structure–function study of gp19K, 13 in-frame virus deletion mutants (four to 12 amino acids deleted) in the ER luminal domain were analyzed for the ability of mutated gp19K to bind to class I antigens, retain them in the ER, and prevent cytolysis by adenovirus-specific CTL (Hermiston et al. 1993b). The mutant proteins were also examined for reaction with the Tw1.3 monoclonal antibody to gp19K, an antibody that recognizes a noncontiguous epitope in the ER luminal domain (HERMISTON et al. 1993b; COX et al. 1991). All but one of the mutants (residues 102–107 deleted) were defective in binding to class I antigens, and all but three mutants (residues 84–107 deleted) were defective in binding to Tw1.3. These experiments, together with the observation that the ER luminal domain has regions that are conserved and diverged among serotypes in different subgroups (FLOMENBERG et al. 1988), led to the proposal that the ER luminal domain of gp19K has three subdomains (HERMISTON et al. 1993b). These domains are termed the ER luminal variable domain (ERL-VD; residues 1 to approximately 77–83), the ER luminal conserved domain (ERL-CD; residues approximately 84–98), and the ER luminal spacer domain (ERL-SD; residues approximately 99–107) (Fig. 2). The ERL-VD, which is only about 22%–25% identical among serotypes in different subgroups, is proposed to bind the variable $\alpha_1$-and $\alpha_2$-domains of class I antigens. The $\alpha_1$- and $\alpha_2$-domains of class I $\alpha$-chain are variable because they form the peptide-binding cleft and they make contacts with the variable T cell receptor. The $\alpha_1$- and $\alpha_2$- domains of $\alpha$-chain have been shown to be important for binding to gp19K (see below). The ERL-CD, which is about 75% identical among serotypes in different subgroups, is proposed to interact with some conserved molecule, perhaps the $\alpha_3$-domain of $\alpha$-chain. [We note, however, that there is no evidence that $\alpha_3$ binds to gp19K (JEFFERIES and BURGERT 1990; BEIER et al. 1994)]. The ERL-VD is proposed to extend the luminal domain away from the membrane. Additional work is required to verify these proposals.

FLOMENBERG et al. (1992) studied the effects of truncations and point mutations in Ad35 gp19K binding to class I antigens in transiently transfected monkey cells. Their results were in general agreement with those of HERMISTON et al. (1993b) with Ad2 gp19K, except that they found that a truncation at Ad35 gp19K residue 127 (equivalent to Ad2 residue 102) abolished binding, whereas such a truncation does not abolish binding in Ad2 gp19K.

Of considerable interest, gp19K of Ad2 and Ad5 binds with different affinities to different class I antigens. As indicated in Table 2, gp19K complexes well with mouse H-2 $K^d$ and $D^b$, weakly with $L^d$, and very poorly or not at all with $K^b$, $D^d$, $K^k$, or $D^k$. Gp19K of Ad2 and Ad5 binds to all human class I antigens tested, but with different affinities. Ad2 gp19K was originally shown to bind about twice as well to HLA A2 than to B7 (SEVERINSSON et al. 1986). Recently, BEIER et al. (1994) carried out a detailed study where a vaccinia virus vector expressing Ad5 gp19K was used to infect a human B lymphoblastoid cell line selected for loss of HLA A and B products and then stably transfected with different HLA A and B alleles. Binding of gp19K to these class I antigens was assayed by coimmunoprecipitation of gp19K using the W6/32 monoclonal antibody, which reacts with all HLA molecules, by endo H sensitivity of HLA molecules, and by flow cytometry to detect HLA molecules on the cell surface. They found that HLA A2.1 and B7 bind very well to gp19K, A3, A1, and Aw69 bind six-to 30-fold less well, and Aw68, B27, and Bw58 bind 50- to 150-fold less well (BEIER et al. 1994).

The widely differing affinities of gp19K to different class I antigens has allowed identification of the regions in class I $\alpha$-chain that are important for

**Table 2.** Interaction of 19-kDa glycoprotein (gp19K) from adenovirus type 2 (Ad2) or Ad5 with different mouse major histocompatibility complex (MHC) class I antigens

| H-2 Allele | Co-IP[a] | Endo H sensitivity[b] | FACS[c] | Anti-Ad CTL[d] | Other CTL[e] |
|---|---|---|---|---|---|
| $K^k$ | –[f,i,k] | –[f,i] | ND | –[h] | –[i] |
| $D^k$ | –[i] | –[i] | ND | ND | –[i] |
| $K^d$ | +[f,i,j,k] | +[f,i,j] | +[j] | +[h] | +[i,j] |
| $D^d$ | ND | –[i] | –[l] | –[h] | –[i,j] |
| $L^d$ | + (weak)[i] | +[i] | ND | ND | +(weak)[i,j] |
| $K^b$ | ND | ND | –[g] | ND | –[g] |
| $D^b$ | ND | ND | +[g] | +[h] | +[g] |

Endo H, endoglycosidase H; FACS, fluorescence-activated cell sorter; CTL, cytotoxic T lymphocytes; ND, not determined.

[a] Coimmunoprecipitation of Ad2 or Ad5 gp19K glycoprotein using antisera to class I antigens.

[b] Inhibition of transport of class I antigens from the endoplasmic reticulum (ER) to the Golgi, as determined by sensitivity to digestion with endo H.

[c] Presence of class I antigens on the cell surface, as determined by flow cytometry.

[d] Adenovirus-specific CTL were generated by infecting C3H/HeJ(H-$2^k$), BALB/c(H-$2^d$), or C57BL/10J(H-$2^b$) mice with Ad5 (priming), followed by secondary stimulation of splenocytes by incubating splenocytes with ultraviolet (UV)-irradiated syngeneic cells infected by mutants that retain or lack gp19K glycoprotein (RAWLE et al. 1989, 1991). Secondary stimulation is necessary for the generation of CTL. The resulting CTL were then examined for their ability to lyse sygeneic target cells infected with mutants in the various E3 genes, using a $^{51}$Cr-release assay (RAWLE et al. 1989). Of all the E3 mutants examined, only mutants in gp19K were defective in inhibiting the generation of CTL in the secondary stimulation step. Also, only mutants in gp19K were defective in inhibiting lysis of infected target cells by CTL. The presence or absence of gp19K did not affect priming.

The CTL were assayed for MHC restriction of cytolysis by using Ad5*dl*327 (deleted in E3)-infected target cells matched at either the K or D end of the H-2 complex (RAWLE et al. 1991). The ability of gp19K to inhibit secondary stimulation of CTL and to inhibit killing of target cells by CTL, together with knowledge of the MHC restriction of the CTL, allowed the determination of which H-2 allele was interacting, or not interacting, with gp19K. The anti-adenovirus CTL observed were as follows: $K^k$(+), $D^k$(–), $K^d$(+), $D^d$(+), $K^b$(–), $D^b$(+) (RAWLE et al. 1991). The $D^b$-restricted CTL were exclusively directed against an immunodominant epitope in E1A, whereas the $K^k$-, $K^d$-, and $D^d$-restricted CTL were not specific to E1A, and they probably were specific to E2 proteins (RAWLE et al. 1991).

[e] Mouse cells were coinfected with vaccinia vectors expressing either gp19K or influenza virus proteins, and lysis was examined using CTL restricted to different H-2 alleles and specific to vaccinia or influenza virus-specific epitopes (COX et al. 1990, 1991). In another study, SV40-transformed monkey cells stably transfected with H-$2K^b$ or H-$2D^b$ were infected with Ad2 or Ad2$^+$ND1 (deleted in E3), then lysis was examined using SV40-specific H-$2K^b$- or H-$2D^b$-restricted CTL clones (TANAKA and TEVETHIA 1988).

[f] BURGERT and KVIST (1987).

[g] TANAKA and TEVETHIA (1988).

[h] RAWLE et al. (1989, 1991).

[i] COX et al. (1990).

[j] COX et al. (1991).

[k] JEFFERIES and BURGERT (1990).

[l] BEIER et al. (1994).

binding gp19K. By experiments using chimeric H-2 $K^d$ (binds gp19K) and $K^k$ (does not bind gp19K) α-chains, it was determined that the $\alpha_1$- and $\alpha_2$-domains of α-chain are important in binding to gp19K (BURGERT and KVIST 1987), in particular the $NH_2$-terminal half of the $\alpha_2$-domain (JEFFERIES and BURGERT 1990). In a similar approach using chimeric HLA A2.1 (binds gp19K well) and H-2 $D^d$ (does not bind gp19K) α-chains, BEIER et al. (1994) concluded that $\alpha_1$ and $\alpha_2$ are involved in binding of α-chain to Ad5 gp19K, but that $\alpha_2$ is more important than $\alpha_1$.

BEIER et al. (1994) also performed pairwise amino acid comparisons of HLA α-chains that bind well to gp19K with α chains that do not bind well. Interestingly, the amino acid differences among strong and weak gp19K binders are located primarily in and around the peptide-binding site and point into the binding site. BEIER et al. (1994) make special note of residue 116, which is tyrosine in strong gp19K binders and aspartic acid in weak binders. This residue is located at the bottom of the peptide-binding cleft, and it is inaccessible when peptide is bound (BEIER et al. 1994). There is indirect evidence that gp19K binds to α-chains that already have peptide in the peptide-binding site (COX et al. 1991). Therefore, it remains to be determined how (and why) differences in the peptide-binding site (as well as sites that make contacts with the T cell receptor) result in differences in the affinity of α-chain for gp19K.

The differences in affinity of gp19K from Ad2 and Ad5 for different class I antigens raise the possibility that humans may be more or less susceptible to prolonged acute and persistent adenovirus infections depending on our particular repertoire of class I antigens and their affinity for gp19K (WOLD and GOODING 1989). As a first step to address this hypothesis, HERMISTON et al. (1993a) sequenced the gp19K gene from 17 very different clinical isolates of Ad7. The expectation was that significant differences would be found among these isolates, especially in the ERL-VD, the domain that is presumed to interact with the variable $\alpha_1$- and $\alpha_2$-domains of α-chain (Fig. 2). Surprisingly, 16 of the 17 Ad7 gp19K proteins were identical, and the 17th isolate had only one conservative substitution (HERMISTON et al. 1993a). Thus, there must be strong selective pressure to maintain the sequence of gp19K in Ad7.

The above paragraphs raise interesting questions regarding gp19K, especially if we assume that the only function for this protein is binding class I antigens and retaining them in the ER. First, gp19K of Ad2 apparently requires a specific tertiary structure for it to bind to class I antigens, inasmuch as binding to class I antigens and to the Tw1.3 anti-gp19K monoclonal antibody is very sensitive to small deletions (HERMISTON et al. 1993b). Second, the ERL-VD, the putative α-chain-binding region, is quite diverged (about 22%–25% identity) among serotypes in different subgroups. Despite this divergence in primary sequence, do gp19K's from all subgroups still assume a similar tertiary structure capable of binding the same class I antigens? Third, the ERL-VD is quite conserved among serotypes in the same subgroup, and it was nearly identical among 17 clinical isolates. What is it about this protein that requires such strong conservation in sequence among serotypes within a subgroup, yet allows such divergence among serotypes in different subgroups? Fourth, gp19K from Ad2 and Ad5 binds with markedly different affinities to different human and mouse class I antigens. How does this affect the ability of gp19K to block class I transport in the human population which has widely different combinations of alleles for the A, B, and C class I antigens? Large differences in binding affinity to class I antigens does not seem consistent with gp19K's presumed function, which is to block transport of *all* class I molecules capable of presenting viral peptides. In this connection it may be useful to consider the results from studies in C57BL/10 (H-$2^b$) mice, showing that there is only one immunodominant epitope for adenovirus proteins and that this epitope is located in the E1A proteins (RAWLE et al. 1991). In contrast, in BALB/c(H-$2^d$) and C3H/HeJ(H-$2^k$) mice the immunodominant epitope is not in E1A, but probably is in E2 (RAWLE et al. 1991). If immunodominant epitopes are also haplotype-specific in humans, then perhaps adenoviruses have evolved such that gp19K from any given serotype binds strongly only to those class I antigens that are capable of presenting the immunodominant epitope? Thus, in future studies it may be instructive to identify CTL epitopes that are dominant in humans. It will also be important to examine the affinities among different class I antigens and gp19K proteins from serotypes in different subgroups.

## 3.2 E3 14.7K Protein and the E3 10.4K/14.5K Complex of Proteins Inhibit the Cytotoxic and Inflammatory Responses Mediated by Tumor Necrosis Factor

Tumor necrosis factor (TNF)-α, a major cytokine secreted by activated macrophages, was first recognized for its ability to lyse certain tumor cell lines. It is now known that TNF is a potent mediator of the immune and inflammatory responses (VASSALLI 1992). As such, TNF should have antiviral effects in vivo. TNF synthesis is induced in mice by several viruses (HENNET et al. 1992; LANE et al. 1992), and it stimulates expression of a large number of genes that function in immunity and inflammation (VILCEK and LEE 1991). TNF also has antiviral activity directly on the infected cell: TNF inhibits the replication of many DNA and RNA viruses in cultured cells, and it lyses cells infected by some viruses (MESTAN et al. 1986; WONG and GOEDDEL 1986; MAYER et al. 1992; PAYA et al. 1988; GOODING et al.1988). TNF-β, which is secreted by activated T lymphocytes, is similar in properties to TNF-α.

The mechanisms by which TNF signal transduction and TNF cytolysis occur are only beginning to be understood. There are two cell surface TNF receptors, p55 and p75, whose extracellular domains are part of the nerve growth factor superfamily which includes the Fas antigen, a receptor that mediates apoptosis (BEUTLER and VAN HUFFEL 1994; C.A. SMITH et al. 1994). The intracellular domains of the p55 and p75 receptors are different, indicating that their mechanism of signal transduction is different. In nonhematopoietic cells, TNF cytolysis as well as TNF induction of many genes are mediated through the p55 receptor (TARTAGLIA et al. 1993).

The pathway through which the p55 receptor signals is uncertain: depending on the cell type, it could be phosphatidylcholine, protein kinase A, or protein kinase C (KRÖNKE et al. 1992; VILCEK and LEE 1991; WIEGMANN et al. 1992). However, recent studies suggest that the sphingosine cycle (DBAIBO et al. 1993; HANNUN and BELL 1993; HANNUN 1994) or pathway (MATHIAS and KOLESNICK 1993; KOLESNICK and GOLDE 1994) is especially important in TNF signaling in general, and TNF-induced apoptosis in particular. TNF stimulation of the p55 receptor activates a phosphatidylcholine-specific phospholipase C to convert phosphatidylcholine to diacylglycerol (SCHÜTZE et al. 1992). Diacylglycerol can then activate a cytosolic acidic sphingomyelinase to convert sphingomyelin to ceramide. Ceramide then acts as a second messenger to activate the serine/threonine-specific protein phosphatase 2A (DOBROWSKY et al. 1993). Ceramide also activates a membrane-bound serine/threonine-specific protein kinase (JOSEPH et al. 1993; LIU et al. 1994). Protein phosphatase 2A and the ceramide-activated protein kinase are believed to transmit downstream signaling events and/or provide feedback control for TNF signaling pathways.

TNF (VAN LINT et al. 1992; VIETOR et al. 1993; KYRIAKIS et al. 1994) and ceramide (RAINES et al. 1993) activate the Ras-mitogen-activated protein (MAP) kinase pathway which leads to phosphorylation and activation of transcription factors such as c-Jun. TNF also activates a recently discovered stress-induced pathway which includes a serine/threonine-specific protein kinase named Jnk (DÉRIJARD et al. 1994; KYRIAKIS et al. 1994). Jnk phosphorylates the N terminus of c-Jun, thereby causing activation of c-Jun (DÉRIJARD et al. 1994).Exogenous sphingomyelinase (which generates intracellular ceramide) activates Jnk (KYRIAKIS et al. 1994).

Multiple transcription factors are activated by TNF, especially NFκB and AP1. The mechanism by which NFκB is activated by TNF (BEG et al. 1993) or ceramide (MACHLEIDT et al. 1994) is via degradation of $I_{\kappa}B$, the protein that interacts with TNF and prevents its migration into the nucleus, where it can stimulate transcription. AP1 can be activated by TNF and ceramide through the Ras-MAP kinase pathway.

Regarding TNF cytolysis, there is good evidence that the 85-kDa cytosolic phospholipase $A_2$ ($cPLA_2$) must be activated (HAYAKAWA et al. 1993). $cPLA_2$ is a latent cytosolic enzyme that is activated by phosphorylation by MAP kinase and, in the presence of micromolar amounts of $Ca^{2+}$, becomes associated with membranes where it releases arachidonic acid (AA) specifically from the *sn*-2 position of membrane phospholipids (CLARK et al. 1991; LIN et al. 1993). $cPLA_2$ is activated by many agonists including TNF (HOECK et al. 1993). TNF kills some cells by necrosis and others by apoptosis (LASTER et al. 1988). Ceramide induces apoptosis (OBEID et al. 1993; JARVIS et al. 1994), suggesting that the sphingomyelin cycle/pathway may mediate TNF cytolysis. Thus, one scenario is that TNF stimulates the synthesis of ceramide which stimulates the Ras-MAP kinase pathway to activate $cPLA_2$. Activated $cPLA_2$ produces AA, which can be further metabolized to leukotrienes through the lipoxygenase pathway and to prostaglandins through the cyclo-oxygenase pathway. Studies with enzyme inhibitors suggest that the lipoxygenase pathway is involved in TNF cytolysis (CHANG et al. 1992; SUFFYS et al. 1991).

TNF apparently is very important in adenovirus infections, because several adenovirus proteins mediate the cellular response to TNF. Most cells are not killed by TNF, nor does TNF kill cells infected with wild-type Ad2 or Ad5. However, when cells are infected with certain adenovirus mutants, then the cells are killed by TNF. In the case of mouse cells, the cells are killed by TNF when infected with mutants in the E3 transcription unit (GOODING et al. 1988). This result indicates two things: first, adenovirus infection must render the cells susceptible to lysis by TNF (the uninfected cells were not lysed, whereas the cells infected with the E3 mutant were lysed), and second, product(s) from the E3 region must inhibit lysis by TNF (cells infected with the E3 mutant were lysed by TNF, whereas cells infected with wild-type Ad5 were not lysed). In the case of human cells, the virus mutant must lack both the E3 and the E1B transcription units in order for the cells to be lysed by TNF (GOODING et al. 1991a), indicating that proteins from both E3 and E1B prevent TNF cytolysis.

The adenovirus proteins that induce susceptibility to TNF cytolysis as well as those that prevent TNF cytolysis (Table 1) have been identified through the use of various virus mutants as well as cell lines stably transfected with adenovirus genes. The E1A 289R and 243R proteins induce susceptibility to TNF in adenovirus-infected cells (AMES et al. 1990; CHEN et al. 1987; DUERKSEN-HUGHES et al. 1989). The mechanism by which this occurs is unknown. The E1A proteins have a modular structure, with specific domains having specific functions. In particular, there are three domains that are highly conserved among adenovirus serotypes; these domains are termed CR1 (conserved region 1; residues 40–80), CR2 (residues 120–140), and CR3 (residues 140–188, lacking in the 243R E1A protein). Adenovirus mutants with large deletions in CR1 are unable to sensitize cells to TNF, whereas mutants with deletions elsewhere in the E1A proteins have a wild-type phenotype (DUERKSEN-HUGHES et al. 1991). In cell lines stably transfected with E1A, both CR1 and CR2 must be deleted in order to abrogate the ability of E1A to sensitize cells to TNF (AMES et al. 1990).

The $NH_2$-terminal region of the E1A proteins and the COOH-terminal portion of CR1 are required for E1A to form a complex with the p300 protein. The $NH_2$-terminal portion of CR1 as well as CR2 is required for E1A to form a complex with the tumor suppressor retinoblastoma protein (RB). Mutants with large deletions in CR1 do not bind either p300 or RB efficiently. The picture that emerges is that E1A can sensitize cells to TNF if E1A can form a complex with either p300 or RB, but if binding to both p300 and RB is knocked out, then the E1A proteins are unable to sensitize cells to lysis by TNF.

A major role for RB is to form a complex with E2F and maintain E2F in an inactive state (NEVINS 1992; MORAN 1993). E2F is a S phase-specific transcription factor. The disruption of the RB–E2F complex by E1A allows free E2F to induce transcription from genes containing E2F sites in their promoters (JOHNSON et al. 1993). p300 (ECKNER et al. 1994) exists in a complex with a variety of cellular proteins, including TBP (the TATA box-binding protein) (ABRAHAM et al. 1993; MORAN 1993). It is possible that E1A disrupts the p300–protein complexes, thereby inducing or repressing synthesis of cellular genes. It is clear that the genes affected by the proteins liberated from the RB and p300 complexes include those involved in DNA synthesis and deregulation of the cell cycle, because studies with E1A mutants have shown that the $NH_2$ terminus plus CR1, and CR1 plus CR2, are required to induce cellular DNA synthesis in quiescent cells, to immortalize primary cells, and to oncogenically transform cells in cooperation with activated *ras* (NEVINS 1992; MORAN 1993).

How does susceptibility to TNF cytolysis fit into this picture? It may be that liberation of the transcription factors that are normally complexed with RB and p300 induces and/or represses genes that play a role in TNF cytolysis.

Alternatively, it may be that unscheduled induction of DNA synthesis and deregulation of the cell cycle activate latent mechanisms that direct the cells down a pathway where they are susceptible to TNF cytolysis. It seems likely that susceptibility to TNF cytolysis is a consequence of the many functions associated with the E1A proteins. Therefore, the virus evolved proteins that inhibit TNF cytolysis.

Studies in which cells were infected with adenovirus mutants lacking various viral genes indicated that there are three "sets" of adenovirus proteins that function independently to inhibit TNF cytolysis. However, the three sets of proteins do not prevent TNF lysis of all cells examined. The E3 14.7K protein prevented TNF cytolysis in 13 of 15 mouse cell lines tested (GOODING et al. 1991b). The E3 10.4K and E3 14.5K proteins, which function as a heterodimer termed 10.4K/14.5K, prevented TNF cytolysis in 11 of 15 mouse cell lines (GOODING et al. 1991b). In human cells there is an E3 function that inhibits TNF cytolysis, but it is not yet known if this is due to 14.7K, 10.4K/14.5K, or both sets of E3 proteins (GOODING et al. 1991a). Remarkably, in human cells, the 19-kDa protein encoded by the E1B transcription unit can also inhibit TNF cytolysis (GOODING et al. 1991a). For unknown reasons, E1B 19K does not inhibit TNF cytolysis of mouse cells.

It is of great interest that of the approximately 25 adenovirus early proteins, four of them inhibit TNF cytolysis. Furthermore, DNA sequencing indicates that these E3 TNF-protecting proteins are conserved among serotypes in all human subgroups (Figs. 4–7). The 14.7K protein has been shown to be expressed in some of these subgroups, and serotypes in subgroups A–E inhibit TNF lysis of infected cells (HORTON et al. 1990). Also, E1B 19K is conserved in serotypes in subgroups A–C and F (WHITE et al. 1992). Such a large commitment of the viral genome by the different adenovirus subgroups indicates that TNF is an important antiviral defense of the host.

Some information is available on the properties of these TNF-protecting proteins. The 14.7K protein of Ad5 is a hydrophilic (26% of the residues are acidic or basic) non-membrane protein of 128 amino acids (Fig. 4) that is localized in the cytoplasm and nucleus as indicated by immunofluorescence and cell fractionation (PERSSON et al. 1978; A. TOLLEFSON and W. WOLD, unpublished results). There are no obvious features in the sequence of 14.7K that give clues to its function.

E3 10.4K (Fig. 5; TOLLEFSON et al. 1990b) and E3 14.5K (Fig. 6; TOLLEFSON et al. 1990a) are integral membrane proteins that localize as a complex to the plasma membrane (HOFFMAN et al. 1992; STEWART et al. 1995; Fig. 8). The 10.4K protein exists as two species joined by a disulfide bond between the cysteine at residue 31 (HOFFMAN et al. 1992; Fig. 8). One species of 10.4K has a cleaved $NH_2$-terminal signal sequence, and the protein has a $N_{exo}$ $C_{cyt}$ orientation in the membrane (KRAJCSI et al. 1992a). With the other species of 10.4K, the $NH_2$-terminal signal is retained as a transmembrane domain, such that the protein spans the membrane twice and has both its $NH_2$ and COOH terminus in the cytoplasm (KRAJCSI et al. 1992a; Fig. 8). The complex between 10.4K and 14.5K is noncovalent. The 14.5K protein has a cleaved $NH_2$-terminal signal sequence, and the protein has an $N_{exo}$ $C_{cyt}$ orientation in the membrane (KRAJCSI et al. 1992b; Fig. 8). The 14.5K

```
Subgroup C Ad5    M T D • • • T L • • • • • D • L E M D G I I T E Q R L L E R R R A A A  26
           Ad2    - - E • • • S - • • • • • - • - - L - - - N - - - - - - - - - K - - S  26
Subgroup B Ad3    - - - P I A - S S T A A K E L - D - - - R A S - - - - I • Q L - I R Q  34
           Ad7    - - E • I L - T S N S A E - L - D - - - R V S - - - - A • Q L - I R Q  32
           Ad11   - - E • T L - T S N S A E - L - D - - - R V S - - - - A • Q L - I R Q  33
Subgroup D Ad9                                      - - - Q A S - - - I - • Q L - V R Q  19
Subgroup A Ad12   - I E P • • • • • • • • • - • - - I - - R - - - - - - - T D - A R R R  24
Subgroup F Ad40   - S - • • • • • • • • • • • Q - - I - - Q C - - - L I - A - - K L K Q  24

Subgroup C Ad5    E Q Q R M N Q E L Q D M V N L H Q C K R G I F C L V K Q A K V T Y D • •  60
           Ad2    - R E - L K - - V E - - - - - - - - - - - - - - V - - - - - L - - E • •  60
Subgroup B Ad3    Q - E - A V K - - R - A I G I - - - - K - - - - - - - - S - I S - E • •  70
           Ad7    Q - E - V T K - - R - V I Q I - - - - K - - - - - - - - - - I S - E • •  66
           Ad11   Q - E - A A K - - R - V I Q I - - - - K - - - - - - - - - - I S - E • •  67
Subgroup D Ad9    Q - E - A A K - - L - A I - I - - - - K - - - - - - - - - - I - - E L V  55
Subgroup A Ad12   Q - D Q K - K - - I - L Q T V - - - - K - L - - - - - - - T L R - E S •  59
Subgroup F Ad40   Q N - E L F N • - - • • • A - - - - - K - L - - - - - - - E L C - - • •  54

Subgroup C Ad5    S N T T G H R L S Y K L P T K R Q K L V V M V G E K P I T I T Q H S   94
           Ad2    K T - - - N - - - - - - - - Q - - - - - L - - - - - - - - V - - - -   94
Subgroup B Ad3    I T A - D - - - - - E - G P Q - - - F T C - - - I N - - V - - - Q -  102
           Ad7    I T A - D - - - - - E - G P Q - - - F T C - - - V N - - V - - - Q -  100
           Ad11   I T A - D - - - - - E - G P Q - - - F T C - - - I N - - V - - - Q -  101
Subgroup D Ad9    - G G K Q - - - T - E - - Q Q K - - F T C - - - V N - - V - - - Q -   89
Subgroup A Ad12   L P G K E - Q - C - T - - - Q - - T F T A - - - S V - - K V S - Q A   93
Subgroup F Ad40   V T Q Q - - E - - - T - N K Q - - S F M T - - - V - - - K V - - Q -   88

Subgroup C Ad5    V E T E G C I H S P C Q G P E D L C T L I K T L C G L K D L I P F N  128
           Ad2    A - - - - - L - F - Y - - - - - - - - - - - - M - - I R - - - - - -  128
Subgroup B Ad3    G D - K - - - Q C S - D S T - C I Y - - L - - - - - - R - - L - M -  136
           Ad7    G D - K - - - - C S - D S I - C T Y - - L - - - - - - R - - L - M -  134
           Ad11   G D - K - - - - C S - D S I - C T Y - - L - - - - - - R - - L - M -  135
Subgroup D Ad9    G - - N - - - - C S - E S - - C I Y S - L - - - - - - R - - L - M -  123
Subgroup A Ad12   G - Q - - S - R C L - D N - - C - Y - - - - - - - - - R N - L - M -  127
Subgroup F Ad40   G P V - - S - L C Q - T N - - C M Y - M V - - - - - - R E - L - - -  122
```

**Fig. 4.** Sequence of the E3 14.7K protein. The DNA sequence was determined in adenovirus type 2 (Ad2; HÉRISSÉ and GALIBERT 1981), Ad5 (CLADARAS and WOLD 1985), Ad3 (SIGNÄS et al. 1986), Ad7 (HONG et al. 1988), Ad11 (MEI and WADELL 1993), Ad9 (L. HAWKINS, T. HERMISTON, and W. WOLD, unpublished results), Ad12 (SPRENGEL et al. 1994), and Ad40 (DAVISON et al. 1993)

protein is O-glycosylated with mucin-type oligosaccharides at a site in the extracellular domain (KRAJCSI et al. 1992c). It is also phosphorylated on serine residues, probably at a single site near the COOH terminus (KRAJCSI and WOLD 1992). With mutants that lack 10.4K, 14.5K is also phosphorylated on threonine residues.

The E1B 19K protein is a membrane-associated protein (probably peripheral rather than integral) that localizes predominantly to the nuclear membrane (WHITE and CIPRIANI 1989, 1990). E1B 19K disrupts vimentin-containing intermediate filaments and the nuclear lamina proteins; the significance of this is not known.

As noted above, these proteins inhibit TNF cytolysis when the cells are rendered sensitive to TNF by expression of E1A. Cells can also be sensitized to TNF by inhibition of protein synthesis using cycloheximide (CHI), by treating cells with cytochalasin E (which disrupts microfilaments), and some transformed cells are "spontaneously" susceptible to TNF. [The 14.7K protein can prevent TNF cytolysis under all these conditions in adenovirus-infected cells (GOODING et al. 1990). [The 14.7K protein did not prevent TNF cytolysis in two cell lines that are

```
                            Signal/Transmembrane
Subgroup C Ad2    M I P R V L I L L T L V A L F C A C S T L A A V A H I E V D   30
           Ad5    - - - - - F - - - - - - - - - - - - - - - - - - S - - - - -   30
Subgroup B Ad3    - - - - N F F F T I - I C A - N V - A - F T - - - T A S P -   30
           Ad7    - - - - N F F F T I - I C A - N V - A - F T - - - T A S P -   30
           Ad35   - - - - N F F F T I - I C A - N V - A - F T - - - T A S P -   30
           Ad11   - V - - N F F F T I - I C A - N V - A - F T - - - T A T P -   30
Subgroup D Ad9    - - - - F F L F N I - F C - - N I - A A F - - - S - A S P -   30
Subgroup A Ad12   - V T V L - - F - C - P V I - S S • - - F - - - S D L D P E   29
Subgroup F Ad40   - V T P L - L - V C - P I I Y A S • T - F - - - S - L D T -   29

                                          transmembrane
Subgroup C Ad2    C I P P F T V Y L L Y G F V T L I L I C S L V T V V I A F I   60
           Ad5    - - - A - - - - - - - - - - - - T - - - - - I - - - - - - -   60
Subgroup B Ad3    - - G - - A S - A - F A - - - C - C V - - I - C L - - N - F   60
           Ad7    - - G - - A S - A - F A - - - C - C V - - I - C L - - N - F   60
           Ad35   - - G - - A S - A - F A - - - C - C V - - I - C L - - N - F   60
           Ad11   - - G A - A S - A - F A - - - C - C V - - I - C L - - N - F   60
Subgroup D Ad9    - L G - - P T - - - F A L L - C T C V - - I - C - - - T - L   60
Subgroup A Ad12   - L A - - A - - - I F T - - - A T C V - - I I - L L - T S L   59
Subgroup F Ad40   - L - A L L T - - I F T S - C C T A - - - I A - F F V - I F   59

Subgroup C Ad2    Q F I D W V C V R I A Y L R H H P Q Y R D R T I A D L L R I L     91
           Ad5    - C - - - - - - - F - - - - - - - - - - - - - - - E - - - - -     91
Subgroup B Ad3    - L V - - I F - - - - - - - - - - E - - N Q N V - A - - - L I     91
           Ad7    - L V - - I F - - - - - - - - - - E - - N Q N V - A - - - L I     91
           Ad35   - L V - - I F - - - - - - - - - - E - - N Q N V - A - - - L I     91
           Ad11   - L - - - I L - - - - - - - - - - E - - N Q N - - A - - - L I     91
Subgroup D Ad9    - L - - - S - A - Y N - - H - S - E - - - E N  83
Subgroup A Ad12   - - F - Y Y Y - - - V - R - - - - R - Q N P Q - - A - - Q L Q P   91
Subgroup F Ad40   - T A - Y L Y - - V - - Y - - - - - - - N H E V - T - - C L S     90
```

**Fig. 5.** Sequence of the E3 10.4K protein. See the legends to Figs. 2 and 4 for references

"spontaneously" susceptible to TNF, so it cannot prevent TNF cytolysis under all conditions (GOODING et al. 1991b)]. The 10.4K/14.5K (GOODING et al. 1991b) and E1B 19K (GOODING et al. 1991a) proteins also can prevent TNF cytolysis of spontaneously sensitive cells as well as cells sensitized to TNF by inhibition of protein synthesis. These results suggest that 14.7K, 10.4K/14.5K, and E1B 19K are more or less general inhibitors of TNF cytolysis, even though there are a limited number of transformed cell lines in which these proteins do not prevent TNF cytolysis.

The 14.7K protein prevents TNF cytolysis when expressed from a vector in stably transfected mouse cell lines (HORTON et al. 1991). This has also been shown for E1B 19K, both in stably (HASHIMOTO et al. 1991) and transiently (WHITE et al. 1992) transfected human cell lines. Thus, these proteins can prevent TNF cytolysis in the absence of other adenovirus proteins. Studies of this type have not yet been done for the 10.4K/14.5K proteins.

A structure-function analysis has been done on E3 14.7K (RANHEIM et al. 1993). Thirteen virus mutants with in-frame deletions throughout the 14.7K gene were constructed, as were six mutants with the six cysteine residues converted to serine. All but two of the deletion mutants and three of the cysteine mutants were defective for preventing TNF cytolysis. This study suggests that the entire 14.7K protein represents the functional moiety and that the protein does not have a subdomain that embodies the TNF-protecting function.

Not a great deal is known about the molecular mechanism by which these adenovirus proteins inhibit TNF cytolysis. One mechanism by which they could

```
                          Signal
Subgroup C Ad5       MKFTVTFLLII•CTLSAFCSPTSKP•QRHISCRFTRIWNIPS   40
           Ad2       --RS-I-V-L-F-A-PVL--Q--A-PK---------Q------   42
Subgroup B Ad3    MQA-LP•-ILI-LLP-•IPLASTA-RATPEQLRK-K-QQP-SFLD   43
           Ad7    MQA-LP•-ILI-LLP-•IALASTA-RATPEQLRK-K-QQP-SFLD   43
           Ad11   ---ILP•IFL--LLPY•AVSTPAAY-T-PEHLRK-K-QQP-SFLA   43
Subgroup D Ad9    --•••••TLLM-LSHL•SP-LA••--A••••DSK-K-AE---FLD   33
Subgroup A Ad12      --••TALV-FFML•IPVWA-SC•••••QLHK•••••P--FLD   29
Subgroup F Ad40      --VPLLC-ILLHK•VL-N-••••••••HLHR•••••PTEFLR   28

                               transmembrane
Subgroup C Ad5    CYNEKSDLSEAWLYAIISVMVFCSTILALAIYPYLDIGWNAIDAM   85
           Ad2    ---KQ---------------------F------------------   87
Subgroup B Ad3    --H----FPTY-IVIVGIINILSC-FFSIT---TFNF---SPN-L   88
           Ad7    --H----FPTY-IVIVGIINILSC-FFSIT---TFNF---SPN-L   88
           Ad11   --R---EIPPNLIMIAGIINII-C--ISFL---LF-F----PN-H   88
Subgroup D Ad9    --Q--I-MPSYY-VIVGI---CSC-FF-IM---CF-L---SVE-F   78
Subgroup A Ad12   --TKET•NYIG-V-G-M-GL--V-SVVS-QL-AR-NFS--KYTDD   73
Subgroup F Ad40   --S••TET-SF---S--FILI-FA-F-G-Q--GC-HL--MHPPNN   71

Subgroup C Ad5    NHPTF•••PAPAMLPLQQVVAGGFVPANQPRPPSPTPTEISYFNLTGGDD   132
           Ad2    -----•••-V--VI-----I-••••-I-----------------------   130
Subgroup B Ad3    GY-QE•••-D•EHI---HIQQPLALVQYENE-QPSL-PA-----------   134
           Ad7    GY-QE•••-D•EHI---HIQQPLALVEYENE-QPSL-PA-----------   134
           Ad11   D--QD•••-E•EHI---NMQHPIALIDYESE-QP-LLPA-----------   134
Subgroup D Ad9    TYTLESSSLSTPPPTPPPRRNQFPLIQYLEE--PRP-STV---HI-----   128
Subgroup A Ad12   LPEYP•••NPQDD---NI-F•••••••••-E-•PRP-SVV---KFT-E--   110
Subgroup F Ad40   LPRFP•••GFLLQP-PPPPA•••••••••••-VQRA-SV----H-NSE-V   107
```

**Fig. 6.** Sequence of the E3 14.5K protein. See the legends to Figs. 2 and 4 for references

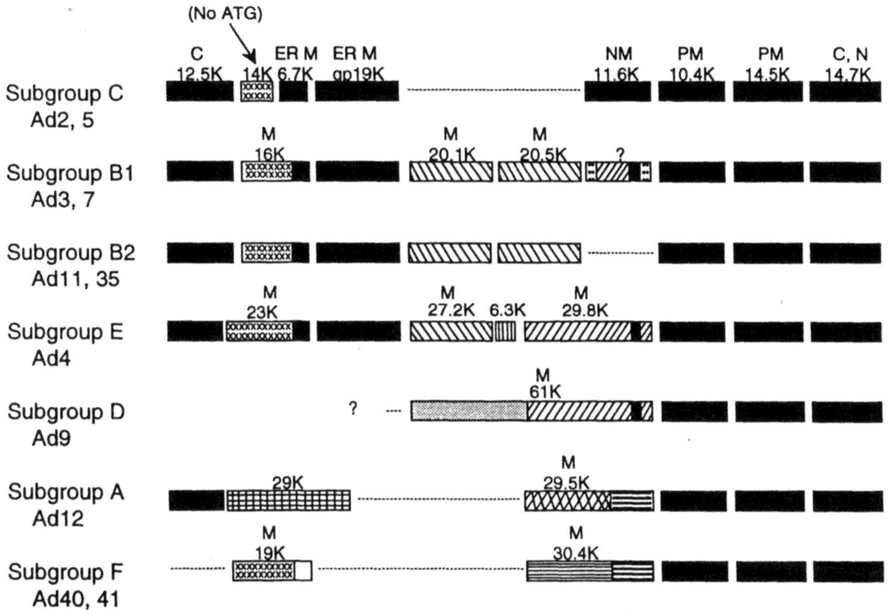


**Fig. 7.** Predicted E3 proteins from serotypes in different human subgroups. Proteins or portions of proteins that are conserved are indicated by identical marking of the *bar* depicting the protein. For example, the 10.4K, 14.5K, and 14.7K proteins are conserved in all subgroups. The 11.6K protein of subgroup C (adenovirus types 2 and 5, Ad2 and Ad5) is conserved (distantly) in subgroup B1, and there is homology (weak) to 11.6K in the 29.8K protein of subgroup E (Ad4) and the 61K protein of subgroup D (Ad9). *Dashed lines* indicate that the protein is absent. *Dashed lines ending in a question mark* indicate that the region has not been sequenced. *C*, localized in the cytoplasm; *N*, localized in the nucleus; *ER M*, localized in the endoplasmic reticulum membrane; *NM*, localized in the nuclear membrane; *PM*, localized in the plasma membrane; *gp*, glycoprotein. References for the DNA se-quences are given in the legends of Figs. 2 and 4. The Ad4 E3 sequence was determined by HERMISTON et al. (manuscript in preparation)

function is by downregulating the p55 TNF receptor. However, no effect on $^{125}$I-TNF binding was observed in transfected cells stably expressing the 14.7K protein (HORTON et al. 1991) or in infected cells expressing the E1B 19K protein (WHITE et al. 1992). Thus, 14.7K and E1B 19K apparently do not downregulate this TNF receptor. It is not yet known whether 10.4K/14.5K affect cell surface expression of the TNF receptors.

An important advance in understanding how these adenovirus proteins may prevent TNF cytolysis was provided by ZILLI et al. (1992). As noted above, TNF activation of $cPLA_2$, which produces AA from membrane phospholipids, appears to be crucial to TNF cytolysis. When ZILLI et al. (1992) treated mouse cells with TNF, AA was not released, nor were the cells lysed. However, when the cells were treated with TNF plus cycloheximide (CHI) AA was released and cell lysis occurred. Apparently CHI is able to sensitize the cells to TNF-induced release of AA, just as it does to TNF cytolysis. The release of AA was presumably due to TNF activation of $cPLA_2$. Of great interest, when mouse cells stably transfected with the E3 14.7K gene (HORTON et al. 1991) were treated with TNF plus CHI, AA was

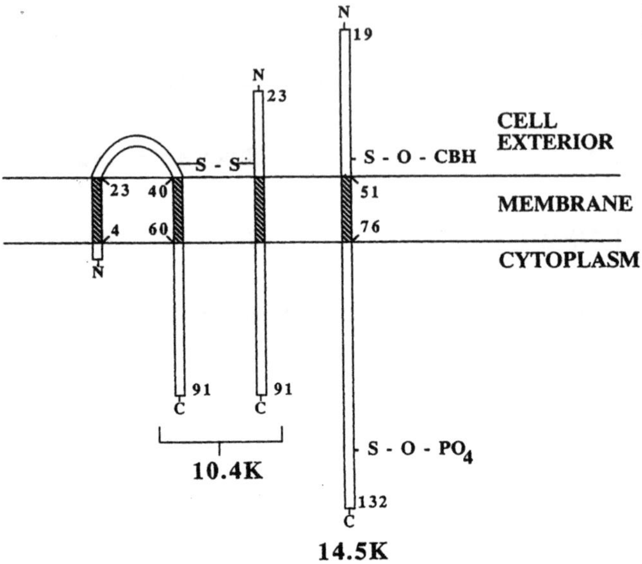


**Fig. 8.** E3 10.4K/14.5K protein complex. See text for details

not released and the cells did not lyse; this suggests that E3 14.7K may prevent TNF cytolysis by preventing TNF activation of $cPLA_2$ (Zilli et al. 1992).

TNF-induced release of AA has also been studied in mouse cells infected with various adenovirus mutants (P. Krajcsi, A. Scaria, S. Laster, and W. Wold, unpublished results). AA was not released when wild-type adenovirus-infected cells were treated with TNF, but it was released from TNF-treated cells infected with an E3 deletion mutant. This observation parallels the lytic response to TNF, i.e., adenovirus infection renders the cells susceptible to TNF-induced release of AA (and TNF cytolysis), and protein(s) from E3 inhibit TNF-induced release of AA (and TNF cytolysis). It is likely that the E1A proteins sensitize the cells to TNF-induced release of AA, although this has not yet been shown. Significantly, the 14.7K protein, and the 10.4K/14.5K complex of proteins, were *independently* able to prevent TNF-induced release of AA, just as they were *independently* able to prevent TNF cytolysis. Of further interest, when human cells were infected with appropriate adenovirus mutants, E1B 19K was found to inhibit TNF-induced release of AA, just as it inhibits TNF cytolysis. Thus, all the adenovirus proteins that inhibit TNF cytolysis also inhibit TNF-induced release of AA. Since TNF-induced release of AA is probably caused by TNF activation of $cPLA_2$, the adenovirus proteins may prevent TNF activation of $cPLA_2$.

Although the 14.7K protein (Zilli et al. 1992) and the E3 10.4K/14.5K proteins (unpublished results) may inhibit activation of $cPLA_2$ by TNF, they do not inhibit activation of $cPLA_2$ by activators of protein kinase C (phorbol ester plus $Ca^{2+}$ ionophore). Thus, these proteins probably act at a step(s) in a signaling pathway that is more or less specific to TNF. Since the 14.7K, 10.4K/14.5K, and E1B 19K proteins are so different in properties, yet they can independently inhibit TNF-induced release of AA, they may act at different critical steps in the signaling pathway between the TNF receptor and $cPLA_2$.

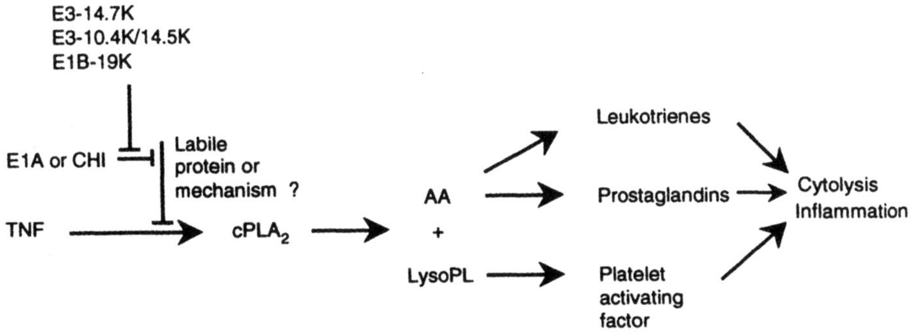


**Fig. 9.** Proposed role for adenovirus proteins in tumor necrosis factor (TNF)-induced cytolysis and inflammation. See text for details. *cPLA$_2$*, cytosolic phospholipase A$_2$; *AA*, arachidonic acid; *Lyso PL*, lysophospholipid

The implications of these findings are illustrated in Fig. 9. In uninfected cells, cPLA$_2$ is not activated by TNF; perhaps a labile protein or mechanism (e.g., phosphorylation) blocks the TNF-cPLA$_2$ signal transduction pathway. However, when the cells are expressing E1A or protein synthesis is inhibited with CHI, the labile protein or mechanism is eliminated and cPLA$_2$ is activated by TNF. This leads to synthesis of AA and lysophospholipid, which are then converted to eicosinoids. Cell lysis then occurs by unknown mechanisms. The 14.7K, 10.4K/14.5K and E1B 19K proteins block one or more steps in the TNF-cPLA$_2$ pathway, so they prevent cytolysis.

The eicosinoids are mediators of the inflammatory response. The adenovirus proteins, by inhibiting the activation of cPLA$_2$ by TNF, would be expected to prevent an inflammatory response due to adenovirus infection. In accord with this prediction, GINSBERG et al. (1989) saw increased infiltration of neutrophils into the lungs of mice infected with a virus mutant that lacks the 10.4K, 14.5K, and 14.7K genes. TUFARIELLO et al. (1994), working with a vaccinia virus vector that expresses 14.7K, TNF, or both 14.7K and TNF, found that 14.7K increased the virulence of the TNF-producing vaccinia vector in mice. The increased virulence was associated with an increase in inflammation, which is the opposite to what is predicted in Fig. 9. However, perhaps the 14.7K effect observed in the vaccinia virus system is due to its ability to inhibit cytolysis by TNF rather than inhibition of eicosinoid synthesis induced by TNF.

As mentioned, Ad2 and Ad5 can establish persistent infections (FOX et al. 1977; MATSUSE et al. 1992), and lymphoid cells, which are less permissive than many other cell types for adenovirus (LAVERY et al. 1987; SILVER and ANDERSON 1988), may be the reservoir for persistent infections. The E3 promoter is expressed in lymphoid cells in an E1A-independent manner (KÖRNER and BURGERT 1994) via the NFκB sites (WILLIAMS et al. 1990). TNF activates NFκB (see above), and TNF activates the E3 promoter, probably via the NFκB sites (KÖRNER et al. 1992). Thus, E1A-independent, NFκB-dependent expression of E3 may maintain persistence in lymphoid cells. Also, expression of E3 may be upregulated under inflammatory conditions associated with high TNF levels.

### 3.3 E3 10.4K/14.5K Protein Complex Downregulates the Epidermal Growth Factor Receptor

The epidermal growth factor receptor (EGF-R), insulin receptor (I-R), and insulin-like growth factor receptor (IGF1-R) are cell surface plasma membrane receptors of the protein tyrosine kinase class. Binding of ligand to the extracellular ligand-binding domain of these receptors activates the cytoplasmic tyrosine kinase; this leads to autophosphorylation of the receptors as well as phosphorylation of proteins that mediate signal transduction. In addition to mitigating responses to TNF, the 10.4K/14.5K complex stimulates endosome-mediated internalization and degradation of cell surface EGF-R (CARLIN et al. 1989; TOLLEFSON et al. 1991). The 10.4K protein also is required to downregulate I-R and IGF1-R, but these receptors are downregulated less efficiently than EGF-R (KUIVINEN et al. 1993); it is not known whether 14.5K is involved, but this seems likely.

The biological significance of downregulation of EGF-R, I-R, and IGF1-R is unknown. The ligands for these receptors stimulate internalization of the receptors, but they initially activate the kinase activity of the receptors. It is possible than 10.4K/14.5K stimulate the kinase activities of these receptors prior to downregulating them; this could activate quiescent infected cells for efficient virus replication. On the other hand, considering that EGF (LIN et al. 1992) and probably insulin and IGF1 stimulate the Ras-MAP kinase pathway to activate $cPLA_2$, perhaps the role of 10.4K/14.5K is to remove these receptors from the cell surface so they cannot activate $cPLA_2$. That is, receptor downregulation could be another mechanism by which 10.4K/14.5K inhibit inflammation.

### 3.4 E3 11.6K Protein from Adenovirus Types 2 and 5 Promotes Cell Death

The 11.6K protein from Ad2 and Ad5 is an integral membrane protein that is modified with complex oligosaccharides at its single N-linked site (SCARIA et al. 1992). It is also O-glycosylated with mucin-type oligosaccharides (A. SCARIA, A. TOLLEFSON, and W. WOLD, unpublished results). Studies with mutants indicate that the internal hydrophobic domain (Fig. 10) serves as the signal to insert 11.6K into the membrane as well to anchor the protein in the membrane (unpublished results). Immunofluorescence indicates that 11.6K initially localizes to the ER and the Golgi complex, but ultimately to the nuclear membrane and Golgi (SCARIA et al. 1992). The NH2-terminal portion of 11.6K extends into the lumen and the COOH-terminal portion into the cytoplasm and/or nucleoplasm (it is not known whether 11.6K localizes to the outer or inner nuclear membrane).

Recent experiments indicate that 11.6K may function to promote cell death (A. TOLLEFSON, A. SCARIA, J. RYERSE, L. WOLD, and W. WOLD, unpublished results). The first clue to this function came from observations that virus mutants that lack 11.6K have small plaques that are slow to develop. Subsequent studies established that 11.6K mutants grow as well as wild-type adenovirus, but the

```
Subgroup C Ad2  M T G S T I A P T T D Y R N T T A T G • • • • • • • • L T S A L • • • • N L P Q  28
           Ad5  - - N T - N - A • • • • • • • A - - - • • • • • • • • - - - T T • • • • - T - -  21
Subgroup B Ad3          M - L F Q S N T T T S Y - Y T • • • • • • • • • • • • • • • • • - I Q P  19
           Ad7          M - L F Q S N T T - - I N V Q T T L N H D M E N H T T S Y A Y I - I Q P  36

                                                |_______________hydrophobic_______________|
Subgroup C Ad2  V H A F V N D W A S L D M W W F S I A L M F V C L I I M W L I C C L K R R R A R  68
           Ad5  - S - - - - N - D N - G - - - - - - - - - - - - - - - - - - - - - - - K - - -  61
Subgroup B Ad3  K Y - M Q L E I T I - I V I G I L - L S V I L Y F - F C R Q - P N V H - N S K -  59
           Ad7  K Y - M H L K I T I - I V I G I L - L S V I L Y F L F S Y D  66
                              |_____________hydrophobic_____________|

Subgroup C Ad2  P P I Y R P I I V L N P H N E K I H R L D G L K P C S L L L Q Y D  101
           Ad5  - - - - S - - - - - H - N - D G - - - - - - - - H M F F S - T V  93
Subgroup B Ad3  R - - - S - M - S R P H M A L N E I  77
```

**Fig. 10.** Sequence of the E3 11.6K protein

mutant virions are released from the infected cells more slowly; thus, it takes longer for the mutant virions to spread from cell to cell, and the plaques are small.

Further experiments showed that cells infected with 11.6K mutants stay alive much longer than do cells infected with wild-type adenovirus. For example, with cells infected with *rec*700 (wild type), cell lysis began at 2–3 days postinfection and continued until 7 days, when nearly all the cells were dead (Fig. 11). In

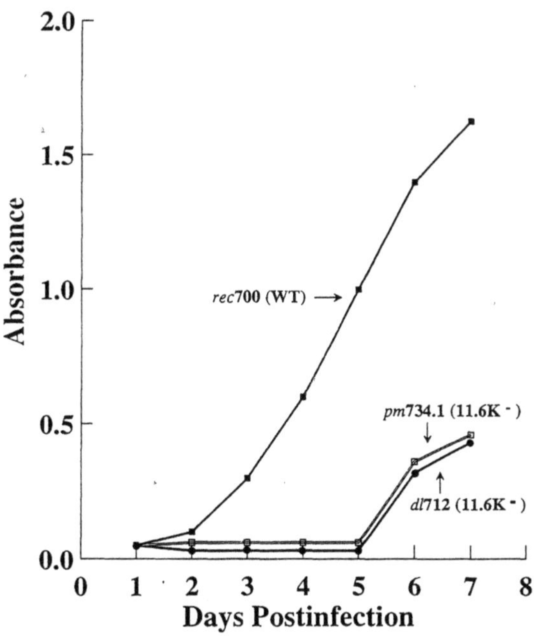


**Fig. 11.** Viability assay for cells infected with wild-type adenovirus (*rec*700) or mutants lacking the E3 11.6K protein. Human A549 cell monolayers were infected and cell lysis was then determined at different days postinfection by measuring the release of lactate dehydrogenase into the culture supernatant

contrast, cells infected with two 11.6K mutants (*dl*712 and *pm*734.1) stayed completely alive until 5 days postinfection, and only by day 6 did cell lysis begin to occur. Similar results have been obtained using, as indicators of cell death, trypan blue exclusion, the 3-(4,5 dimethylthiazol-2-yl)-2,5-diphenyltetrazolium bromide (MTT) assay for mitochondrial activity, microscopy, and the release of virus from infected cell monolayers.

We suggest that the function of 11.6K in adenovirus biology is to promote cell lysis and allow virus to be released from the infected cell. The 11.6K protein is not synthesized abundantly until very late stages of infection (> 20 h; TOLLEFSON et al. 1992), which is when virions are beginning to accumulate in the cell nucleus. This is what would be expected for a protein whose function is to promote virus release.

Is this proposed function of 11.6K in keeping with the theme that E3 proteins block host responses? Possibly it is, because CTL, TNF, and phagocytic cells eventually should kill infected cells. Also, the infected cell may synthesize AA (eventually) and other cytokines that could stimulate the inflammatory response. The 11.6K protein, by promoting rapid cell death and the release of virus, could preclude these events.

A further interesting aspect of 11.6K is that the gene is located in a region of E3 that is highly diverged among serotypes in different subgroups and even among serotypes in the same subgroup. The sequence of 11.6K in Ad2 and Ad5 (subgroup C) is quite different (CLADARAS and WOLD 1985; HÉRISSÉ et al. 1980), and the putative 11.6K proteins of Ad3 and Ad7 (subgroup B1) (SIGNÄS et al. 1986; HONG et al. 1988) are highly diverged (Fig. 10). The 11.6K gene is completely lacking in Ad35 (FLOMENBERG et al. 1988) and Ad11 (MEI and WADELL 1992), serotypes in subgroup B2. Versions of 11.6K may be expressed in Ad12 (subgroup A) (SPRENGEL et al. 1994) and Ad40 (subgroup F) (DAVISON et al. 1993), where putative proteins of 29.5K and 30.4K, respectively, contain a small region with patchy homology to 11.6K of Ad2 and Ad5 (see Fig. 7). If 11.6K does in fact promote the release of Ad2 and Ad5 virions from the infected cell, as we postulate, then how are the virions from these other serotypes released from the infected cell?

## 3.5 Functions of the E3 6.7K and 12.5K Proteins Are Unknown

The 6.7K protein is an integral membrane glycoprotein that is localized in the ER (WILSON-RAWLS et al. 1990; WILSON-RAWLS and WOLD 1993). Considering that 6.7K is translated from the same mRNA as gp19K and that both proteins are ER integral membrane proteins, it seems intuitive that the function of 6.7K should be related to that of gp19K. For example, 6.7K could block the transport of peptides into the ER so that they cannot bind to class I antigens. However, studies with 6.7K mutants have not revealed an effect on export of class I antigens from the ER or on killing of infected cells by CTL (HERMISTON et al. 1993b). Perhaps the effect of gp19K on the metabolism of class I antigens masks any effect that 6.7K may have.

The 12.5K protein (HAWKINS and WOLD 1992) is a nonmembrane protein that is localized in the nucleus and cytoplasm. It is a well conserved protein, so it probably has an important function.

# 4 Region E3 in Adenovirus Subgroups A, B, D, and F

Much less is known about the molecular biology of adenovirus serotypes in subgroups other than subgroup C. However, a fascinating aspect is that the subgroups appear to differ considerably in the E3 proteins expressed (Fig. 7). Although most serotypes can probably infect many tissues, there is a decided tendency in the more common infections (HORWITZ 1990). Subgroup C types cause respiratory infections in children and can form persistent infections in lymphoid tissues. Subgroup B types cause more severe infections of the respiratory tract and the kidney of children and adults. Ad4 (subgroup E) is a respiratory virus of adults. Subgroup D types cause epidemic keratoconjunctivitis. Subgroup F types cause about 5% of GI disease in infants. Subgroup A types also infect the GI tract. If it is true that E3 governs the interaction between the virus and the host's antiviral defenses, then it is possible that the E3 proteins of the different subgroups are responsible for the unique pathogenic properties of the subgroups.

Subgroups B1 (Ad3, Ad7) and B2 (Ad11, Ad35) have ORF for two unique proteins of 20.1K and 20.5K (SIGNÄS et al. 1986; HONG et al. 1988; FLOMENBERG et al. 1988; MEI and WADELL 1992). Both proteins are predicted to be type I integral membrane glycoproteins. The 20.1K and 20.5K proteins are 34% identical, suggesting that they arose by a gene duplication event (SIGNÄS et al. 1986). Using antipeptide antisera, we have shown that the 20.5K protein is expressed at early and late stages of infection and that it is a membrane glycoprotein modified with O-linked oligosaccharides and complex N-linked oligosaccharides (HAWKINS and WOLD 1995; L. HAWKINS and W. WOLD, unpublished results).

Subgroup B1 and B2 types also have an ORF for a 16K protein that is located in the same position in E3 as the 6.7K protein (Fig. 7). The 16K is predicted to be an integral membrane protein. Using antipeptide antisera, we have shown that 16K is expressed in Ad3 and Ad7 as a membrane protein (HAWKINS et al. 1995). The COOH-terminal region of 16K has patchy homology to 6.7K, suggesting that the proteins have related functions. The $NH_2$-terminal part of 16K has homology to a potential subgroup C 14K protein (with no ATG); this Ad2 ORF does not have an ATG (Fig. 7), and there is no evidence that it is expressed as protein. However, given the homology to 16K and considering that adenovirus does not contain useless DNA, it is possible that the Ad2 14K protein is expressed and has the function embodied in the $NH_2$-terminal part of 16K. The $NH_2$-terminal region of 16K and putative 14K has homology to proteins at the equivalent position in the other subgroups.

Ad4 (subgroup E) has four ORFs for potential unique proteins of 23K, 27.2K, 6.3K, and 29.8K. Again, 23K, 27.2K, and 29.8K are predicted to be membrane proteins (Hermiston et al. manuscript in preparation). The 19K protein is related to the 16K protein of subgroup B. The Ad4 27.2K protein is 35% identical to the Ad3 20.1K and 20.5K proteins, suggesting that Ad4 27.2K may have a similar function to these Ad3 proteins.

The Ad4 29.8K protein is related to the COOH-terminal half of a putative Ad9 (subgroup D) 61K membrane glycoprotein (Fig. 7). The putative transmembrane domains of Ad4 29.8K and Ad9 61K are similar in sequence to a putative Ad3 9.0K protein (Signäs et al. 1986), which is in the same position in E3 as the subgroup C 11.6K protein. These proteins have a small patch of homology to a 12 residue domain immediately downstream from the transmembrane domain of 11.6K. It will be interesting to determine whether the Ad4 29.8K and Ad9 61K function in cell death.

Subgroup A Ad12 has two putative unique proteins, 29.0K and 29.5K (Sprengel et al. 1994). Subgroup F Ad40 (Davison et al. 1993) has a putative unique 30.4K protein as well as a 19K protein that is related to the 16K protein of Ad3/7 and the 23K protein of Ad4. The COOH terminus of Ad12 29.5K is related to the COOH terminus of Ad40 30.4K. As mentioned earlier, Ad12 and Ad40 do not encode gp19K. Ad40 also lacks 12.5K (the 12.5K region has not been sequenced in subgroups B2 or D).

*Acknowledgments.* The research by the authors that was described in this article was supported by grants CA24710, CA58538, CA49540, and AI29492 from the National Institutes of Health.

## References

Abraham SE, Lobo S, Yaciuk P, Wang H-GHW, Moran E (1993) p300, and p300-associated proteins, are components of TATA-binding protein (TBP) complexes. Oncogene 8: 1639–1647
Ames RS, Holskin B, Mitcho M, Shalloway D, Chen M-J (1990) Induction of sensitivity to the cytotoxic action of tumor necrosis factor alpha by adenovirus E1A is independent of transformation and transcriptional activation. J Virol 64: 4115–4122
Andersson M, Pääbo S, Nilsson T, Peterson PA (1985) Impaired intracellular transport of class I MHC antigens as a possible means for adenoviruses to evade immune surveillance, Cell 43: 215–222
Andersson M, McMichael A, Peterson PA (1987) Reduced allorecognition of adenovirus-2 infected cells. J Immunol 138: 3960–3966
Baker CC, Ziff EB (1981) Promoters and heterogeneous 5' termini of the messenger RNAs of adenovirus serotype 2. J Mol Biol 149: 189–221
Beg AA, Finco TS, Nantermet PV, Baldwin AS (1993) Tumor necrosis factor and interleukin-1 lead to phosphorylation and loss of IκBα: a mechanism for NF-κβ activation. Mol Cell Biol 13: 3301–3310
Beier DC, Cox JH, Vining DR, Cresswell P, Engelhard VH (1994) Association of human class I MHC alleles with the adenovirus E3/19K protein. J Immunol 152: 3862–3872
Berk AJ, Sharp PA (1978) Structure of the adenovirus 2 early mRNAs. Cell 14: 695–711
Berk AJ, Lee F, Harrison T, Williams J, Sharp P (1979) Pre-early adenovirus 5 gene product regulates synthesis of early viral messenger RNAs. Cell 17: 935–944
Bernards R, Schrier PI, Houweling A, Bos JL, van der Eb AJ, Zijlstra M, Melief CJM (1983) Tumorigenicity of cells transformed by adenovirus type 12 by evasion of T-cell immunity. Nature 305: 776–779
Beutler B, van Huffel C (1994) Unraveling function in the TNF ligand and receptor families. Science 264: 667–668

Bhat BM, Wold WSM (1985) ATTAAA as well as downstream sequences are required for RNA 3'-end formation in the E3 complex transcription unit of adenovirus. Mol Cell Biol 5: 3183–3193
Bhat BM, Wold WSM (1986a) Genetic analysis of mRNA synthesis in region E3 of adenovirus at different stages of infection by RNA processing mutants. J Virol 60: 54–63
Bhat BM, Wold WSM (1986b) Adenovirus mutants with splice-enhancing mutations in the E3 complex transcription unit are also defective in E3A RNA 3'-end formation. J Virol 57: 1155–1158
Bhat BM, Wold WSM (1987) A small deletion distant from a splice or polyadenylation site dramatically alters pre-mRNA processing in region E3 of adenovirus. J Virol 61: 3938–3945
Bhat BM, Brady HA, Pursley MH, Wold WSM (1986) Deletion mutants that alter differential RNA processing in the E3 complex transcription unit of adenovirus. J Mol Biol 190: 543–557
Binger MH, Flint SJ (1984) Accumulation of early and intermediate mRNA species during subgroup C adenovirus productive infections. Virology 136: 387–403
Brady HA, Wold WSM (1987) Identification of a novel sequence that governs both polyadenylation and alternative splicing in region E3 of adenovirus. Nucleic Acids Res. 15: 9397–9416
Brady HA, Wold WSM (1988) Competition between splicing and polyadenylation reactions determines which adenovirus region E3 mRNAs are synthesized. Mol Cell Biol 8: 3291–3297
Brady HA, Scaria A, Wold WSM (1992) Map of *cis*-acting sequences that determine alternative pre-mRNA processing in the E3 complex transcription unit of adenovirus. J Virol 66: 5914–5923
Brandt CD, Kim HW, Vargosko AJ, Jeffries BC, Arrobio JO, Rindge B, Parrott RH, Chanock RM (1969) Infections in 18,000 infants and children in a controlled study of respiratory tract disease. I. Adenovirus pathogenicity in relation to serologic type and illness syndrome. Am J Epidemiol 90: 484–500
Buckbinder L, Miralles VJ, Reinberg D (1989) TPA can overcome the requirement for E1a and together act synergistically in stimulating expression of the adenovirus E3 promoter. EMBO J 8: 4239–4250
Burgert H-G, Kvist S (1985) An adenovirus type 2 glycoprotein blocks cell surface expression of human histocompatibility class I antigens. Cell 41: 987–997
Burgert H-G, Kvist S (1987) The E3/19K protein of adenovirus type 2 binds to the domains of histocompatibility antigens required for CTL recognition. EMBO J 6: 2019–2026
Burgert H-G, Maryanski JL, Kvist S (1987) "E3/19K" protein of adenovirus type 2 inhibits lysis of cytolytic T lymphocytes by blocking cell-surface expression of histocompatibility class I antigens. Proc Natl Acad Sci USA 84: 1356–1360
Carlin CR, Tollefson AE, Brady HA, Hoffman BL, Wold WSM (1989) Epidermal growth factor receptor is down-regulated by a 10,400 MW protein encoded bythe E3 region of adenovirus. Cell 57: 135–144
Chang DJ, Ringold GM, Heller RA (1992) Cell killing and induction of manganous superoxide dismutase by tumor necrosis factor-$\alpha$ is mediated by lipoxygenase metabolites of arachidonic acid. Biochem Biophys Res Commun 188: 538–546
Chen M-J, Holskin B, Strickler J, Gorniak J, Clark MA, Johnson PJ, Mitcho M, Shalloway D (1987) Induction by *E1A* oncogene expression of cellular susceptibility to lysis by TNF. Nature 330: 581–583
Chow LT, Broker TR (1978) The spliced structures of adenovirus 2 fiber message and the other late mRNAs. Cell 15: 497–510
Chow LT, Roberts JM, Lewis JB, Broker TR (1977) A map of cytoplasmic RNA transcripts from lytic adenovirus type 2, determined by electron microscopy of RNA:DNA hybrids. Cell 11: 819–836
Chow LT, Lewis JB, Borker TR (1979a) RNA transcription and splicing at early and intermediate times after adenovirus-2 infection. Cold Spring Harbor Symp Quant Biol 44: 401–414
Chow LT, Broker TR, Lewis JB (1979b) Complex splicing patterns of RNAs from the early regions of adenovirus-2. J Mol Biol 134: 265–303
Cladaras C, Wold WSM (1985) DNA sequence of the early E3 transcription unit of adenovirus 5. Virology 140: 28–43
Cladaras C, Bhat BM, Wold WSM (1985) Mapping the 5'ends, 3' ends, and splice sites of mRNAs from the E3 transcription unit of adenovirus 5. Virology 140: 44–54
Clark JD, Lin L-L, Kriz RW, Ramesha CS, Sultzman LA, Lin AY, Milona N, Knopf JL (1991) A novel arachidonic acid-selective cytosolic $PLA_2$ contains a $Ca^{2+}$-dependent translocation domain with homology to PKC and GAP. Cell 65: 1043–1051
Cousin C, Winter N, Gomes SA, D'Halluin JC (1991) Cellular transformation of E1 genes of enteric adenoviruses. Virology 181: 277–287
Cox JH, Yewdell JW, Eisenlohr LC, Johnson PR, Bennink JR (1990) Antigen presentation requires transport of MHC class I molecules from the endoplasmic reticulum. Science 247: 715–718
Cox JH, Bennink JR, Yewdell JW (1991) Retention of adenovirus E19 glycoprotein in the endoplasmic reticulum is essential to its ability to block antigen presentation. J Exp Med 174:1629–1637

Dahllöf B, Wallin M, Kvist S (1991) The endoplasmic reticulum retention signal of the E3/19K protein of adenovirus-2 is microtubule binding. J Biol Chem 266: 1804–1808
Davison AJ, Telford AR, Watson MS, McBride K, Mautner V (1993) The DNA sequence of adenovirus type 40. J Mol Biol 234: 1308–1316
Dbaibo GS, Obeid LM, Hannun YA (1993) Tumor necrosis factor-α (TNF-α) signal transduction through ceramide. J Biol Chem 268: 17762–17766
Dérijard B, Hibi M, Wu I-H, Barrett T, Su B, Deng T, Karin M, Davis RJ (1994) JNK1: a protein kinase stimulated by UV light and Ha-Ras that binds and phosphorylates the c-Jun activation domain. Cell 76: 1025–1037
Deutscher SL, Bhat BM, Pursley MH, Cladaras C, Wold WSM (1985) Novel deletion mutants that enhance a distant upstream 5' splice in the E3 transcription unit of adenovirus 2. Nucleic Acids Res 13: 5771–5788
Dobrowsky RT, Kamibayashi C, Mumby MC, Hannun YA (1993) Ceramide activates heterotrimeric protein phosphatase2A. J Biol Chem 268: 15523–15530
Duerksen-Hughes P, Wold WSM, Gooding LR (1989) Adenovirus E1A renders infected cells sensitive to cytolysis by tumor necrosis factor. J Immunol 143: 4193–4200
Duerksen-Hughes PJ, Hermiston TW, Wold WSM, Gooding LR (1991) The amino-terminal portion of CD1 of the adenovirus E1A proteins is required to induce susceptibility to tumor necrosis factor cytolysis in adenovirus-infected mouse cells. J Virol 65: 1236–1244
Eckner R, Ewen ME, Newsome D, Gerdes M, DeCaprio JA, Lawrence JB, Livingston DM (1994) Molecular cloning and functional analysis of the adenovirus E1A-associated 300-kD protein (p300) reveals a protein with properties of a transcriptional adaptor. Genes Dev 8: 869–884
Flomenberg PR, Chen M, Horwitz MS (1987) Characterization of a major histocompatibility complex class I antigen-binding glycoprotein from adenovirus type 35, a type associated with immunocompromised hosts. J Virol 61: 3665–3671
Flomenberg PR, Chen M, Horwitz MS (1988) Sequence and genetic organization of adenovirus type 35 early region 3. J Virol 62: 4431–4437
Flomenberg P, Szmulewicz J, Gutierrez E, Lupatkin H (1992) Role of the adenovirus E3-19K conserved region in binding major histocompatibility complex class I molecules. J Virol 66: 4778–4783
Fox JP, Hall CE, Cooney MK (1977) The Seattle virus watch: observations of adenovirus infections. Am J Epidemiol 105: 362–386
Friedman DJ, Ricciardi RP (1988) Adenovirus type 12 E1A gene represses accumulation of MHC class I mRNAs at the level of transcription. Virology 165: 303–305
Gabathuler R, Levy F, Kvist S (1990) Requirements for the association of adenovirus type 2 E3/19K wild-type and mutant proteins with HLA antigens. J Virol 64: 3679–3685
Garcia J, Wu F, Gaynor R (1987) Upstream regulatory regions required to stabilize binding to the TATA sequence in an adenovirus early promoter. Nucleic Acids Res 15: 8367–8385
Ginsberg HS, Lundholm-Beauchamp U, Horswood RL, Pernis B, Wold WSM, Chanock RM, Prince GA (1989) Role of early region 3 (E3) in pathogenesis of adenovirus disease. Proc Natl Acad Sci USA 86: 3823–3827
Gooding LR (1992) Virus proteins that counteract hose immune defenses. Cell 71: 5–7
Gooding LR, Wold WSM (1990) Molecular mechanisms by which adenovirus counteract antiviral immune defenses. CRC Crit Rev Immunol 10: 53–71
Gooding LR, Elmore LW, Tollefson AE, Brady HA, Wold WSM (1988) A 14,700 MW protein from the E3 region of adenovirus inhibits cytolysis by tumor necrosis factor. Cell 53: 341–346
Gooding LR, Sofola IO, Tollefson AE, Duerksen-Hughes P, Wold WSM (1990) The adenovirus E3-14. 7K protein is a general inhibitor of tumor necrosis factor-mediated cytolysis. J Immunol 145: 3080–3086
Gooding LR, Aquino L, Duerksen-Hughes PJ, Day D, Horton TM, Yei S, Wold WSM (1991a) The E1B 19,000-molecular-weight protein of group C adenoviruses prevents tumor necrosis factor cytolysis of human cells but not of mouse cells. J Virol 65: 3083–3094
Gooding LR, Ranheim TS, Tollefson AE, Aquino L, Duerksen-Hughes P, Horton TM, Wold WSM (1991b) The 10,400- and 14,500-dalton proteins encoded by region E3 of adenovirus function together to protect many but not all mouse cell lines against lysis by tumor necrosis factor. J Virol 65: 4114–4123
Grunhaus A, Horwitz MS (1992) Adenoviruses as cloning vectors. Semin Virol 3: 237–252
Grunhaus A, Cho S, Horwitz MS (1994) Association of vaccinia virus-expressed adenovirus E3-19K glycoprotein with class I MHC and its effects on virulence in a murine pneumonia model. Virology 200: 535–546
Hannun YA (1994) The sphingomyelin cycle and the second messenger function of ceramide. J Biol Chem 269: 3125–3128

Hannun YA, Bell RM (1993) The sphingomyelin cycle: a prototypic sphingolipid signaling pathway. Adv Lipid Res 25: 27–41
Hashimoto S, Ishii A, Yonehara S (1991) The E1b oncogene of adenovirus confers cellular resistance to cytotoxicity of tumor necrosis factor and monoclonal anti-Fas antibody. Int Immunol 3: 343–351
Hawkins LK, Wold WSM (1992) A 12,500 MW protein is coded by region E3 of adenovirus. Virology 188: 486–494
Hawkins LK, Wold WSM (1995) A 20 500-dalton protein is coded by region E3 of subgroup B but not subgroup C human adenoviruses. Virology (in press)
Hawkins LK, Wilson-Rawls J, Wold WSM (1995) Region E3 of subgroup B human adenoviruses encodes a 16,000 MW membrane protein that may be a distant analog of the E3-6.7K protein of subgroup C adenoviruses. J Virol (in press)
Hayakawa M, Ishida N, Takeuchi K, Shibamoto S, Hori T, Oku N, Ito F, Tsujimoto M (1993) Arachidonic acid-selective cytosolic phospholipase A2 is crucial in the cytotoxic action of tumor necrosis factor. J Biol Chem 268: 11290–11295
Hérissé J, Galibert F (1981) Nucleotide sequence of the EcoRI E fragment of adenovirus 2 genome. Nucleic Acids Res 9: 1229–1240
Hérissé J, Courtois G, Galibert F (1980) Nucleotide sequence of the EcoRI D fragment of adenovirus 2 genome. Nucleic Acids Res 8: 2173–2192
Hennet T, Ziltener HJ, Frei K, Peterhans E (1992) A kinetic study of immune mediators in the lungs of mice infected with influenza A virus. J Immunol 149: 932–939
Hermiston TW, Hellwig R, Hierholzer JC, Wold WSM (1993a) Sequence and functional analysis of the human adenovirus type 7 E3-gp19K protein from 17 clinical isolates. Virology 197: 593–600
Hermiston TW, Tripp RA, Sparer T, Gooding LR, Wold WSM (1993b) Deletion mutant analysis of the adenovirus type 2 E3-gp19K protein: identification of sequences within the endoplasmic reticulum lumenal domain that are required for class I antigen binding and protection from adenovirus-specific cytotoxic T lymphocytes. J Virol 67: 5289–5298
Hermiston TW, Tollefson AE, Bhat BM, Skdheer S, Davis A, Hung P, Wold WSM. DNA sequence of the E3 transcription unit of human adenovirus type 4 (manuscript in preparation)
Hoeck WG, Ramesha CS, Chang DJ, Fan N, Heller RA (1993) Cytoplasmic phospholipase A2 activity and gene expression are stimulated by tumor necrosis factor: dexamethasone blocks the induced synthesis. Proc Natl Acad Sci USA 90: 4475–4479
Hoffman P, Yaffe MB, Hoffman BL, Yei SP, Wold WSM, Carlin C (1992) Characterization of the adenovirus E3 protein that down-regulates the epidermal growth factor receptor. J Biol Chem 267: 13480–13487
Hong JS, Mullis KG, Engler JA (1988) Characterization of the early region 3 and fiber genes of Ad7. Virology 167: 545–553
Horton TM, Tollefson AE, Wold WSM, Gooding LR (1990) A protein serologically and functionally related to the group C E3 14,700-kilodalton protein is found in multiple adenovirus serotypes. J Virol 64: 1250–1255
Horton TM, Ranheim TS, Aquino L, Kusher DI, Saha SK, Ware CF, Wold WSM, Gooding LR (1991) Adenovirus E3 14.7K protein functions in the absence of other adenovirus proteins to protect transfected cells from tumor necrosis factor cytolysis. J Virol 65: 2629–2639
Horwitz MS (1990) Adenovirus. In: Fields BN, Knipe DM (eds) Virology, 2nd edn. Raven, New York, pp 1679–1721
Hurst HC, Jones NC (1987) Identification of factors that interact with the E1A-inducible adenovirus E3 promoter. Genes Dev 1: 1132–1146
Jackson MR, Nilsson T, Peterson PA (1990) Identification of a consensus motif for retention of transmembrane proteins in the endoplasmic reticulum. EMBO J 9: 3153–3162
Jackson MR, Nilsson T, Peterson PA (1993) Retrieval of transmembrane proteins to the endoplasmic reticulum. J Cell Biol 121: 317–333
Jarvis WD, Kolesnick RN, Fornari FA, Traylor RS, Gewirtz DA, Grant S (1994) Induction of apoptotic DNA damage and cell death by activation of the sphingomyelin pathway. Proc Natl Acad Sci USA 91: 73–77
Jefferies WA, Burgert H-G (1990) E3/19K from adenovirus 2 is an immunosubversive protein that binds to a structural motif regulating the intracellular transport of major histocompatibility complex class I proteins. J Exp Med 172: 1653–1664
Johnson DG, Schwarz JK, Cress WD, Nevins JR (1993) Expression of transcription factor E2F1 induces quiescent cells to enter S phase. Nature 365: 349–352
Jones N, Shenk T (1979) Isolation of adenovirus type 5 host range deletion mutants defective for transformation of rat embryo cells. Cell 17: 683–689

Joseph CK, Byun H-S, Bittman R, Kolesnick RN (1993) Substrate recognition by ceramide-activated protein kinase. J Biol Chem 268: 20002–20006
Kämpe O, Bellgrau D, Hammerling U, Lind P, Pääbo S, Ṡeverinsson L, Peterson PA (1983) Complex formation of class I transplantation antigens and a viral glycoprotein. J Biol Chem 258: 10594–10598
Kapoor QS, Wold WSM, Chinnadurai G (1981) A non-essential glycoprotein is coded by early region E3 of adenovirus type 7. Virology 112: 780–784
Kelly Jr TJ, Lewis Jr AM (1973) Use of nondefective adenovirus-simian virus 40 hybrids for mapping the simian virus 40 genome. J Virol 12: 643–652
Kitchingman GR, Westphal H (1980) The structure of adenovirus 2 early nuclear and cytoplasmic RNAs. J Mol Biol 137: 23–48
Kolesnick R, Golde DW (1994) The sphingomyelin pathway in tumor necrosis factor and interleukin-1 signaling. Cell 77: 325–328
Körner H, Burgert H-G (1994) Down-regulation of HLA antigens by the adenovirus type 2 E3/19K protein in a T-lymphoma cell line. J Virol 68: 1442–1448
Körner H, Fritzsche U, Burgert H-G (1992) Tumor necrosis factor $\alpha$ stimulates expression of adenovirus early region 3 proteins: implications for viral persistence. Proc Natl Acad Sci USA 89: 11857–11861
Kornfeld R, Wold WSM (1981) Structures of the oligosaccharides of the glycoprotein coded by early region E3 of adenovirus 2. J Virol 40: 440–449
Kornuc M, Kliewer S, Garcia J, Harrich D, Li C, Gaynor R (1990) Adenovirus early region 3 promoter regulation by E1A/E1B is independent of alterations in DNA binding and gene activation of CREB/ATF and AP1. J Virol 64: 2004–2013
Kozak M (1987) An analysis of 5'-noncoding sequences from 699 vertebrate messenger RNAs. Nucleic Acids Res 15: 8125–8132
Krajcsi P, Wold WSM (1992) The adenovirus E3-14.5K protein which is required for prevention of TNF cytolysis and for down-regulation of the EGF receptor contains phosphoserine. Virology 187: 492–498
Krajcsi P, Tollefson AE, Anderson CW, Stewart AR, Carlin CR, Wold WSM (1992a) The E3-10.4K protein of adenovirus is an integral membrane protein that is partially cleaved between $Ala_{22}$ and $Ala_{23}$ and has a $C_{cyt}$ orientation. Virology 187: 131–144
Krajcsi P, Tollefson AE, Anderson CW, Wold WSM (1992b) The adenovirus E3 14.5-kilodalton protein, which is required for down-regulation of the epidermal growth factor receptor and prevention of tumor necrosis factor cytolysis, is an integral membrane protein oriented with its C terminus in the cytoplasm. J Virol 66: 1665–1673
Krajcsi P, Tollefson AE, Wold WSM (1992c) The E3-14.5K integral membrane protein of adenovirus that is required for down-regulation of the EGF receptor and for prevention of TNF cytolysis is O-glycosylated but not N-glycosylated. Virology 188: 570–579
Krönke M, Schütze S, Scheurich P, Pfizenmaier K (1992) TNF signal transduction and TNF-responsive genes. Immunol Ser 56: 189–216
Kuivinen E, Hoffman BL, Hoffman PA, Carlin CR (1993) Structurally related class I and Class II receptor protein tyrosine kinases are down-regulated by the same E3 protein coded for by human group C adenoviruses. J Cell Biol 120: 1271–1279
Kvist S, Ostberg L, Persson H, Philipson L, Peterson PA (1978) Molecular association between transplantation antigens and cell surface antigen in adenovirus-transformed cell line. Proc Natl Acad Sci USA 75: 5674–5678
Kyriakis JM, Banerjee P, Nikolakaki E, Dai T, Rubie EA, Ahmad MF, Avruch J, Woodgett JR (1994) The stress-activated protein kinase subfamily of c-Jun kinases. Nature 369: 156–160
Lane JR, Neumann DA, Lafond-Walker A, Herskowitz A, Rose NR (1992) Interleukin 1 or tumor necrosis factor can promote coxsackie B3-induced myocarditis in resistant B10.A mice. J Exp Med 175: 1123–1129
Laster SM, Wood JG, Gooding LR (1988) Tumor necrosis factor can induce both apoptic and necrotic forms of cell lysis. J Immunol 141: 2629–2634
Lavery DJ, Fu SM, Lufkin T, Chen-Kiang S (1987) Productive infection of cultured human lymphoid cells by adenovirus. J Virol 61: 1466–1472
Lin L-L, Lin AY, Knopf JL (1992) Cytosolic phospholipase A2 is coupled to hormonally regulated release of arachidonic acid. Proc Natl Acad Sci USA 89: 6147–6151
Lin L-L, Wartmann M, Lin AY, Knopf JL, Seth A, Davis RJ (1993) cPLA2 is phosphorylated and activated by MAP kinase. Cell 72: 269–278
Lippé R, Luke E, Kuah YT, Lomas C, Jefferies WA (1991) Adenovirus infection inhibits the phosphorylation of major histocompatibility complex class I proteins. J Exp Med 174: 1159–1166

Liu J, Mathias S, Yang Z, Kolesnick RN (1994) Renaturation and tumor necrosis factor-α stimulation of a 97-kDa ceramide-activated protein kinase. J Biol Chem 269: 3047–3052
Machleidt T, Wiegmann K, Henkel T, Schütze S, Baeuerle P, Krönke M (1994) Sphingomyelinase activates proteolytic IκB-α degradation in a cell-free system. J Biol Chem 269: 13760–13765
Mathias S, Kolesnick R (1993) Ceramide: a novel second messenger. Adv Lipid Res 25: 65–90
Matsuse T, Hayashi S, Kuwano K, Keunecke H, Jefferies WA, Hogg JC (1992) Latent adenoviral infection in the pathogenesis of chronic airways obstruction. Am Rev Respir Dis 146: 177–184
Mayer A, Gelderblom H, Kümel G, Jungwirth C (1992) Interferon-γ-induced assembly block in the replication cycle of adenovirus 2: augmentation by tumor necrosis factor-α. Virology 187: 372–376
Mei Y-F, Wadell G (1992) The nucleotide sequence of adenovirus type 11 early 3 region: comparison of genome type Ad11p and Ad11a. Virology 191: 125–133
Mei Y-F, Wadell G (1993) Hemagglutination properties and nucleotide sequence analysis of the fiber gene of adenovirus genome types 11p and 11a. Virology 194: 453–462
Mestan J, Digel W, Mittnacht S, Hillen H, Blohm D, Möller A, Jacobsen H, Kirchner H (1986) Antiviral effects of recombinant tumour necrosis factor in vitro. Nature 323: 816–819
Moran E (1993) Interaction of adenoviral proteins with pRB and p53. FASEB J 7: 880–885
Morin JE, Lubeck MD, Barton JE, Conley AJ, Davis AR, Hung PP (1987) Recombinant adenovirus induces antibody response to hepatitis B virus surface antigen in hamsters. Proc Natl Acad Sci USA 84: 4626–4630
Nevins JR (1992) E2F: a link between the Rb tumor supressor protein and viral oncoproteins. Science 258: 424–429
Nevins JR, Ginsberg HS, Blanchard J-M, Wilson MC, Darnell Jr JE (1979) Regulation of the primary expression of the early adenovirus transcription units. J Virol 32: 727–733
Nilsson T, Jackson M, Peterson PA (1989) Short cytoplasmic sequences serve as retention signals for transmembrane proteins in the endoplasmic reticulum. Cell 58: 707–718
Obeid LM, Linardic CM, Karolak LA, Hannun YA (1993) Programmed cell death induced by ceramide. Science 259: 1769–1771
Ozawa K, Hagiwara H, Tang X, Saka F, Kitabayashi I, Shiroki K, Fujinaga K, Israël A, Gachelin G, Yokoyama K (1993) Negative regulation of the gene for H-$2K_b$ class I antigen by adenovirus 12-E1A is mediated by a CAA repeated element. J Biol Chem 268: 27258–27268
Pääbo S,Weber F, Kämpe O, Schaffner W, Peterson PA (1983) Association between transplantation antigens and a viral membrane protein synthesized from a mammalian expression vector. Cell 35: 445–453
Pääbo S, Nilsson T, Peterson PA (1986a) Adenoviruses of subgenera B, C, D, and E modulate cell-surface expression of major histocompatibility complex class I antigens. Proc Natl Acad Sci USA 83: 9665–9669
Pääbo S, Weber F, Nilsson T, Schaffner W, Peterson PA (1986b) Structural and functional dissection of an MHC class I antigen-binding adenovirus glycoprotein. EMBO J 5:1921–1927
Pääbo S, Bhat BM, Wold WSM, Peterson PA (1987) A short sequence in the COOH terminus makes an adenovirus membrane glycoprotein a resident of the endoplasmic reticulum. Cell 50: 311–317
Pääbo S, Severinsson L, Andersson M, Martens I, Nilsson T, Peterson PA (1989) Adenovirus proteins and MHC expression. Adv Cancer Res 52: 151–163
Paya CV, Kenmotsu N, Schoon RA, Leibson PJ (1988) Tumor necrosis factor and lymphotoxin secretion by human natural killer cells leads to antiviral cytotoxicity. J Immunol 141: 1989–1995
Persson H, Öberg B, Philipson L (1978) Purification and characterization of an early protein (E14K) from adenovirus type 2-infected cells. J Virol 28: 119–139
Persson H, Signäs C, Philipson L (1979) Purification and characterization of an early glycoprotein from adenovirus type 2-infected cells. J Virol 29: 938–948
Persson H, Jansson M, Philipson L (1980a) Synthesis and genomic site for an adenovirus type 2 early glycoprotein. J Mol Biol 136: 375–394
Persson H, Jornväll H, Zabielski J (1980b) Multiple mRNA species for the precursor to an adenovirus-encoded glycoprotein: identification and structure of the signal sequence. Proc Natl Acad Sci USA 77: 6349–6353
Raines MA, Kolesnick RN, Golde DW (1993) Sphingomyelinase and ceramide activate mitogen-activated protein kinase in myeloid HL-60 cells. J Biol Chem 268: 14572–14575
Rammensee H-G, Falk K, Rötzschke O (1993) Peptides naturally presented by MHC class I molecules. Annu Rev Immunol 11: 213–244
Ranheim TS, Shisler J, Horton TM, Wold LJ, Gooding LR, Wold WSM (1993) Characterization of mutants within the gene for the adenovirus E3 14.7-kilodalton protein which prevents cytolysis by tumor necrosis factor. J Virol 67: 2159–2167

Rawle FC, Tollefson AE, Wold WSM, Gooding LR (1989) Mouse anti-adenovirus cytotoxic T lymphocytes. Inhibition of lysis by E3 gp19K but not E3 14.7K. J Immunol 143: 2031–2037

Rawle FC, Knowles BB, Ricciardi RP, Brahmacheri V, Duerksen-Hughes P, Wold WSM (1991) Specificity of the mouse cytotoxic T lymphocyte response to adenovirus 5. E1A is immunodominant in H-$2^b$, but not in H-$2^d$ or H-$2^k$ mice. J Immunol 146: 3977–3984

Roberts RJ, Akusjärvi G, Aleström P, Gelinas RE, Gingeras TR, Sciaky D, Pettersson U (1986) A consensus sequence for the adenovirus-2 genome. In: Doerfler WE (ed) Adenovirus DNA, the viral genome and its expression. Nijhoff, Boston, pp 1–51

Ross S, Levine AJ (1979) The genomic map position of the adenovirus type 2 glycoprotein. Virology 99: 427–430

Routes JM, Metz BA, Cook JL (1993) Endogenous expression of E1A in human cells enhances the effect of adenovirus E3 on class I major histocompatibility complex antigen expression. J Virol 67: 3176–3181

Scaria A, Wold WSM (1994) Fine-mapping of sequences that suppress splicing in the E3 complex transcription unit of adenovirus. Virology 205: 406–416

Scaria A, Tollefson AE, Saha SK, Wold WSM (1992) The E3-11.6K protein of adenovirus is an Asn-glycosylated integral membrane protein that localizes to the nuclear membrane. Virology 191: 743–753

Schütze S, Potthoff K, Machleidt T, Berkovic D, Wiegmann K, Krönke M (1992) TNF activates NF-κB by phosphatidylcholine-specific phospholipase C-induced "acidic" sphingomyelin breakdown. Cell 71: 765–776

Schrier PI, Bernards R, Vaessen RTMJ, Houweling A, van der Eb AJ (1983) Expression of class I major histocompatibility antigens switched off by highly oncogenic adenovirus 12 in transformed rat cells. Nature 305: 771–775

Severinsson L, Martens I, Peterson PA (1986) Differential association between two human MHC class I antigens and an adenoviral glycoprotein. J Immunol 137: 1003–1009

Shaw AR, Ziff EB (1980) Transcripts from the adenovirus-2 major late promoter yield a single early family of 3' coterminal mRNAs and five late families. Cell 22: 905–916

Shemesh J, Rotem-Yehudar R, Ehrlich R (1991) Transcriptional and postranscriptional regulation of class I major histocompatibility complex genes following transformation with human adenoviruses. J Virol 65: 5544–5548

Signäs C, Katze MG, Persson H, Philipson L (1982) An adenovirus glycoprotein binds heavy chains of class I transplantation antigens from man and mouse. Nature 299: 175–178

Signäs C, Akusjärvi G, Pettersson U (1986) Region E3 of human adenoviruses; differences between the oncogenic adenovirus-3 and the non-oncogenic adenovirus-2. Gene 50: 173–184

Silver J, Anderson CW (1988) Interaction of human adenovirus serotype 2 with human lymphoid cells. Virology 165: 377–387

Smith CA, Farrah T, Goodwin RG (1994) The TNF receptor superfamily of cellular and viral proteins: activation, costimulation, and death. Cell 76: 959–962

Smith GL (1994) Virus strategies for evasion of the host response to infection. Trends Microbiol 2: 81–88

Sprengel J, Schmitz B, Heuss-Neitzel D, Zock C, Doerfler W (1994) Nucleotide sequence of human adenovirus type 12 DNA: comparative functional analysis. J Virol 68: 379–389

Stålhandske P, Persson H, Perricaudet M, Philipson L, Pettersson U (1983) Structure of three spliced mRNAs from region E3 of adenovirus type 2. Gene 22: 157–165

Stein R, Ziff EB (1984) HeLa cell beta-tubulin gene transcription is stimulated by adenovirus 5 in parallel with viral early genes by an E1a-dependent mechanism. Mol Cell Biol 4: 2792–2801

Stewart AR, Tollefson AE, Krajcsi P, Yei S-P, Wold WSM (1995) The adenovirus E3 10.4K and 14.5K proteins, which function to prevent cytolysis by tumor necrosis factor and to downregulate the epidermal growth factor receptor, are localized in the plasma membrane. J Virol 69: 172–181

Suffys P, Beyaert R, DeValck D, Vanhaesebroeck B, Van Roy F, Fiers W (1991) Tumour-necrosis-factor-mediated cytotoxicity is correlated with phospholipase-$A_2$ activity, but not with arachidonic acid release *per se*. Eur J Biochem 195: 465–475

Tanaka Y, Tevethia SS (1988) Differential effect of adenovirus 2 E3/19K glycoprotein on the expression of H-$2K^b$ and H-$2D^b$ class I antigens and H-$2K^b$-and H-$2D^b$-restricted SV40-specific CTL-mediated lysis. Virology 165: 357-366

Tartaglia LA, Ayres TM, Wong GHW, Goeddel DV (1993) A novel domain within the 55 kd TNF receptor signals cell death. Cell 74: 845–853

Tollefson AE, Wold WSM (1988) Identification and gene mapping of a 14,700-molecular-weight protein encoded by region E3 of group C adenoviruses. J Virol 62: 33–39
Tollefson AE, Krajcsi P, Pursley MH, Gooding LR, Wold WSM (1990a) A 14,500 MW protein is coded by region E3 of group C human adenoviruses. Virology 175: 19–29
Tollefson AE, Krajcsi P, Yei S, Carlin CR, Wold WSM (1990b) A 10,400-molecular-weight membrane protein is coded by region E3 of adenovirus. J Virol 64: 794–801
Tollefson AE, Stewart AR, Yei S, Saha SK, Wold WSM (1991) The 10,400- and 14,500-dalton proteins encoded by region E3 of adenovirus form a complex and function together to down-regulate the epidermal growth factor receptor. J Virol 65: 3095–3105
Tollefson AE, Scaria A, Saha SK, Wold WSM (1992) The 11,600-*Mw* protein encoded by region E3 of adenovirus is expressed early but is greatly amplified at late stages of infection. J Virol 66: 3633–3642
Tufariello J, Cho S, Horwitz MS (1994) The adenovirus E3 14.7-kilodalton protein which inhibits cytolysis by tumor necrosis factor increases the virulence of vaccinia virus in a murine pneumonia model. J Virol 68: 453–462
Uhlen M, Svensson C, Josephson S, Aleström P, Chattapadhyaya JB, Pettersson U, Philipson L (1982) Leader arrangement in the adenovirus fiber mRNA. EMBO J 1: 249–254
Van Lint J, Agostinis P, Vandevoorde V, Haegeman G, Fiers W, Merlevede W, Vandenheede JR (1992) Tumor necrosis factor stimulates multiple serine/threonine protein kinases in Swiss 3T3 and L929 cells. J Biol Chem 267: 25916–25921
Vasavada R, Eager KB, Barbanti-Brodano G, Caputo A, Ricciardi RP (1986) Adenovirus type 12 early region 1A proteins repress class I HLA expression in transformed human cells. Proc Natl Acad Sci USA 83: 5257–5261
Vassalli P (1992) The pathophysiology of tumor necrosis factors. Annu Rev Immunol 10: 411–452
Vietor I, Schwenger P, Li W, Schlessinger J, Vilcek J (1993) Tumor necrosis factor-induced activation and increased tyrosine phosphorylation of mitogen-activated protein (MAP) kinase in human fibroblasts. J Biol Chem 268: 18994–18999
Vilcek J, Lee TH (1991) Tumor necrosis factor. J Biol Chem 266: 7313–7316
Wang EW, Scott MO, Ricciardi RP (1988) An adenovirus mRNA which encodes a 14,700-Mr protein that maps to the last open reading frame in region E3 is expressed during infection. J Virol 62: 1456–1459
White E, Cipriani R (1989) Specific disruption of intermediate filaments and the nuclear lamina by the 19-kDa product of the adenovirus E1B oncogene. Proc Natl Acad Sci USA 86: 9886–9890
White E, Cipriani R (1990) Role of adenovirus E1B proteins in transformation: altered organization of intermediate filaments in transformed cells that express the 19-kilodalton protein. Mol Cell Biol 10: 120–130
White E, Sabbatini P, Debbas M, Wold WSM, Kusher DI, Gooding LR (1992) The 19-kilodalton adenovirus E1B transforming protein inhibits programmed cell death and prevents cytolysis by tumor necrosis factor α. Mol Cell Biol 12: 2570–2580
Wiegmann K, Schütze S, Kampen E, Himmler A, Machleidt T, Krönke M (1992) Human 55-kDa receptor for tumor necrosis factor coupled to signal transduction cascades. J Biol Chem 267: 17997–18001
Williams JL, Garcia J, Harrich D, Pearson L, Wu F, Gaynor R (1990) Lymphoid specific gene expression of the adenovirus early region 3 promoter is mediated by NF-κ binding motifs. EMBO J 9: 4435–4442
Wilson-Rawls J, Wold WSM (1993) The E3-6.7K protein of adenovirus is an Asn-linked integral membrane glycoprotein localized in the endoplasmic reticulum. Virology 195: 6–15
Wilson-Rawls J, Saha SK, Krajcsi P, Tollefson AE, Gooding LR, Wold WSM (1990) A 6700 MW membrane protein is encoded by region E3 of adenovirus type 2. Virology 178: 204–212
Wilson-Rawls J, Deutscher SL, Wold WSM (1994) The signal-anchor domain of adenovirus E3-6.7K, a type III integral membrane protein, can direct adenovirus E3-gp19K, a type I integral membrane protein, into the membrane of the endoplasmic reticulum. Virology 201: 66–76
Wold WSM (1993) Adenovirus genes that modulate the sensitivity of virus-infected cells to lysis by TNF. J Cell Biochem 53: 329–335
Wold WSM, Gooding LR (1989) Adenovirus region E3 proteins that prevent cytolysis by cytotoxic T cells and tumor necrosis factor. Mol Biol Med 6: 433–452
Wold WSM, Gooding LR (1991) Region E3 of adenovirus: a cassette of genes involved in host immunosurveillance and virus-cell interactions. Virology 184: 1–8
Wold WSM, Cladaras C, Magie SC, Yacoub N (1984) Mapping a new gene that encodes an 11,600-molecular-weight protein in the E3 transcription unit of adenovirus 2. J Virol 52: 307–313

Wold WSM, Cladaras C, Deutscher SL, Kapoor QS (1985) The 19-kDa glycoprotein coded by region E3 of adenovirus: Purification, characterization, and structural analysis. J Biol Chem 260: 2424–2431
Wold WSM, Deutscher SL, Takemori N, Bhat BM, Magie SC (1986) Evidence that AG*U*A*UAUG*A and CC*A*AG*AUG*A initiate translation in the same mRNA in region E3 of adenovirus. Virology 148: 168–180
Wong GHW, Goeddel DV (1986) Tumour necrosis factors α and β inhibit virus replication and synergize with interferons. Nature 323: 819–822
Yewdell JW, Bennink JR, Eager KB, Ricciardi RP (1988) CTL recognition of adenovirus-transformed cells infected with influenza virus: lysis by anti-influenza CTL parallels adenovirus-12-induced suppression of class I MHC molecules. Virology 162: 236–238
Zilli D, Voelkel-Johnson C, Skinner T, Laster SM (1992) The adenovirus E3 region 14.7 kDa protein, heat and sodium arsenite inhibit the TNF-induced release of arachidonic acid. Biochem Biophys Res Commun 188: 177–183

# Subject Index

# Current Topics in Microbiology and Immunology

Volumes published since 1989 (and still available)

Vol. 157: **Swanstrom, Roland; Vogt, Peter K. (Ed.):** Retroviruses. Strategies of Replication. 1990. 40 figs. XII, 260 pp. ISBN 3-540-51895-9

Vol. 158: **Muzyczka, Nicholas (Ed.):** Viral Expression Vectors. 1992. 20 figs. IX, 176 pp. ISBN 3-540-52431-2

Vol. 159: **Gray, David; Sprent, Jonathan (Ed.):** Immunological Memory. 1990. 38 figs. XII, 156 pp. ISBN 3-540-51921-1

Vol. 160: **Oldstone, Michael B. A.; Koprowski, Hilary (Eds.):** Retrovirus Infections of the Nervous System. 1990. 16 figs. XII, 176 pp. ISBN 3-540-51939-4

Vol. 161: **Racaniello, Vincent R. (Ed.):** Picornaviruses. 1990. 12 figs. X, 194 pp. ISBN 3-540-52429-0

Vol. 162: **Roy, Polly; Gorman, Barry M. (Eds.):** Bluetongue Viruses. 1990. 37 figs. X, 200 pp. ISBN 3-540-51922-X

Vol. 163: **Turner, Peter C.; Moyer, Richard W. (Eds.):** Poxviruses. 1990. 23 figs. X, 210 pp. ISBN 3-540-52430-4

Vol. 164: **Bækkeskov, Steinnun; Hansen, Bruno (Eds.):** Human Diabetes. 1990. 9 figs. X, 198 pp. ISBN 3-540-52652-8

Vol. 165: **Bothwell, Mark (Ed.):** Neuronal Growth Factors. 1991. 14 figs. IX, 173 pp. ISBN 3-540-52654-4

Vol. 166: **Potter, Michael; Melchers, Fritz (Eds.):** Mechanisms in B-Cell Neoplasia. 1990. 143 figs. XIX, 380 pp. ISBN 3-540-52886-5

Vol. 167: **Kaufmann, Stefan H. E. (Ed.):** Heat Shock Proteins and Immune Response. 1991. 18 figs. IX, 214 pp. ISBN 3-540-52857-1

Vol. 168: **Mason, William S.; Seeger, Christoph (Eds.):** Hepadnaviruses. Molecular Biology and Pathogenesis. 1991. 21 figs. X, 206 pp. ISBN 3-540-53060-6

Vol. 169: **Kolakofsky, Daniel (Ed.):** Bunyaviridae. 1991. 34 figs. X, 256 pp. ISBN 3-540-53061-4

Vol. 170: **Compans, Richard W. (Ed.):** Protein Traffic in Eukaryotic Cells. Selected Reviews. 1991. 14 figs. X, 186 pp. ISBN 3-540-53631-0

Vol. 171: **Kung, Hsing-Jien; Vogt, Peter K. (Eds.):** Retroviral Insertion and Oncogene Activation. 1991. 18 figs. X, 179 pp. ISBN 3-540-53857-7

Vol. 172: **Chesebro, Bruce W. (Ed.):** Transmissible Spongiform Encephalopathies. 1991. 48 figs. X, 288 pp. ISBN 3-540-53883-6

Vol. 173: **Pfeffer, Klaus; Heeg, Klaus; Wagner, Hermann; Riethmüller, Gert (Eds.):** Function and Specificity of γ/δ TCells. 1991. 41 figs. XII, 296 pp. ISBN 3-540-53781-3

Vol. 174: **Fleischer, Bernhard; Sjögren, Hans Olov (Eds.):** Superantigens. 1991. 13 figs. IX, 137 pp. ISBN 3-540-54205-1

Vol. 175: **Aktories, Klaus (Ed.):** ADP-Ribosylating Toxins. 1992. 23 figs. IX, 148 pp. ISBN 3-540-54598-0

Vol. 176: **Holland, John J. (Ed.):** Genetic Diversity of RNA Viruses. 1992. 34 figs. IX, 226 pp. ISBN 3-540-54652-9

Vol. 177: **Müller-Sieburg, Christa; Torok-Storb, Beverly; Visser, Jan; Storb, Rainer (Eds.):** Hematopoietic Stem Cells. 1992. 18 figs. XIII, 143 pp. ISBN 3-540-54531-X

Vol. 178: **Parker, Charles J. (Ed.):** Membrane Defenses Against Attack by Complement and Perforins. 1992. 26 figs. VIII, 188 pp. ISBN 3-540-54653-7

Vol. 179: **Rouse, Barry T. (Ed.):** Herpes Simplex Virus. 1992. 9 figs. X, 180 pp. ISBN 3-540-55066-6

Vol. 180: **Sansonetti, P. J. (Ed.):** Pathogenesis of Shigellosis. 1992. 15 figs. X, 143 pp. ISBN 3-540-55058-5

Vol. 181: **Russell, Stephen W.; Gordon, Siamon (Eds.):** Macrophage Biology and Activation. 1992. 42 figs. IX, 299 pp. ISBN 3-540-55293-6

Vol. 182: **Potter, Michael; Melchers, Fritz (Eds.):** Mechanisms in B-Cell Neoplasia. 1992. 188 figs. XX, 499 pp. ISBN 3-540-55658-3

Vol. 183: **Dimmock, Nigel J.:** Neutralization of Animal Viruses. 1993. 10 figs. VII, 149 pp. ISBN 3-540-56030-0

Vol. 184: **Dunon, Dominique; Mackay, Charles R.; Imhof, Beat A. (Eds.):** Adhesion in Leukocyte Homing and Differentiation. 1993. 37 figs. IX, 260 pp. ISBN 3-540-56756-9

Vol. 185: **Ramig, Robert F. (Ed.):** Rotaviruses. 1994. 37 figs. X, 380 pp. ISBN 3-540-56761-5

Vol. 186: **zur Hausen, Harald (Ed.):** Human Pathogenic Papillomaviruses. 1994. 37 figs. XIII, 274 pp. ISBN 3-540-57193-0

Vol. 187: **Rupprecht, Charles E.; Dietzschold, Bernhard; Koprowski, Hilary (Eds.):** Lyssaviruses. 1994. 50 figs. IX, 352 pp. ISBN 3-540-57194-9

Vol. 188: **Letvin, Norman L.; Desrosiers, Ronald C. (Eds.):** Simian Immunodeficiency Virus. 1994. 37 figs. X, 240 pp. ISBN 3-540-57274-0

Vol. 189: **Oldstone, Michael B. A. (Ed.):** Cytotoxic T-Lymphocytes in Human Viral and Malaria Infections. 1994. 37 figs. IX, 210 pp. ISBN 3-540-57259-7

Vol. 190: **Koprowski, Hilary; Lipkin, W. Ian (Eds.):** Borna Disease. 1995. 33 figs. IX, 134 pp. ISBN 3-540-57388-7

Vol. 191: **ter Meulen, Volker; Billeter, Martin A. (Eds.):** Measles Virus. 1995. 23 figs. IX, 196 pp. ISBN 3-540-57389-5

Vol. 192: **Dangl, Jeffrey L. (Ed.):** Bacterial Pathogenesis of Plants and Animals. 1994. 41 figs. IX, 343 pp. ISBN 3-540-57391-7

Vol. 193: **Chen, Irvin S. Y.; Koprowski, Hilary; Srinivasan, Alagarsamy; Vogt, Peter K. (Eds.):** Transacting Functions of Human Retroviruses. 1995. 49 figs. IX, 240 pp. ISBN 3-540-57901-X

Vol. 194: **Potter, Michael; Melchers, Fritz (Eds.):** Mechanisms in B-cell Neoplasia. 1995. 152 figs. XXV, 458 pp. ISBN 3-540-58447-1

Vol. 195: **Montecucco, Cesare (Ed.):** Clostridial Neurotoxins. 1995. 28 figs. XI, 278 pp. ISBN 3-540-58452-8

Vol. 196: **Koprowski, Hilary; Maeda, Hiroshi (Eds.):** The Role of Nitric Oxide in Physiology and Pathophysiology. 1995. 21 figs. IX, 90 pp. ISBN 3-540-58214-2

Vol. 197: **Meyer, Peter (Ed.):** Gene Silencing in Higher Plants and Related Phenomena in Other Eukaryotes. 1995. 17 figs. IX, 232 pp. ISBN 3-540-58236-3

Vol. 198: **Griffiths, Gillian M.; Tschopp, Jürg (Eds.):** Pathways for Cytolysis. 1995. 45 figs. IX, 224 pp. ISBN 3-540-58725-X

Vol. 199/II: **Doerfler, Walter; Böhm, Petra (Eds.):** The Molecular Repertoire of Adenoviruses II. 1995. Approx. 36 figs. Approx. XIII, 296 pp. ISBN 3-540-58829-9

Vol. 199/III: **Doerfler, Walter; Böhm, Petra (Eds.):** The Molecular Repertoire of Adenoviruses III. 1995. Approx. 50 figs. Approx. XIII, 324 pp. ISBN 3-540-58987-2

MIX
Papier aus verantwortungsvollen Quellen
Paper from responsible sources
FSC® C105338

If you have any concerns about our products,
you can contact us on
**ProductSafety@springernature.com**

In case Publisher is established outside the EU,
the EU authorized representative is:
**Springer Nature Customer Service Center GmbH**
**Europaplatz 3, 69115 Heidelberg, Germany**

Printed by Libri Plureos GmbH
in Hamburg, Germany